Tributes
Volume 49

Logically Speaking
A Festschrift for Marie Duží

Tributes Series Editor
Dov Gabbay

dov.gabbay@kcl.ac.u

Logically Speaking
A Festschrift for Marie Duží

edited by

Pavel Materna

Bjørn Jespersen

ISBN 978-1-84890-419-4

College Publications
Scientific Director: Dov Gabbay
Managing Director: Jane Spurr

http://www.collegepublications.co.uk

Cover design by Laraine Welch

CONTENTS

v

Foreword

Marie was one of the invited speakers at an international conference on propositions in Barcelona back in the summer of 2015, and it was my pleasure to introduce Marie to a mixed audience, not all of whom might have known who she was. How to cram a multi-faceted personality with a multi-faceted career into one handy phrase? This is what came out: "Marie is originally a mathematician, who works in a department of computer science, who teaches mathematical logic and who publishes on philosophical logic."

This summary leaves out much, no doubt too much, not least Marie's prior career in industry and the private sector. However, it is within the areas of philosophical logic and multi-agent systems that Marie has made a name for herself within the last twenty-odd years. In 2008 the Rector of her home university, the VŠB-Technical University of Ostrava, granted Marie the award Outstanding Results in the Development of Science and Research. In 2011 the Council of the Academy of Sciences of the Czech Republic granted Marie and her co-authors its prestigious Award for Outstanding Results of Major Scientific Importance for the monograph *Procedural Semantics for Hyperintensional Logic*. This is still the authoritative tome on Transparent Intensional Logic; that is, of course, until the next generation reshuffles and reinvents TIL, taking account of new objections and new insights and thereby keeping TIL as fresh as ever.

In 2016 Marie was made a Full Professor in the Department of Computer Science, allowing her to continue her academic career for as long as she decides. The major achievements of TIL that Marie has been involved in over the last decade have been compiled into the volume *Transparent Intensional Logic: Selected Recent Essays*, also published by College Publications. The editors span the three generations TIL has spawned so far: Marie from the first generation, Bjørn Jespersen from the next generation, and Miloš Kosterec and Daniela Vacek (*née* Glavaničová) from the current generation.

Marie is literally a globetrotter, who has presented her research on six continents and spent research stays in some of her favourite countries, such as Japan and Finland, but who also remains rooted in her native Silesia, in the village of Háj ve Slezsku. This seems to fit a pattern, because Marie has a keen eye both for the global and the local, both the big picture and the minute detail.

When we as editors approached a string of friends and colleagues of Marie about a year ago, our opening line was that a Festschrift in Marie's honour was long overdue. The fact that the invitation to contribute was so well-received only confirms that the twenty-four contributors to this volume thought so as well. It is far from an exaggeration that Marie has, more than anyone else, been the driving force behind the development of TIL the last quarter of a century. And not only as a researcher in her own right, but also with respect to, for instance, grant capture, workshops, supervision, and guest-editing of special issues of journals. Furthermore, and perhaps most importantly, we all know Marie as an exceptionally committed, engaged, knowledgeable, witty and humorous person who brightens up any company she is in. We, the editors and the contributors, are more than happy that we are able to present Marie with this Festschrift as a small token of our appreciation of Marie both as a researcher and as a person.

Pavel Materna and Bjørn Jespersen

A Personal Tribute to Marie

PAVEL MATERNA
Czech Academy of Sciences, Prague
MaternaPavel@seznam.cz

As a young philosopher, I would often take part in various conferences. One remarkable such conference was SOFSEM. It consisted of a series of seminars oriented to computer science and also broadly to various philosophical topics. This combination of philosophy and algorithms was an ideal combination for me, because it was Bolzano, Frege, Church and others who were attracting my attention at the time. I began to strongly accept the views of my friend Pavel Tichý, the best philosophical logician in what was then Czechoslovakia. He would later emigrate to New Zealand, where he continued building his remarkable system of Transparent Intensional Logic (TIL) in steep opposition to the formalist orientation of most systems of modern logic then.

My friends told me about Marie, who was successfully developing several computer programs at the time. She knew neither TIL, nor Pavel Tichý, but I soon realized that she might find TIL interesting, because she was interested in logic and its applications in computer science. We started to discuss, discovering that it was not only TIL (and, of course, anti-Marxist ideology) that connected us with Tichý. Moreover, good music was a strong motive for our developing friendship. Both Marie and I played piano as capable amateurs, also four-handed, and we even performed in the evening at the SOFSEM conferences; our masterpieces were (together with another friend of ours, the mathematician Jaroslav Pokorný) Dvořák's *Legends* and Bach's *Brandenburg Concertos*. Thus, our meetings became a regular part of our lives. When Tichý visited the Czech Republic after the Velvet Revolution, he appreciated Marie; "One of the few who understands me", as he said.

The more I came to appreciate the depth of Tichý's ideas, the more curious I became: Is there perhaps some possibility to combine my rather good knowledge of TIL with Marie's great skills to navigate in the world of computer programs? I started to persuade Marie that this would be an idea worth pursuing. At first, I was surprised: She hesitated, saying that she was not sure that she would find a way to carry out such a collaboration. I did not give up, but continued my propaganda, and after a while Marie recognized the main ideas of Tichý's system and started to write some common papers with me. After some of the papers were published, TIL started to be better known both at home and abroad. Many friends of mine were convinced that the main purpose of this collaboration was that I simply wanted to help Marie, because my publishing activity was well known. Nothing could be farther from the truth, however. Soon it was admitted that it was Marie who was bringing new interesting ideas to the table, and that my part was just a welcome philosophical commentary.

The period after Marie got involved, or entangled, in TIL was no more one of underestimating TIL. Many deep thoughts rooted in Tichý's work began to gain respect. An outstanding Danish philosopher (now based in the Netherlands) earned his doctorate at Masaryk University of Brno and became an active member of the TIL community. And due to the fact that Marie continues to teach TIL at, for instance, the VŠB-Technical University of Ostrava and the Masaryk University of Brno, a young generation of TILians have begun to regularly publish in prominent venues and get quoted also internationally.

In conclusion, I would also like to mention Marie's sense of humour; here is an example: "What is the difference between an optimist and a pessimist? The optimist believes that the actual world is the best possible one. The pessimist knows that it is."

With best wishes for many more years of creative activities, Marie, many people like you very much!

Intensional Proof-Theoretic Semantics and the Rule of Contraction

Peter Schroeder-Heister
University of Tübingen
psh@uni-tuebingen.de

Abstract

The structural rule of contraction, which allows the identification of two occurrences of the same formula, is discussed from the viewpoint of intensional proof-theoretic semantics. It is argued that sometimes such occurrences have different meanings and therefore should not be identified. This suggests a restriction of contraction based on a definitional (and thus intensional) ordering of formulas within proofs, which is related to ramification in type theories such as the one proposed by Marie Duží.

In her impressive scientific work, Marie Duží has put forward and established *Transparent Intensional Logic* as a comprehensive approach to semantics, according to which the meaning of an expression is viewed as a construction that generates its denotation.[1] Here a construction is not understood as a set-theoretical function but as a procedure in a fine-grained sense. This construction-oriented stance makes Duží's approach particularly attractive to proof-theoretic semantics, even though the starting points differ in many respects. In what

---

[1] The *Opus Magnum* is Duží, Jespersen and Materna [6], but there is a great number of further papers by Duží propagating this topic. I just mention [3, 5].

follows I would like to make some remarks on what may be called "intensional proof-theoretic semantics" and discuss the structural rule of contraction as a typical example. This rule, which allows one to identify two occurrences of the same formula, gives rise to intensional considerations from a procedural point of view. In the context of paradoxical reasoning I suggest a definitional (and thus intensional) ordering of formula occurrences within proofs corresponding to the idea of ramification in Duží-style type theory.[2]

1   The standard approach to intensional proof-theoretic semantics

What might be called "intensional" proof-theoretic semantics goes back to the development of *general proof theory*, which was proclaimed at the beginning of the 1970s, most notably by Dag Prawitz [12, 13, 14] — he coined this term — and Georg Kreisel [10]. The topic was present also in the writings of Girard, Martin-Löf and others, and in a sense already in Gentzen [9]. Its idea was that in proof theory we should not only be interested in the power of deductive systems and the reduction of deductive systems to others with the aim to establish their consistency (in the spirit of Hilbert's programme), but also in the form and structure of proofs as a topic in its own right. In this sense, proof theory goes way beyond provability theory: beyond our interest in what we can we prove, we should be interested in how something provable is proved. It was only natural that, when proofs as such were put forward as genuine objects of study, the question of proof identity became central. It was explicitly, and under this heading, put on the proof-theoretic agenda by Prawitz [12] and Kreisel [10]. One might relate it to Quine's [15] slogan "no entity without identity": if we want to talk about proofs as objects we must give identity criteria for them. This is definitely an intensional question: if one considers proofs to be extensionally identical when they prove the same theorem, the

---

[2]I have mentioned certain aspects of these ideas in [16], [17, in particular Supplement "Definitional Reflection and Paradoxes"] and [18], but never from the intensional perspective.

wider problem of when they are different, though proving the same, is intensional. It is obvious that Euclid's and Euler's proofs establishing that there are infinitely many prime numbers are different, as they are based on different proof ideas (see [1]). Mathematicians often have a very clear intuition of whether two proofs are based on the same proof idea and are essentially identical. However, it is not clear at all how to make this precise.[3]

At the very elementary level of proof theory, and in particular in the neighbouring discipline of categorial proof theory, several proposals have been made. The most prominent one is that two proofs are identical if they can be transformed into each other by certain proof-theoretic reductions along the lines of normalization procedures as given by Prawitz [11]. For example, the proofs

$$
\begin{array}{cc}
\vdots \quad | \\
\dfrac{A \quad B}{\dfrac{A \wedge B}{A}} \quad \text{and} \quad \begin{array}{c} \vdots \\ A \end{array}
\end{array}
\qquad \text{as well as} \qquad
\begin{array}{cc}
\vdots \quad | \\
\dfrac{A \quad B}{\dfrac{A \wedge B}{B}} \quad \text{and} \quad \begin{array}{c} | \\ B \end{array}
\end{array}
$$

are intensionally identical, as they result from removing the redundancy of introduction followed by elimination of conjunction. This very elementary example already shows that we must be very careful in our understanding of proofs and rules. If we replace the proof $\begin{array}{c} | \\ B \end{array}$ with an arbitrary proof $\begin{array}{c} | \\ A \end{array}$ of A, then the proof

$$
\dfrac{\dfrac{\begin{array}{cc} \vdots & | \\ A & A \end{array}}{A \wedge A}}{A} \tag{1}
$$

is both identical to $\begin{array}{c} \vdots \\ A \end{array}$ and to $\begin{array}{c} | \\ A \end{array}$, which implies that any two proofs of A are identical. This *intensional collapse* of all proofs of

---

[3]I am using the term "proof" in an ambiguous way to denote both a concrete proof figure (a derivation) and the abstract proof represented by this proof figure. Therefore, when I say that two different proofs are identical I mean that two syntactically different proof figures represent the same abstract proof.

A into a single one is not a contradiction, but something that may be called an *intensional paradox* that we want to avoid if we want to talk about possibly different proofs of a provable sentence A. In the present case it can easily be remedied by distinguishing between the right and left projection of conjunction elimination, which means that, without further annotation, the proof figure (1) is ambiguous. This is not a trivial matter. It implies that the annotation of a proof telling which rule has been applied at which step must be considered a part of the proof itself, something which is immediately relevant for logic teaching and often not mentioned there. When we annotate proofs with terms, this is, of course, automatically satisfied.

In general, the situation is not always that simple. If we have a

proof of A which after further steps arrives again at A, that is, $\begin{array}{c} \vdots \\ A \\ | \\ A \end{array}$,

we do not want that it reduces to $\begin{array}{c} \vdots \\ A \end{array}$ as long as we do not know what is happening between the upper and the lower A. It is not easy to tell what exactly should be allowed as a reduction generating intensional identity, apart from the negative criterion that we want to avoid an intensional collapse. Ekman's paradox is a very instructive example pointing to these issues (see [20]). It is related to the issue of harmony, that is, the relation between introduction and elimination rules in natural deduction, as harmonious rules give rise to identifications between proofs[4]. I do not want to further elaborate on this approach to intensional proof-theoretic semantics, which puts the identity of proofs in the foreground. Tranchini [21] gives a thorough discussion of it. In the following I would like to draw the reader's attention to another aspect of intensional proof-theoretic semantics.

---

[4]For a general discussion see [19].

8

2 Contraction: a case of intensionality in proof-theoretic semantics

The question of identity of proofs extends the intensional identity problem: from that of objects, functions, concepts and propositions discussed in philosophical semantics since Frege, to another category of entities, namely proofs. However, the identity problem for the common categories still remains present in proof-theoretic semantics and needs to be approached there. I would like to focus on an aspect of propositional identity, that is, of identity between sentences or formulas, that has not yet found adequate consideration. At least four senses of propositional identity can be distinguished (in ascending order from extensional towards intensional):

1. Material equivalence: A is identical to B if A and B have the same truth value. We would nevertheless intensionally distinguish between A and B as they may not be *logically* equivalent.

2. Logical equivalence: A is identical to B if A and B are logically equivalent (with respect to some logical system such as intuitionistic or classical logic). We would nevertheless intensionally distinguish between A and B as they may not be *isomorphic*.

3. Logical isomorphism: A is identical to B if there are proofs between A and B such that the composition of these proofs is the identity proof (with respect to some concept of proof identity). We would neverthelessss intensionally distinguish between A and B as they may not be *syntactically identical*.

4. Syntactic identity: A is identical to A.

The second and third notions are discussed in categorial proof theory (see, e.g., [2]). An example of logical equivalence lacking isomorphism is the relation between $p \wedge q$ and $p \wedge (p \to q)$, whereas an example of isomorphism is the relation between $p \wedge q$ and $q \wedge p$. The fourth notion is not normally considered explicitly, as it is always taken for granted.

However, I would like to go one step further by arguing that syntactic identity is not the finest-possible semantic identity relation,

and that different occurrences of the same sentence or formula may
be semantically distinguished. If we have two occurrences of the same
sentence A, they point, of course, to the same content, as the content
is tied to the sentence itself and not to its occurrences. However, in a
non-perfect language, A in one context may mean something different
from A in another context, which in the semantic formalism we would
disambiguate by different annotations in evaluating A. Therefore I
propose to rephrase the fourth identity notion and add a fifth one:

4. Syntactic identity: A is identical to A. We would nevertheless
   intensionally distinguish between two occurrences of A as they
   may have a different semantic status.

5. Intensional identity of occurrences: Two occurences of A are
   identical, if they have the same semantic status.

It turns out that in proof theory this problem — when can two
occurrences of formulas be identified and in which context do they
denote something different? — manifests itself explicitly in the rule
of contraction.

In the sequent calculus the rule of contraction allows one to identify two occurrences of a formula A:

$$\frac{\ldots, A, A, \ldots \vdash \ldots}{\ldots, A, \ldots \vdash \ldots}$$

In the classical variant of the sequent calculus — here I only consider
the intuitionistic case —, we have an analogous rule also for the right
side of the turnstile. The rule of contraction is often concealed by
the fact that the antecedents and succedents of sequents are taken as
sets, in which the multiplicity of occurrences of the same formula does
not count. In natural deduction contraction occurs in the form that
more than one occurrence of A may be discharged as assumption. For
example, when introducing $A \to B$ from a derivation of B which uses

the assumption A at more than one place

$$
\begin{array}{ccc}
{\scriptstyle(1)} & & {\scriptstyle(1)} \\
[A] & \cdots & [A] \\
& | & \\
& B & \\
\hline
& A \to B & {\scriptstyle(1)}
\end{array}
$$

then these occurrences become a single occurrence in the implication introduced.[5]

Contraction is normally seen as absolutely unproblematic. The validity of an argument is not affected by using an assumption twice rather than once. However, in substructural logics, in particular in the context of relevance logic and entailment as well as in linear logic, the rule of contraction has become an issue of discussion. Putting the taxonomy of possible formal systems aside, the arguments for taking contraction-free logics seriously are not very convincing, at least not to me. Mathematical and other argumentation is full of contractions.

If we want to argue against the identification of two occurrences of A, we must argue that these two occurrences are semantically different. Semantical difference here means that they belong to different semantical contexts, so that they can be disambiguated by distinguishing these contexts, perhaps by attaching corresponding context-labels to them. This is definitely a topic of intensional proof-theoretic semantics, as we would argue that the formula A presented by the two occurrences of A is given to us *in two different ways*. In other words, the *way of being given* or *mode of presentation* of the formula A differs between the two occurrences of A. "Way of being given" or "mode of presentation" ("Art des Gegebenseins") is actually Frege's term for 'sense' in contradistinction to 'reference'. The planet Venus is given to us in two different ways when described as 'the morning/evening star'. In the deductive context, if we wanted to abolish contraction, we would have first to argue for the difference in meaning (sense) of

---

[5] As a side remark, I should mention that the notion of occurrence, though appearing in every logic textbook, is still an underexposed concept that deserves further considerations. A detailed elaboration has been provided by Gazzari [8].

the two occurrences of A and then argue that this difference counts in the given surrounding. That there is such an argument for *any* contraction candidate A can be doubted.

Actually, there is one area where the absence of contraction would solve big problems: that of logical and semantical paradoxes. Rather than criticizing their internal construction (such as negative self-referentiality etc.), one may attack the logical derivation of a contradiction from these constructions. As has been pointed out already by Fitch [7], this derivation uses contraction at an essential place. Certainly, it also uses other logical means (at essential places). However, one can at least say that abolishing contraction blocks the derivation of inconsistency.

I do not want to go into the construction details of any paradoxes here. For the point to be made it suffices to assume that there is a sentence R, which by definition has the same meaning as its opposite $\neg R$. Thus I assume that in the sequent calculus there are definitional rules of the following form, which govern the meaning of R ("R" should remind one of "**R**ussell").

$$\frac{\dots, \neg R, \dots \vdash \dots}{\dots, R, \dots \vdash \dots} \text{ DEF} \qquad\qquad \frac{\dots \vdash \neg R}{\dots \vdash R} \text{ DEF}$$

In a natural deduction system, we would, as definitional rules, postulate introduction and elimination rules for R, which allow one to pass from $\neg R$ to R and from R to $\neg R$.

Then the following derivations generate $\neg R$ and R, respectively, and thus a contradiction. The right derivation is nothing more than the left derivation extended with an additional step. Thus, when discussing the issue of contraction, we can focus on the left one.

$$\frac{\dfrac{\dfrac{R \vdash R}{R, \neg R \vdash} \text{ DEF}}{\dfrac{R, R \vdash}{R \vdash} \text{ Contr}}}{\vdash \neg R} \qquad\qquad \frac{\dfrac{\dfrac{\dfrac{R \vdash R}{R, \neg R \vdash} \text{ DEF}}{\dfrac{R, R \vdash}{R \vdash} \text{ Contr}}}{\vdash \neg R} \text{ DEF}}{\vdash R} \text{ DEF}$$

The natural deduction formulation would be as follows.

$$
\cfrac{\cfrac{[R]^{(1)}}{\neg R}\text{ DEF} \quad [R]^{(1)}}{\neg R}{}^{(1)}
\qquad\qquad
\cfrac{\cfrac{\cfrac{[R]^{(1)}}{\neg R}\text{ DEF} \quad [R]^{(1)}}{\neg R}{}^{(1)}}{R}\text{ DEF}
$$

The critical use of contraction is obvious: in the sequent calculus formulation as the application of an explicit rule, in the natural deduction formulation as the joint cancellation of two occurrences of R when applying the rule of constructive reductio (negation introduction).

However, should one regard this observation as a proper argument against contraction? The argument is of a consequentialist nature. Assuming contraction, we get into trouble: thus abolish contraction. This looks like it would fit well into Popperian falsificationist methodology. However, many other assumptions may be responsible for the trouble of paradoxicality, such as the above-mentioned internal construction of the paradoxical sentence, something we have deliberately ignored. There should be additional arguments that speak for the selection of contraction as the critical feature (rather than any other feature) to be rejected. These arguments are not very strong. One does not get very far when trying to do mathematics (or other argumentation) in a logic without contraction. Rejecting a logical principle which has proved almost indispensible in the context of mathematics should be our last resort after testing other, more specific principles, which are crucial for the paradoxes but do not affect mathematical reasoning. Actually, formulating principles as specific as possible and testing those would be really in line with Popperian methodology.

Can such specific principles be found in our framework? What if we can find a plausible restriction of contraction, which disallows contraction in the case of paradoxical reasoning, but would otherwise, in standard mathematical contexts, allow for the use of contraction? I will argue that such a restriction can indeed be formulated by intensional considerations. In the case of the paradoxes, the two occurrences of R can be claimed to have different semantical contents. This is not arguing ad hoc, saying that contraction should be disallowed when it has a paradoxical outcome, but arguing by general semantical considerations for a difference in meaning of the two occurrences of R, which, when ignored, has paradoxical consequences, and that this difference plays no role in standard mathematical reasoning.

3 Levels of definitional evaluation (*order of meaning*)

Consider the upper part of the the derivation of inconsistency down to the application of contraction.

$$
\cfrac{\cfrac{\cfrac{\boxed{R} \vdash R}{\boxed{R}, \neg R \vdash}}{\boxed{R}, \textcircled{R} \vdash} \text{ DEF}}{R \vdash} \text{ Contr}
$$

Certain occurrences of R are annotated by boxing or encircling them. Contraction is applied to a boxed and an encircled occurrence of R. It is obvious that these two occurrences have a different 'derivational history'. The boxed occurrence of R is the repetition of an R occurring in an initial identity sequent. It does not result from a definitional rule, that is, a meaning rule. Contrary to that, the encircled occurrence of R is the result of such a rule, leading from $\neg R$ to R on the left side of the turnstile, thus using the definition that defines R in terms of $\neg R$. In this sense the encircled R is a *specific* or evaluated occurrence of R, whereas the boxed R is completely *unspecific*, as it has just been laid down as an assumption within an initial sequent.

In the corresponding part of the natural deduction formulation we have the analogous situation.

$$
\cfrac{\cfrac{\boxed{\textcircled{R}}^{(1)}}{\neg R} \text{ DEF} \qquad \boxed{\boxed{R}}^{(1)}}{\neg R}\,{}^{(1)}
$$

The encircled R is the major (actually the only) premiss of a meaning inference and thus understood according to its definition, while the boxed R, which is discharged simultaneously and thus identified with the encircled R, is just laid down as an assumption, without presupposing any meaning rule associated with it.

Now we argue that it makes a semantical difference of whether, in the sequent calculus, an occurrence of a formula A is specific, that is, results from a meaning rule, or whether it is unspecific, that is, does not result from such a rule. In the first case, establishing it (either in the antecedent or in the succedent of a sequent) requires knowledge of its definition, whereas in the second case no definitional knowledge is presupposed, as the formula just remains as it has been stated, without any reference to its content. In the case of natural deduction, an occurrence A of a formula is specific, if it is involved in a definitional rule, either as the conclusion of an introduction of A, or as the major premiss of an elimination of A, and unspecific, if it is not of this form. If this difference is considered intensionally significant, which we argue it is, contraction should be disallowed in these cases, even though the two occurrences have the same shape. Understanding A as a sentence *defined* in a certain way, that is, having a specific meaning, is different from regarding A just as an arbitrary sentence whose semantic content does not play any role. These are two different ways in which A is given to us, that is, two senses in Frege's terminology (the reference here being the sentence A). Using another terminology, we can understand the meaning of A as the *construction* embodied by its meaning rules. A specific use of A is one in which A has actually been construed according to this meaning, whereas in the unspecific use such a construction is not taken into account. This might be viewed as a *procedural distinction* between A as the dynamic result/output of a specific operation (or, in natural deduction, as its argument/input), and A as being static.

To formally deal with this situation, I propose to introduce an indexing discipline. With every formula occurrence in a sequent calculus derivation, a natural number is associated as a *meaning index*, which is increased if a meaning rule (left- or right-introduction rule in the sequent calculus) is applied. Contraction is then prohibited, if the meaning indices of the sentences involved differ. In the above example, the encircled R undergoing contraction receives a higher meaning index under this discipline than the boxed one, as it results from an application of a definitional rule, whereas the boxed R is available without any such application. This procedure can be iterated in that any application of a meaning rule adds to the index. For example, using an upper index to denote the meaning level, the following derivation ends with two occurrences of R in the antecedent (or as assumptions in natural deduction), which due to the different meaning index must not be contracted[6].

$$\frac{\dfrac{\dfrac{\dfrac{\dfrac{\dfrac{R^0 \vdash R^0}{R^0, \neg R^0 \vdash}}{\dfrac{\neg R^0 \vdash \neg R^0}{\neg R^0 \vdash R^1} \text{ DEF}} \text{ DEF}}{\dfrac{R^1 \vdash R^1}{R^1, \neg R^1 \vdash}}}{R^1, R^2 \vdash} \text{ DEF}}{R^? \vdash} \text{ NOT ALLOWED}}{\vdash \neg R^?}$$

In natural deduction:

$$\frac{\dfrac{\dfrac{R^2}{\neg R^1} \text{ DEF} \quad \dfrac{\dfrac{R^1}{\neg R^0} \text{ DEF} \quad R^0 \, {}^{(1)}}{\dfrac{\neg R^0}{R^1} \text{ DEF}}}{\neg R^?}}{} \, {}^{(2)} \text{ NOT ALLOWED}$$

Due to the non-local features of natural deduction the indexing regime is not so straightforward as in the sequent calculus, something that speaks for the sequent calculus in this situation (which is anyway the system making the contraction rule fully explicit).

---

[6]For simplicity, we do not apply the indexing regime to the negation rules. In a properly worked-out setting, indexing would be applied to all logical constants, as the rules governing their meaning have a definitional status.

The general result is a kind of higher-order theory with respect to meaning, according to which the application of a meaning rule increases the order. If in the underlying theory a sort of typing is already under way, this yields a ramified type theory, which represents another link to Duží's work (see, e.g., [4]). Ramification essentially means classification by definitional order, which is exactly what we are doing. We distinguish items according to their definition. The definiendum is of higher order than its definiens. In the current situation, R is defined by $\neg R$ and thus occurs in its definiens, which in the context of contraction has the effect that the defined R is of higher order than the R from which it is defined. In this way one may look at the proposal made here as a way of evading impredicativity by ramification. The crucial issue, however, is that we carry out this ramification at the deduction level and not at the definitional level. As far as definitions are concerned, we do not impose any restrictions and thus allow R to be defined by $\neg R$ (or being obtained from $\neg R$ by other means).

Our claim is that in ordinary mathematical reasoning, this situation does not show up. Our proposal to disallow contraction, but only when the occurrences to be contracted differ in their order of meaning, can be seen as a well-founded semantical approach which at the same time is precisely targeted at the paradoxes. This shows that a definitional hierarchy according to the order of meaning is not only a conceptually well-founded, but also a useful device. At least there is no need for a Russellian reducibility postulate to enable proper mathematical reasoning.

Note that in such a system we do not obtain full admissibility of cut, but have a restriction for the cut formula analogous to that for contraction. The cut rule

$$\frac{\Gamma \vdash A \quad A, \Delta \vdash C}{\Gamma, \Delta \vdash C}$$

can be shown to be admissible, but only if the occurrence of A in the left premiss and the occurrence of A in the right premiss, that is, the two occurrences of the cut formula, are of the same order of meaning. With an unrestricted cut rule we could easily override our restriction for contraction, as we could generate sequents such as $R^1 \vdash R^0$ and thus transform a sequent of the form $\ldots, R^1, R^0, \ldots \vdash \ldots$ into the sequent $\ldots, R^1, R^1, \ldots \vdash \ldots$ However, this outcome is only natural. When performing a cut, we are identifying two occurrences of the cut formula, and we should expect that they have the same meaning apart from representing the same formula.

These considerations, which still need to be spelled out in detail, show that intensionality in proof-theoretic semantics goes way beyond the question of identity of proofs. They support a ramified approach, which may be seen as an inferentialist extension of the constructive approach found in Duží's procedural semantics.

Acknowledgements

I would like to thank Andrzej Indrzejczak and Michał Zawidzki for inviting me to the 10th International Conference "Non-Classical Logics: Theory and Applications" (NCL'22, Łódź, Poland, March 14–18, 2022), where basic ideas of this paper were presented. Moreover, I am extremely grateful to Bjørn Jespersen and Luca Tranchini for their very careful reading of an earlier version of this contribution and for many comments and suggestions for improvement.

References

[1] Martin Aigner and Günter M. Ziegler. *Proofs from THE BOOK*. 6th edition. Berlin: Springer, 2018. DOI: 10.1007/978-3-662-57265-8.

[2] Kosta Došen. Models of deduction. *Synthese* 148 (2006). Special issue *Proof-Theoretic Semantics*, edited by R. Kahle and P. Schroeder-Heister, 639–657. DOI: 10.1007/s11229-004-6290-7.

[3] Marie Duží. Extensional logic of hyperintensions. In: *Conceptual Modelling and Its Theoretical Foundations: Essays Dedicated to Bernhard Thalheim on the Occasion of His 60th Birthday*. Ed. by Antje Düsterhöft, Meike Klettke and Klaus-Dieter Schewe. Berlin: Springer Berlin Heidelberg, 2012, 268–290. DOI: 10.1007/978-3-642-28279-9_19.

[4] Marie Duží. Deduction in TIL: From simple to ramified hierarchy of types. *Organon F* 20 (Suppl. Issue 2) (2013), 5–36.

[5] Marie Duží. Negation and presupposition, truth and falsity. *Studies in Logic, Grammar and Rhetoric* 54 (2018), 15–46. DOI: 10.2478/slgr-2018-0014.

[6] Marie Duží, Bjørn Jespersen and Pavel Materna. *Procedural Semantics for Hyperintensional Logic: Foundations and Applications of Transparent Intensional Logic*. Berlin: Springer, 2010. DOI: 10.1007/978-90-481-8812-3.

[7] Frederic B. Fitch. A system of formal logic without an analogue to the Curry W operator. *Journal of Symbolic Logic* 1 (1936), 92–100. DOI: 10.2307/2269029.

[8] René Gazzari. *Formal Theories of Occurrences and Substitutions*. University of Tübingen: Doctoral Dissertation, 2020. DOI: 10.15496/publikation-47553.

[9] Gerhard Gentzen. Untersuchungen über das logische Schließen. *Mathematische Zeitschrift* 39 (1934–1935), 176–210, 405–431 (DOI: 10.1007/BF01201353, 10.1007/BF01201363). English translation in: *The Collected Papers of Gerhard Gentzen*. Ed. by M. E. Szabo. Amsterdam: North Holland 1969, 68–131.

[10] Georg Kreisel. A survey of proof theory II. In: *Proceedings of the Second Scandinavian Logic Symposium*. Ed. by Jens Erik Fenstad. Amsterdam: North-Holland, 1971, 109–170. DOI: 10.1016/S0049-237X(08)70845-0.

[11] Dag Prawitz. *Natural Deduction: A Proof-Theoretical Study*. Stockholm: Almqvist & Wiksell, 1965. Reprinted Mineola, NY: Dover Publ., 2006.

[12] Dag Prawitz. Ideas and results in proof theory. In: *Proceedings of the Second Scandinavian Logic Symposium (Oslo 1970)*. Ed. by Jens E. Fenstad. Amsterdam: North-Holland, 1971, 235–308. DOI: 10.1016/S0049-237X(08)70849-8.

[13] Dag Prawitz. The philosophical position of proof theory. In: *Contemporary Philosophy in Scandinavia*. Ed. by R. E. Olson and A. M. Paul. Baltimore: Johns Hopkins Press, 1972, 123–134.

[14] Dag Prawitz. On the idea of a general proof theory. *Synthese* 27 (1974), 63–77. DOI: 10.1007/BF00660889.

[15] W. V. Quine. Speaking of objects. In: *Ontological Relativity and Other Essays*. New York: Columbia University Press, 1969, 1–25.

[16] Peter Schroeder-Heister. Paradoxes and structural rules. In: *Insolubles and Consequences: Essays in Honour of Stephen Read*. Ed. by C. Dutilh Novaes and O. Hjortland. London: College Publications, 2012, 203–211. DOI: 10.15496/publikation-70824.

[17] Peter Schroeder-Heister. Proof-Theoretic Semantics. In: *Stanford Encyclopedia of Philosophy*. Ed. by Edward N. Zalta. Metaphysics Research Lab, Stanford University, 2012. URL: https://plato.stanford.edu/entries/proof-theoretic-semantics/. Revised 2018 and 2023.

[18] Peter Schroeder-Heister. Restricting initial sequents: the trade-offs between identity, contraction and cut. In: *Advances in Proof Theory*. Ed. by Reinhard Kahle, Thomas Strahm and Thomas Studer. Basel: Birkhäuser, 2016, 339–351. DOI: 10.1007/978-3-319-29198-7_10.

[19] Peter Schroeder-Heister. Axiomatic thinking, identity of proofs and the quest for an intensional proof-theoretic semantics. In: *Axiomatic Thinking I*. Ed. by F. Ferreira, R. Kahle and G. Sommaruga. Berlin, 2022, 145–163. DOI: 10.1007/978-3-030-77657-2_8.

[20]   Peter Schroeder-Heister and Luca Tranchini. Ekman's paradox. *Notre Dame Journal of Formal Logic* 58 (2017), 567–581. DOI: 10.1215/00294527-2017-0017.

[21]   Luca Tranchini. *Harmony and Paradox: Intensional Aspects of Proof-Theoretic Semantics*. Berlin: Springer, 2023.

TWO MEANINGS OF *IF-THEN-ELSE* AND NORMATIVE CONATIVE FACTS

FRANTIŠEK GAHÉR
Department of Logic and Methodology of Sciences, Faculty of Arts,
Comenius University in Bratislava, Slovak Republic
`frantisek.gaher@uniba.sk`

Abstract

In computer science, the "if-then-else" function is well established and has a stable meaning. Duží, Jespersen and Materna [7] have convincingly resolved the question of its strictness on the basis of partial type theory. However, in legal theory and practice, the expression "if-then-else" is used in two different senses. In neither of those does it refer to program instructions. Rather, what is at stake are instructions for persons to act on. The first sense is fundamentally identical with the one common in computer science. This is evident especially in the chaining of the so-called perfect legal norm. In the second sense, the distinction between the components bound by the operator "if-then-else" comes to the fore. Some of these components are brute facts, appearing in the role of empirical conditions. Others are legal regulations (commands, prohibitions, permissions) which cannot be said to be true or false in the usual sense. Adherents of logical empiricism in legal theory have viewed this as a fundamental obstacle to the application of classical logic to the domain of norms.

In the second part of the paper, I propose a conservative solution to this problem that does not require a special deontic logic. I propose an extension of the domain of facts by the domain of social facts of two distinct types: a) *institutional*

This work was supported by the Science Grants Agency (VEGA 1/0197/20) and the Research and Development Agency under contract No. APVV-21-0405.

facts; b) normative *conative* facts. Institutional social facts are the various distributions of *social roles* and social *relations* (as opposed to physical or biological properties and relations) *onto individuals*. They are akin to brute facts in that we have no difficulty with viewing them as indicators of truth or falsity.

Normative conative facts are the various distributions of *legal regulations* which stimulate action (commands, prohibitions, permissions) onto individuals. Using this conceptual basis, we can abstract from the pragmatic component of the meaning of legal regulations, as expressed, e.g., in an imperative, and focus only on their semantic component. In this way, normative conative facts can be viewed as true or false, and can thus become components of inferences that can be manipulated using logic. Since conative facts stimulate action with a particular goal, they are best viewed as explicit attitudes of a hyperintensional nature. The constraints resulting from their hyperintensional nature are in practice compensated for by a system of praxeological principles.

If-then-else in computer science

The *if-then-else* function, as used in computer science, has a stable meaning. "If A, then B, else C" — that is, if A is true, do B; if A is false, do C. The analysis of this function using propositional logic encounters two obstacles. First, the components B and C are not typical statements, but instructions to act. Second, classical logic is strictly compositional and evaluates both components B and C. If even one of the components B, C were not a defined (executable) statement, then the *if-then-else* function would not be strict. It would violate the principle of compositionality.

A solution to the first problem for a field outside computer science — namely, for law — will be proposed in the second part of this paper. A solution to the second problem was proposed by the Marie Duží, Bjørn Jespersen and Pavel Materna [7, 260ff]. In the case of Transparent intensional logic, i.e., a hyperintensional, partial, typed lambda calculus, the authors studied the *if-then-else* function and its strictness or "non-strictness" in the context of β-reduction and the strictness requirement of composition. The procedure[1] If A, then B, else C is explained as consisting of two stages:

> "First, select a construction to be executed on the basis of a specific condition. Second, execute the selected construction." [7, 263]

Duží *et al.* explain the preservation of the *if-then-else* function's strictness by the fact that in the first step, condition A is evaluated and neither of the subprocedures B, C are executed. The latter are only mentioned: they occur in *de dicto* supposition and need not be restricted to their empirically identifiable value. In the second stage, e.g., component C does not need to be evaluated. Therefore, the evaluation itself is not strict and it is sufficient to evaluate one of the procedures — procedure B. The problem of the β-reduction rule and the correct definition of substitution was also addressed by Duží in her other works, e.g. [9].

The first meaning of *if-then-else* in law

In the provisions of legal norms, the *if-then-else* function is used in two different senses. The first type of use coincides with the meaning of the function common in computer science. However, there is an important difference: in law, the expression does not specify instructions for a step in a computer program. Rather, it specifies instructions for actions of natural or legal persons in space-time. *Pars pro toto* we can mention the following provisions of the legal norms of the Criminal Code[2]:

> §52 (1)
> **If the** *parolee leads an orderly life during the probationary period, complies with the conditions of probation supervision, fulfills the imposed restrictions and meets the obligations imposed,* **[then]** *the court shall issue a ruling*

---

[1]For more on the notion of procedure or construction see [31, 63ff] and [7, 42ff].

[2]https://www.slov-lex.sk/pravne-predpisy/SK/ZZ/2005/300/   CIVIL LAW.

declaring that the parolee has proved themselves; otherwise
[i.e., else] *the court may issue an unconditional sentence,*
including while the probationary period is in progress.

§68 (2)
If the *parolee has led an orderly life during the probation-*
ary period and has complied with the restrictions and obli-
gations imposed, *[***then***]* *the court shall issue a ruling that*
the parolee has proved themselves; otherwise **[i.e., else]** *the*
court may rule, while the parole period is still in progress,
that the remainder of the sentence be served.

The bifurcation point in the execution of this complex procedure is the complex empirical condition A (its truth-value): namely, the parolee's compliance with probationary supervision and with the other restrictions and obligations imposed. If these conditions are satisfied, the court shall rule the parolee to have proved themselves. If they are not, the court shall issue an unconditional sentence or that the remainder of the sentence be served. Schematically,

$$A \to B; \neg A \to C$$
$$(A_1 \land A_2) \to B; \neg(A_1 \land A_2) \to C,$$

where B and C represent legal regulations.

The second meaning of *if-then-else* in law: the case of a perfect legal norm

In legal theory, however, the *if-then-else* function also has a second meaning. The latter serves to capture the logical structure of the so-called perfect legal norm[3] and those provisions where the bifurcation point in the execution of the complex procedure is not an empirical condition, but the execution (or its absence) of the legal regulation conditioned by it. I will describe this using the case of a perfect legal norm, and then turn to a more general form.

---

[3]The perfect legal norm itself is not commonly identical with a particular provision of the legal norm, but can usually be reconstructed from the extensive text of the law.

The procedure "If Hypothesis, then Disposition, else Sanction" (If
H, then D, else S) expresses the logical structure of the so-called
perfect legal norm. This complex procedure has several stages of
execution, whose steps and inner connections are different from the
execution of the *if-then-else* procedure in the first meaning.

The procedure *If H, then D, else S* starts by establishing whether
the empirical condition H is true for a particular addressee. Its truth
is a necessary condition for the very application of the legal norm's
provision as a procedure. If H is false, the entire procedure remains
inactive, and therefore neither of the subprocedures D, S are selected
nor executed. The entire procedure remains a potential course of
action in a *de dicto* supposition.

After determining that H is true for a particular addressee, the
addressee has to decide whether or not to act on the regulation D
within the time period specified. If they choose to execute D and do
so successfully (efficiently) within the time limit (ExeD), the complex
procedure is terminated and the sanction S remains inactive.

If the addressee chooses not to act on D[4] and does not actually
execute D[5] within the time period specified, then this non-execution
of D becomes a condition — a secondary disposition for the execution
of a secondary legal regulation — for the sanction S. As an example
of a provision that roughly exhibits such a logical structure, consider
the following provisions of the Civil Code:

§577 (1)
*The debtor shall notify the creditor without undue delay
after the debtor becomes aware of a circumstance which
renders the performance impossible. If they fail to do so,
they shall be liable for any damage which the creditor incurs due to the failure to duly notify the creditor of such
impossibility*

After rephrasing into an explicit *if-then-else* structure, it reads:

---

[4]It can also be a prohibition of doing something or a more permanent exercise
of the power of attorney.

[5]They may also decide to execute D, but they do not do so successfully, efficiently, within the time period specified.

(1) **If** *the debtor becomes aware of a circumstance which renders the performance impossible, [then] the debtor shall notify the creditor without undue delay.* **Otherwise [i.e., else]**, *they shall be liable for any damage incurred by the creditor due to the failure to notify the creditor of such impossibility.*

Another similar example is in the third provision of §695:

The lessor is only entitled to undertake any construction work in the flat and any other material changes to the flat with the lessee's consent. Such consent may be only withheld for serious reasons. **If** *the lessor undertakes such work under the order of a competent authority of state administration,* **[then]** *the lessee is obliged to allow that such work be undertaken, otherwise* **[i.e., else]** *the lessee shall be liable for any damage incurred due to the failure to fulfil this obligation.*

Here, the distinction between empirical conditions (empirical facts; H) and legal regulations (rules for action; D, S) comes to the fore. A legal regulation D is activated by an empirical condition H being satisfied, and it stimulates the execution of the subprocedure D ($\text{Exe}D$) by the addressee of the norm:

$$H \to D$$

The execution of D by the addressee within the time limit specified results in the desired state. The primary goal of the norm is achieved. By executing the legal regulation ($\text{Exe}D$), a new empirical fact is created and the relevant legal regulation D ceases to be valid. The sanction S remains inactive.

On the other hand, by failing to execute D ($\text{non}(\text{Exe}D)$), a new empirical condition is created that is a secondary hypothesis for the secondary disposition S activated by it:

$$(H \wedge \neg(\text{Exe}D)) \to S$$

By executing the sanction S ((ExeS)), the secondary objective of the norm's provision is satisfied, upon which the S sanction as a regulation ceases to apply.

A general look at the second meaning of *if-then-else* in law

"Sanction" designates a specific type of a secondary disposition. More generally, we can distinguish primary, secondary, tertiary, etc. dispositions, where the criterion is the varying degree of their desirability, commonness or pragmatic convenience. Simply put, in legal systems, some legal regulations or rules may be preferred over others based on their utility (or legal laxity) with respect to this or that party in a legal relationship. An example of a sequence of legal regulations is the following provision:

> *(1)* **If** *the defect may not be rectified and the property may not be used duly or in the agreed manner due to the defect,* *[then]* *the acquirer is entitled to demand termination of the contract.* **Otherwise [i.e., else],** *the acquirer may demand either a reasonable discount from the price or the repair or supplementation of a missing part.*

$$(\neg H_1 \wedge (\neg H_2 \vee \neg H_3)) \rightarrow \mathrm{Prim}D;$$
$$((\neg H_1 \wedge (\neg H_2 \vee \neg H_3)) \wedge \neg(\mathrm{ExePrim}D)) \rightarrow$$
$$(\mathrm{Sek}D_1 \vee \vee (\mathrm{Sek}D_2 \vee \mathrm{Sek}D_3)^6$$

The acquirer's demand that the contract be terminated if a complex empirical condition is satisfied is the primary disposition (PrimD). If the acquirer does not take advantage of it ($\neg$(ExePrimD)), they may take advantage of one (and only one) of the alternative secondary dispositions:

1. demand a reasonable discount on the price ($\mathrm{Sec}D_1$);

2. demand repair or supplementation of a missing part (which is an inclusive disjunctive disposition — the acquirer may demand

---

[6]"∨" indicates inclusive disjunction, while "∨∨" indicates exclusive disjunction.

the first ($\mathrm{Sec}D_2$), the second ($\mathrm{Sec}D_3$) or both of its components ($\mathrm{Sec}D_2$ and $\mathrm{Sec}D_3$)).

Either $((\neg H_1 \wedge \neg H_2 \vee \neg H_3) \wedge \neg(\mathrm{ExePrim}D)) \rightarrow \mathrm{ExeSek}D_1$ or $((\neg H_1 \wedge \neg H_2 \vee \neg H_3) \wedge \neg(\mathrm{ExePrim}D)) \rightarrow \mathrm{Exe}(\mathrm{Sek}D_2 \vee \mathrm{Sek}D_3)$.

Thus, the second scenario also applies to complex procedures where the subprocedure that follows after the *else* connective is not a sanction, but a secondary disposition for the addressee — e.g., a less preferable disposition. Examples of such provisions are frequent:

Section 154 **If** another legal person is a member of an elected body of a legal person, [**then**][7] the former shall authorise a natural person to represent it in the body; otherwise [**else**], the legal person is represented by a member of its governing body.[8]

This secondary disposition may contain alternative legal regulations. The addressee then chooses only one of them and can subsequently execute it.

In computer science, logic and mathematics, the *if-then-else* function has abstract objects — numbers, concepts, procedures, etc. - as its main operands. The executor is an impersonal program that has no free will and never violates rules. In law, both abstract meanings and empirical facts, events, legal regulations and human actions in space-time are in play. The executors are people or institutions that can break the rules. The analysis presented here assumes that propositional logic or logics such as the system of Transparent Intensional Logic are applicable to legal regulations. However, this does not come naturally to many.

Challenging the applicability of logic in normative disciplines

There were no serious doubts about the possibility of an effective application of logic in law and other normative disciplines until the

---

[7]The word *"then"* is usually omitted from the *if-then* expression, and the condition is separated from the legal regulation simply by the comma.

[8]§154 of the Czech Civil Code https://www.refworld.org/docid/5da57dd04.html.

1920s. In legal theory as the paradigmatic normative discipline, the
role of logic was not questioned at all. What was up for debate was
only the degree of importance of logic to law. European law schools of
the time still produced textbooks in logic which were dominated by the
traditional approach — focusing on critical thinking and the theory
of argumentation and building upon the foundations of legal methodology laid down by the German expert on Roman law, Friedrich Carl
von Savigny.[9]

A turning point was the penetration of logical positivism into the
circle of researchers of the application of logic in law: Jørgen Jørgensen, Walter Dubislav, and Alfred Ross. The use of the principle
of verification as the sole criterion for identifying scientific knowledge
(apart from formal truths) led to a questioning of the very possibility
of a scientific approach to commands as a kind of legal regulations,
and thus to legal norms in general.

Since legal practice as well as everyday life involve reasoning from
assumptions of *ought* (*sollen*), explanations were sought as to how such
inference is possible. In his 1937 article, the German logician Walter
Dubislav proposed a clever solution — the idea of *parallelism*. The
relation of entailment between two obligations holds if it holds between
their roots, i.e., their descriptive parts. But how can *Dubislav's trick*
be used to deal with a situation where the premises are mixed?

The Danish logician Jørgen Jørgensen [19] showed that there is a
conflict between the generally accepted definition of (logical) inference
and the common practice of reasoning using imperative sentences.
The crux of the contradiction lay in the question of the truth-value
of normative sentences. In logic, entailment is defined as a relation
from sets of propositions to propositions, while an inference is a function from sets of premises to conclusions that transfers truth from
the premises to the conclusion. In inferences where the some of the
premises or the conclusion express commands or prohibitions in the
form of imperatives, one cannot speak of entailment, since imperatives
cannot be true or false. Hence, such inferences, even if they seem intuitively correct, would not by definition be correct. None of the

---

[9]Savigny [29, 213ff].

solutions to the so-called *Jørgensen dilemma* proposed by Jørgensen himself were accepted.[10]

Knud Grue-Sørensen [16] responded to Jørgensen's article by defending the idea of a science-based ethics. He argued that although propositional sentences do not have a truth-value in the ordinary sense, they nevertheless have "logoid" values (ethical validity or invalidity). In reasoning, the latter play the role of truth-values. This view was consistent with the intuitive use of logic in normative practice.

Deontic logic — the mainstream

Due to space constraints, I cannot discuss the development of mainstream deontic logic initiated by von Wright (1951) in any detail. Let me note, though, that the original objects of commands were actions. The consensus that the object of commands are states-of-affairs (empirical facts) was only established later, under the influence of Arthur Prior's critique. Deontic modalities thus became sentence operators [34]. This reversal, made almost in unison by logicians, was later viewed by Peter Geach [13, 35] as a fatal move. Nicholas Rescher, already in his *Logic of Commands* [26, 7], argued that in the field of deontic logic, there was in fact no single point on which a lasting consensus had been reached. He himself insisted that the objects of commands are actions.

Social facts as a new kind of facts

Let us now turn back more than a hundred years. For a long time, normativity had been studied by sociologists independently of logicians, using their own methods. At the end of the 19th century, Émile Durkheim introduced a new kind of facts: *social facts*. His sociological investigations were based on a positivist theory of science. Man lives not only in a natural environment, but also — and especially — in a society that acts upon him and enforces certain kinds of behavior. When I perform my duties as a brother, husband or citizen

---

[10]Gahér [12, 32].

and honor the commitments I have entered into, I fulfil obligations
defined by law and custom that are external to myself and my actions
[6, 50]. When Durkheim characterizes social facts in more detail, he
emphasizes, on the one hand, their specific nature: they are the beliefs, inclinations and habits of a group. That is, they are not tangible
or directly observable. On the other hand, Durkheim describes their
function in motivating social action — the realization of such facts:
"What constitutes social facts are the beliefs, tendencies and practices
of the group taken collectively." [6, 54].

Understood in this way, social facts are instructions (or abbreviations of instructions) for action combined with a motivating attitude
towards the execution of these instructions. In the resulting definition
of a social fact, Durkheim reiterates precisely this procedural aspect
of action which exerts external pressure on the individual. Here, it
is important to distinguish between the fact itself (as a way of acting) and the realization (execution) of this way of action itself: "A
social fact is any way of acting, whether ruled or not, capable of
exerting-over the individual an external constraint; or: which is general over the whole of a given society whilst having an existence of us
own, independent of its individual manifestations." [6, 59]. A specific
characteristic of social facts is that they exert pressure on or *coerce*
individuals. They are objective, independent of individuals, though
they do depend on society. An individual cannot change social facts
by personal decision. In Durkheim, a new kind of facts invaded the
field of scientific inquiry. However, logical empiricists disregarded social facts, and some researchers refuse to this day to acknowledge it
as a realm of facts *sui generis*.

Institutional social facts versus brute facts

In the tradition of analytic philosophy and the theory of meaning,
John Searle has paid particular attention to social facts. He proposed
the distinction between brute facts which do not depend in any way on
social reality and mental facts (e.g., "I am in pain") which, although
ontologically subjective, are nevertheless epistemically as objective as
brute facts. A brute fact, for example, is that the object in front

33

of me is composed of wood and metal. In relation to the intention (intentionality) of the observer, however, it is a screwdriver. On the basis of an intentionalist vocabulary, Searle singles out among mental intentional facts those that are collectively intentional, and he labels these social facts [27, 26]. For example, two people going on a walk together is a social fact. A specific type of social facts are institutional facts. The latter presuppose social institutions. For example, the fact that the Sun is 150 million km away from the Earth is a brute fact because it does not depend on any social institutions. In contrast, the fact that this piece of paper is a 20 Euro banknote, or the fact that Bratislava is the capital of the Slovak Republic, are institutional social facts, since they require the existence of social institutions, including the institution of language [27, 27].

Unlike Durkheim, Searle does not explicitly state that institutional facts can act coercively on the actions of individuals. At first glance, their main role seems to be to objectively describe or declare ("institute") social reality or register the results of actions of individuals and institutions, rather than to reflect any stimulating instructions for action. In his later work, Searle does speak of social pressure: "...we are all subject to 'social pressures' " [28, 159]. However, the nature of this "subjection" is not that of an explicitly normative fact. He refers to this power as Background/Network power, i.e., "a type of power in society that is not codified, is seldom explicit, and may even be largely unconscious." [28, 155]. On the other hand, already in his earlier work, he highlighted the role of duties and obligations when speaking about the role of deontic powers in the creation of rights and obligations [27, 100]. Similarly, in his later work, "Deontic powers are rights, duties, obligations, authorizations, permissions, privileges, authority, and the like" [28, 164]. Searle's examples of performative declarations — e.g., "*The meeting is adjourned*", "*War is hereby declared*" [27, 34] — could, on a charitable reading, be viewed as the results of actions stimulated by imperative utterances by a legitimate authority: "*Adjourn the meeting!*" or "*Declare War!*" However, this is not found explicitly in Searle.

Normativity and normative facts

In today's scholarly circles, few would doubt the existence of social facts and specifically of normative social facts. Numerous papers at the intersection of law, ethics, and philosophy by, for example, Jeremy Randel Koons [20, 21], Mark Greenberg [14], Ram Neta [24], José Noguera [25], Jaap Hage [17, 18], and others confirm that the contemporary debate is not about whether there is such a thing as normative facts, but rather about their exact implications, as well as the question of their naturalistic reduction. To cite Koons's view, "First, it seems absurd to deny that terms of moral and epistemic appraisal have normative meaning." [21, 266]. This view is then re-emphasized:

"Normative facts are normative facts, and an attempt to reduce them without remainder to natural facts (as the attempt to reduce moral normativity to the ability to motivate) strips these normative facts of their normativity." [21, 267]. In the field of (meta)ethics, the situation seems to be similar.

A naturalistic account of normative facts was put forward by Stephen Turner in *Explaining the Normative*. Turner criticizes the version of normativism that views normative facts as in some sense special or transcendental,[11] beyond the realm of causal interactions:

"The normative, however, arises out of ordinary facts: meanings, obligations, rationality and so forth, come into existence through actions, learning, and the like, but have the special added properties of norms: of binding, constraining and the rest. Once the norms are established, they have consequences for behavior. They do not directly cause behavior, but they regulate it normatively, by specifying what is the right way to say something, what obligations one has, what one owes to others as a result of one's meaningful actions, and what is justified for others to do in response to your actions." [32, 9]

---

[11]The 'normativist' philosophers call 'normative facts' are not special or transcendental in any meaningful sense, but can be fully grasped by social science standard explanations

According to Turner, normative facts can be fully grasped by standard social science explanations and are naturalistically reducible. He admits that norms do not directly (causally) cause behavior, but regulate it "normatively". What is meant by this "normative regulating", however, is not entirely clear.

Bicchieri's widely discussed book *The Grammar of Society* (2006) argues that "social norms are not internalized generic simple imperatives such as 'reciprocate,' 'commit,' or 'cooperate' " [2, 79]. The work does not discuss any possible relation between norms and imperatives. In Explaining Norms, Geoffrey Brennan, Lina Eriksson, Robert E. Goodin, and Nicholas Southwood reject some competing approaches to norms, such as the explanation of norms as a kind of practices, desires or wishes. Their own proposal is to explicate norms as "clusters of normative attitudes plus knowledge of those attitudes" [3, 15; 28ff]. The knowledge of attitudes is indeed an essential conceptual feature of norms.

Normative conative facts as distributions of legal regulations onto individuals

A possible worlds semantics assumes that different facts are different **distributions of empirical properties and relations on individuals** (elements of the universe of discourse) and a possible world is any maximal logically conceivable (consistent) set of facts. A possible world or state of affairs may change in time, i.e., the distribution of properties onto individuals may be different at different points in time. We will call the temporal sequence of states of affairs (maximal logically conceivable sets of facts) a world-history. This enables a thinking of possible worlds as temporally determined (see Tichý [31, 177 ff], Duží, Jespersen and Materna [8, 212 ff].

In some worlds, one and the same individual is a young boy. In others, he is a grown man, and in still others a bald man. Since all of these possibilities are consistent with biological human nature, this sequence may represent the history of the actual world. Conversely, in some worlds, the same individual may be white, while in others he is black and in still others he is brown. Since such transformations are unlikely vis-à-vis biological human nature, these worlds would be best thought of not as a history of the actual world, but rather as several alternative worlds.

Since humans are social creatures, his various social statuses, determined by their position in society, participate in the specification of institutional social facts. For example, in some worlds, the same individual is a taxi driver, in others an actor, and in still others the president of a republic. In some cases, such may be the history of the actual world, since a temporal succession of such social roles is realistically possible. In other cases, it may be a history of some possible worlds or the various histories of several alternative worlds. Physical apparatuses used in the study of the natural world are not sufficient to verify institutional social facts. Tests are required which can detect the state of society and the parameters of its bearers. Official records, archives and statistical databases are replete with evidence for such tests or with results of such tests. All this is made possible by the extension of the conceptual superstructure or intensional base. Besides "natural" properties and relations, we study, as a matter of course, the **distribution of *social roles and social relations onto individuals*.** Despite this difference in the methods of testing, institutional social facts can be said to be true or false, even though no physical apparatus can directly verify them. What is of utmost importance for our present analysis, however, is that we can easily use such facts in making inferences.

Over time, people age and change their social roles. More importantly in our context, their duties and responsibilities also change, based on the legal system of the society in which they live. In one state of affairs, a person has the obligation to pay the property tax within a certain time frame (if they are the owner of some property), they have an obligation to pay the income tax within another time frame, they have to pay health and social insurance contributions (if they had had any income), they have to pay a fine (if they were speeding), they can vote in elections (if they fulfil the conditions for passive suffrage), they can stand for elections (if they fulfill the conditions for active suffrage) and so on. The set of all these actual duties and rights, enshrined in a given society's legal system, cannot even be enumerated — it is an extensive and changing list.

To fulfil a legal regulation that applies to an agent (i.e., one that is objectively valid for the agent), the regulation must become part of the agent's beliefs. The agent should know about it and accept it, i.e., adopt it as subjectively valid. Then, the regulation can have a stimulating effect on the agent's actions. If the agent fulfils an effective legal regulation, it ceases to act coercively on them and ceased to be effective. The agent may, knowingly or due to omission or negligence, fail to comply with or violate it. This may result in the creation of new obligations or penalties for the agent.

The relation between a certain obligation and the acting on it (its realization) is not causal. The naturalistic explanation is not very convincing. Noguera, for example, believes that normative facts, as a kind of "propositional mental states", can be investigated using empirical methods:

> "...social norms, conventions, normative beliefs, ethical or political beliefs and attitudes, and the like, are considered as propositional mental states individuals may hold ('beliefs' and 'desires'), whose causes and effects may in principle be investigated empirically. As such, normative beliefs do not pose more epistemological or ontological problems than factual beliefs or other cognitive mental states." [25, 203]

In my view, propositions and propositional procedures are objective semantic entities that are not themselves mental states. Only their public disclosure and understanding by the addressee can have a stimulating effect on the latter.

Beyond the normatively enshrined duties, rights, etc., which are effective and which have been understood and accepted (adopted as one's beliefs) by the agent, morally justified rules of behavior — which need not be codified or explicitly established — also exercise a stimulating effect on human action. Of course, the most frequent stimuli for action range from the basic bodily (physiological) and social needs

to the higher-level needs — social, spiritual, or aesthetic. The latter
kinds of stimuli exhibit varying, but often non-negligible, degrees of
subjectivity. Hence, a complete, objective register of such stimuli is
highly questionable. On the other hand, normative facts that stimulate action have a high degree of objectivity and can be recorded.
In the following, I shall focus on the latter. I will refer to them as
normative conative facts or simply *conative facts*.

The vast majority of the provisions of legal norms of a given legal
system do not act as direct stimuli for a particular person. They only
contain potential legal regulations that only "kick in" when the condition for their activation is met. In this generality, the provisions of
legal norms resemble laws of nature. They do not specify the state of
affairs fully; instead, they specify the actual legal system. However,
each member of society is associated with a specific set of actual duties, permissions, prohibitions and authorizations. This universe of
discourse includes not only individuals as natural persons, but also
legal persons as defined by law. Associations of natural persons are a
typical example of these.[12] The various **distributions of *legal regulations* onto individuals** are the various **normative conative
facts**.

The knowledge or awareness of an effective legal regulation by its
addressee is necessary condition for its acceptance and for acting in
compliance with it. An agent who forgets that he or she has to pay
the tax within a given period of time fails to fulfil the obligation and
may be sanctioned by the tax authority. Since the age of Roman
law, it has been assumed that ignorance of the law cannot excuse its
violation or non-compliance (*ignorantia juris non excusat*). This is the
so-called irrefutable presumption which establishes objective liability
of the addressee of the norm for acting or not acting in accordance
with the norm.[13] Subjective liability of the addressee presupposes an

---

[12]In a ramified type theory, such as that of TIL, legal entities can be easy
defined as objects of a different type than individuals — e.g., as functions that
assign truth values to sets of individuals.

[13]Act on the Collection of Laws of the Slovak Republic No.1/1992 §2: "Anything
published in the Collection of Laws shall be presumed to have become known to
everyone concerned on the date of publication."

understanding and awareness of the actual legal regulation and its subsequent acceptance or rejection. In this sense, the fulfilment of the condition that activates the legal regulation, together with the acceptance of this regulation, should belong to the set of beliefs of the addressee — the agent. We can indirectly infer whether the regulation in question does indeed belong to that set of beliefs based on, for example, the agent's actions or testimonies.

The introduction of of institutional social facts and normative conative facts requires a change in the basic definition of a possible world: a possible world is any *maximal logically consistent set of empirical,* institutional *social, and normative conative facts.*[14] Of course, the epistemic framework and conceptual system of such a theory must be extended not only to include the concepts of social role, valid and effective legal regulations, and of the incorporation of legal regulations into the sets of beliefs of their addressees.

The difference between a statement and a command: is it (also) semantic or purely pragmatic?

By showing that distributions of legal regulations onto individuals are a kind of facts —namely, normative conative facts — we have overcome the seemingly insurmountable obstacle that was supposed to have prevented logic from operating on them as it does on other kinds of facts. The fact that a person is obliged to pay taxes within a given period is true or false. Hence, a sentence which states this obligation and whose authority (that of a competent institution etc.) is implied, e.g.,

(K1) XY has to pay z amount of taxes within period t,

---

[14]Stones, trees or animals have no duties, but their properties co-determine the particular world. However, some animals may acquire certain rights and protections, as can the environment in general (including trees as it) or cultural or natural monuments. In many legal systems, cruelty toward some animal species is criminal. https://www.najpravo.sk/clanky/novela-trestneho-zakona-ucinna-od-1-novembra-2020-prinasa.html?print=1

can be a regulative component of an inference (e.g., its conclusion) in which the formulation of a conditional legal regulation acts as one premise. However, note that the same applies to a sentence which expresses this obligation in an explicitly stimulating way and which is uttered by the same authority, e.g.:

(I1) XY, pay the amount of z within period t!

Once we accept the nature of normative conative facts, we can tackle the obstacle that prevents us from associating truth values with the semantic meaning of the imperative (I1). To truly overcome that obstacle, we must ignore the misleading *prima facie* impression produced by the pragmatic superstructure. That impression is as follows: the imperative (I1) carries a syntactically and formally accentuated *pragmatic charge* responsible for stimulating action, whereas the declarative statement (K1) does not. However, from the point of view of an objective register of all obligations, the two statements represent one and the same item, rather than two different items.

From the point of view of the addressee (XY), the only difference between the two is that in (K1), the authority issuing the command is implicit. Hence the declarative mode of the sentence. On the other hand, in the imperative formulation (I1), the speaker plays the role of the authority and the presence of the addressee is simply assumed. Semantically, there is no difference for the addressee between the meaning (K1) and (I1). The obligation, i.e., the normative conative fact, remains the same. We can attribute truth-values to the proposition that this particular obligation belongs to the set of actual obligations of person XY. If it does, then the presupposition (I1), uttered by a legitimate authority, expresses a true conative fact, just like the statement (K1). If it does not, then neither the statement nor the imperative are true — the speaker has uttered either of them with no justification.

However, the truth of a sentence, which expresses conative fact must not be confused with the reality of its fulfilment. A sentence, which expresses conative fact, can be true if it is (or will be) fulfilled, but it can also be true if it is not fulfilled and never will be. From this perspective, replacing imperatives by descriptions of their fulfilment or their descriptive roots (the so-called *Dubislav's trick*) is a categorial mistake. When a conative fact is fulfilled — insofar as it stimulates action that immediately achieves the goal — the fact ceases to be an actual conative fact and hence ceases to be an item on the list of actual obligations.

Paradoxically, it was not until the beginning of the 20th century that this intuition, which had been viewed as self-evident for millennia, was fundamentally challenged. However, the challengers failed to offer a sufficiently justified explanation of how logic would then be applied in law. Quite the contrary: at the heart of this challenge was a call for the construction of a *special* logic of commands, prohibitions, permissions, and empowerments. Positions stating that "we do not need a special 'logic' of imperative statement, 'logic' of perfomative statement, and so on, as logic over and beyond, or basically different from standard logic of propositions" [11, 40], or that "The need for a special logic of questions, it will be argued, is not greater than the need for special logic of beliefs, for a special logic of conjectures, of wishes, prayers, prejudices, promises, or insults." [30, 275], were ignored. This despite that fact that the special logics offered to legal theorists and practitioners were, and still are, of no real help. In the logical empiricists' laudable effort to purge philosophy and scientific knowledge of speculation and, later, of subjectivism, even some viable shoots were removed from the tree of knowledge. The damage done was to be repaired later — by the ever more elaborate systems of logics with no real applications in law. Deontic logicians wanted to solve the problem by introducing extremely sophisticated systems of deontic logics. However, most legal theorists, and perhaps all practical lawyers, ignored such efforts. In practice, they were able to apply, mostly correctly, intuitive concretization rules of *modus ponens* and *modus tollens*, where the conclusions were normative conative facts.[15]

Written systems of law have long been founded on the convention that legal regulations should not be expressed using explicit imperatives, but by sentences in the indicative mode. On the one hand, this

---

[15]See [12, 27 ff] for more details.

allows the seamless pairing of sentences expressing empirical facts or institutional social facts (in the roles of activation conditions for legal regulations) with sentences expressing conative facts (the legal regulations themselves). On the other hand, when reading the provisions of legal norms, one must carefully identify the syntactic-semantic signals indicating conative facts. The clearest indicators of this kind are *modal verbs* (*shall, may, shall, shall not*, etc.), but other forms of verbs can also be used:

(K2)   The President of the Slovak Republic appoints and dismisses the Prime Minister and other members of the Government of the Slovak Republic.[16]

(K3)   Upon a proposal by the Prime Minister, the President of the Slovak Republic appoints and dismisses other members of the Government and entrusts them with the management of ministries.[17]

Restrictions on conclusions inferred from explicit conative facts are compensated by praxeological principles

One lesson from the famous Ross's paradox [23] leads to an important restriction. From commands, one can only seamlessly logically derive their explicit constituents. For example, there are no good logical reasons for disjunctively connecting an arbitrary command with another command. From the command "Put the letter in the mailbox!", Peter cannot logically infer the disjunction "Put a letter in the mailbox or burn it!". Similarly, one can distinguish between strong (explicit) and weak (implicit) permissions. From a simple explicit permission, we cannot infer a disjunction containing the original permission and another, arbitrary permission. Normative attitudes expressed by commands, prohibitions, etc., thus resemble explicit propositional attitudes and indicate a hyperintensional context.[18]

---

[16]Constitution of the Slovak Republic, Article 102, letter g).
[17]Constitution of the Slovak Republic, Article 111.
[18]For a more detailed discussion of hyperintensionality in deontic logic see, e.g., [10] and of propositional attitudes, see, e.g., [7, 421 ff].

When a command is being fulfilled, the activated conative fact had stimulated the addressee to act in a certain way. However, the conative fact itself does not contain any detailed instructions for carrying out all the subsequent steps of the procedure. Often, it merely expresses the goal, the desired result of the action. With an uncooperative addressee, such a command might not be enforceable. The over-restriction of the logical consequences of such explicit facts is compensated for by *praxeological principles* that are inseparable from practical normative reasoning. Otherwise, a "Good Soldier Svejk" phenomenon could occur: the addressee of a command would not do anything that he or she had not been explicitly ordered to do. If someone is ordered to open the window, we assume that they will take all the steps necessary for achieving that goal. This paves the way to *implicit attitudes*.

Logic alone is not sufficient for specifying command-based action. An *analytic theory of action* capable of addressing the enforceability of commands, prohibitions, and permissions is also needed. If we restrict inference in normative disciplines to deductive rules that only allow reasoning from explicitly stated legal regulations, the set of allowed inferences becomes too limited, jeopardizing the executability of commands.

In light of this requirement of supplementing the logic of declarative sentences with what is specific to the *practice* of executing imperatives, we might understand Nuel Belnap's advocacy of the autonomy of the meanings of imperatives vis-à-vis the meanings of descriptive sentences. His criticism stating that *the logical grammar of imperatives* is still in its infancy [1, 17], as well as his doubts about us ever learning anything from how an imperative can figure as a premise or from some Gentzen rule of disjunction for imperatives, can be viewed as a demand for supplementing logic with praxeological principles.

Action is a practical activity. It should therefore be regulated so that it is effective and achieves its objectives. It is clear that in such a theory of action, certain "rational" (in terms of practical reason) principles or meta-principles will be valid, which we shall formulate for the specific case of commands:

44

1. The principle of feasibility: The target state of the regulation *realistically (practically) attainable* by actions of the subject (addressee). The desired state is realistic and the agent is capable of achieving it by their own action within a given time frame and using the given means. A command that would require sunsets to cease is not practically meaningful, although the requirement is logically conceivable. Of course, the real feasibility of the target state depends on technological development. What had not been feasible just a few years ago may be commonplace today.

2. The principle of *doing all that is necessary* to fulfill the command: Do whatever is necessary to fulfill the command, without violating any other — equivalent or stronger — command! The addressee of the command is simply asked to perform all the *prerequisite activities* that, although not explicitly stated, are necessary to achieve the target state described in the command.

3. The principle of *abstaining from doing anything that would hinder* the execution of the command: Do nothing that would hinder the execution of the command, while not violating another (equivalent or stronger) command! The *auxiliary activities* of the subject *must not impede* the execution of the explicit regulation. If one is commanded to throw a letter in the mailbox, one must not burn it first.

Such praxeological "rationality" requirements implicitly presuppose a "rational" addressee of the command. Every legal regulation implicitly presupposes — and some even state this explicitly — that the addressee of the command (or prohibition, permission) is a human being endowed with average rationality.

The above list of praxeological assumptions or principles is by no means exhaustive. Since the particular situations in the application of legal regulations are determined by a myriad of circumstances, their concrete execution always depends on these contingent conditions. Hence, it is not fully predictable and describable in advance. We simply have to assume a rational subject as the executor of the command (prohibition, permission) and a certain level of not only

practical rationality but also charity in interpreting the generally formulated regulations.

Conclusion

In this paper, I have presented an explanation of the second, alternative meaning of the 'if-then-else' function in law, which contrasts with
its usual meaning in computer science. In this second meaning, the
distinction between the components that the operator binds comes to
the fore. On the one hand, these components are brute facts in the role
of empirical conditions; on the other hand, they are legal regulations
that are not commonly viewed as being true or false. The followers
of logical empiricism in legal theory considered this as a fundamental
obstacle to the application of classical logic in the domain of norms. I
propose a conservative solution to the problem with no recourse to a
special deontic logic. The solution is based on extending the realm of
facts with two new classes: a) the domain of institutional social facts;
b) the domain of normative conative facts. Another element of the
solution is the supplementation of the theoretical framework with the
concepts of social roles and legal regulations. As items of the actual
set of obligations of a given agent, normative conative facts are true.
They seamlessly fit into inferences that can be justifiably operated on
by logic, as demonstrated in everyday legal practice. Because normative conative facts stimulate action with a particular goal in mind,
they are best viewed as explicit attitudes of a hyperintensional nature.
When such attitudes are effectively executed, the constraints imposed
by their hyperintensional nature are compensated for by compliance
with praxeological principles and rules.

References

[1] Belnap, N. (1990): Declarative Are Not Enough.*Philosophical studies*,
vol. 59, 1 - 30.

[2] Bicchieri, C. (2006) *The Grammar of Society: The Nature and Dynamics of Social Norms.* Cambridge, IX.

[3] Brennan, G., Erikson, L., Goodin, R., and Southwood, N. (2013): *Explaining Norms*. Oxford University Press, Oxford.

[4] Carnap, R. (1935): *Philosophy and Logical Syntax*, London: Kegan Paul, Trench, Trubner & Co.

[5] Dubislav, W. (1938): Zur Unbegründbarkeit der Forderungsatze. *Theoria* 3, 330 - 342.

[6] Durkheim, É. (1982): *The Rules of Sociological Method*. New York: The Free Press.

[7] Duží, M., Jespersen, B., and Materna, P. (2010). *Procedural Semantics for Hyperintensional Logic:Foundations and Applications of Transparent Intensional Logic*. Dordrecht: Springer.

[8] Duží, M. and Materna P. (2012): TIL jako procedurálni logika. aleph, Bratislava.

[9] Duží, M. and Kosterec, M. (2017): A Valid Rule of β-conversion for the Logic of Partial Functions. *Organon* F 24 (1), 10 - 36.

[10] Faroldi, F. (2019): Deontic Modals and Hyperintensionality, *Logic Journal of the IGPL*, vol. 27, 387 - 410.

[11] Fitch, F. (1971): On Kinds of Utterance. in Kurtz, Paul: Language and Human Nature, 1971, St. Louis, Warren H. Green, 39 - 40.

[12] Gahér, F. (2014). Aká logika by sa mohla používa' v práve? (What kind of logic could be used in law?) *Organon* F 20 (Suppl.) 20-41

[13] Geach, P. (1991): Whatever Happened to Deontic Logic? In. Geach, P. (ed.). Dordrecht Boston London: Kluwer Academic Publishers.

[14] Greenberg, M. (2004): *How Facts Make Law*. Legal Theory, vol. 10, Cambridge University Press, 157-198.

[15] Greenberg, M. (2006): Hartian Positivism and Normative Facts: How Facts Make Law II. The Social Science Research Network Electronic Paper Collection: http://ssrn.com/abstract=887418

[16] Grue-Sørensen, K. (1939): Imperativsätze und Logik. Begegnung einer Kritik. *Theoria*, 5, 195 - 202.

[17] Hage, J. (2016): Facts and meaning. in Jerzy Stelmach, Bartosz Brozek and Lukasz Kurek (eds.), The Emergence of Normative Orders. Krakow, Copernicus Center, 13 - 42.

[18] Hage, J. (2016b): Facts, Values and Norms. In: Sanne Taekema, Bart van Klink and Wouter de Been (eds.), *Facts and Norms in Law*. Cheltenham, UK - Northampton, MA, USA, Edward Elgar 23 - 49.

[19] Jørgensen, J. (1937/38): Imperatives and Logic. *Erkenntnis*, 7, 288 -

296:

[20] Koons, J. (2000): Do Normative Facts Need to Explain? *Pacific Philosophical Quarterly* 81, 246 - 272.

[21] Koons, J. (2006): An Argument Against Reduction in Morality and Epistemology. *Philosophical Investigations* 29, 3, 256 - 274.

[22] Kurtz, Paul (1971): *Language and Human Nature*, St. Louis, Warren H. Green.

[23] Ross, A. (1944): Imperatives and Logic.*Philosophy of Science*, Vol. 11, No. 1, 30 - 46.

[24] Neta, R. (2004): On the Normative Significance of Brute Facts. In: *Legal Theory*, 10, 199 - 214.

[25] Noguera, J. A. (2013): Social Science and 'Normative Facts.' What's the big deal? RIS, VOL.71. No. 1, ENERO-ABRIL, 200 - 212, ISSN: 0034-9712. DOI: 10.3989.

[26] Rescher, N. (1966): *The Logic of Commands*. London: Routlage & Kegan Paul.

[27] Searl, J. (1995): *The Construction of Social Reality*. New York: The Free Press Simon and Schuster.

[28] Searle, J. (2010): *Making the Social World: The Structure of Human Civilization*. New York: Oxford University Press.

[29] Savigny, C. F. von (2004): *Vorlesungen über juristische Methodologie 1802 - 1842*. Mazzacane, A. (Hrsg.). Franfurkt am Main: Vittorio Klostermann.

[30] Tichý, P. (1978): Questions, Answer, and Logic. In: *American Philosophical Quarterly*, Vol. 15, No. 4., 275 - 284.

[31] Tichý, P. (1988): *The Foundations of Frege's Logic*. Berlin — New York, W. de Gruyter.. Berlin - New York.

[32] Turner, S. P. (2010). *Explaining the Normative*. Cambridge: Polity Press.

[33] Wright, G. H. (1951): Deontic Logic: *Mind*, 60, No. 237, 1-15.

[34] Wright, G. H. (1963): *Norm and Action*, Routledge & Kegan Paul, London.

HYPERINTENSIONALITY OF DEONTIC MODALS AND THE PROCEDURAL THEORY OF INFORMATION

DOMINIK SADLOŇ AND DANIELA VACEK
Department of Logic and Methodology of Sciences, Faculty of Arts,
Comenius University in Bratislava, Slovakia
dominik.sadlon@gmail.com, daniela.vacek@gmail.com

Abstract

The paper explores hyperintensionality of deontic modals in a novel way. In particular, the paper seeks to explain and motivate hyperintensionality of deontic modals in terms of a procedural theory of information. Such a theory was suggested by Marie Duží in her [7] article 'The Paradox of Inference and the Non-Triviality of Analytic Information' as a solution of the paradox of inference, but also as a semantic theory of information. It distinguishes empirical information from analytic information. The present paper applies these two notions in the context of implicit and explicit deontic modalities and examines their potential for the debate on hyperintensionality in deontic logic.

Keywords: analytic and empirical information, deontic logic, failure of substitution, hyperintensionality, procedural theory of information.

1 Introduction

Our agent, Emma, once took part in a geocaching game (an outdoor activity that consists mostly in searching for "treasures" hidden in various places marked by coordinates). This particular treasure was hard to find. To get the coordinates, she had to walk through different spots in her hometown and count various things. She followed the instructions to the letter, from the first line to the last. She was surprised to find out that the next task was to count the number of bricks on an impossibly large brick wall. She observed a pattern and divided the wall into parts of the same shape and size. Serious counting then ensued. She arrived at the enormous number and was confident in its correctness; after a while, she had all the numbers needed to calculate the final coordinates. Curiously, however, the final formula had no variable for the number of bricks. You can imagine her astonishment. But things like this happen all the time. A more straightforward example is described by Fagin, Halpern, Moses and Vardi (see [10, p. 364]):

> A computer program that can determine in time T whether $\phi \vee \psi$ follows from some initial premises might not be able to determine in time T whether $\psi \vee \phi$ follows from those premises. (...) And people do *not* necessarily identify formulas such as $\phi \vee \psi$ and $\psi \vee \phi$. The reader can validate the idea that the order matters by computing the product $1 \times 2 \times 3 \times 4 \times 5 \times 6 \times 7 \times 8 \times 9 \times 0$.

It is true that, from the point of view of a logically omniscient being, the order is irrelevant: for instance, if such a being knows $A \vee B$, then such a being also knows $B \vee A$. Ordinary agents have their limits, however, whether the agent is a computer program with its predictable boundaries or a human being reasoning unpredictably as well as they can. In light of the well-known observations briefly highlighted above, it seems that we cannot always substitute a formula ϕ for a formula ψ in the context of a knowledge operator, even if the two are provably equivalent and differ only to the extent that their subformulas appear in a different order.[1] This is a clear indication that the phenomenon of epistemic attitudes is hyperintensional.

---

[1] In the following sections, which focus on deontic logic construed as a (classical) propositional modal logic, the embedded formulas are provably equivalent in the classical propositional logic; that is, they are classically equivalent, and in turn,

Some have argued that deontic modals are hyperintensional in this sense and have offered various arguments in favour of this claim (see, for instance, [1, 12, 13, 17, 18, 21]), some disagree (see, for instance, [30, 31]). The present paper will provide a broad range of direct arguments for the hyperintensionality of deontic modals. Direct arguments for the hyperintensionality of a given context are standardly formulated in terms of examples in which two formulas are provably equivalent but one cannot be substituted for another within that context. In particular, a modal operator is hyperintensional if and only if we cannot substitute one provably equivalent formula for another within the scope of that operator. Subsequently, to show that deontic modals, such as the operator of obligation, O, or an operator of permission, P, are hyperintensional, we need to find examples where we cannot substitute one provably equivalent formula for another within the scope of a deontic operator. Indeed, this strategy will be the starting point of the present paper. Note that much depends on the default reading of the given deontic operator. For instance, one can read O as an obligation given explicitly by some normative authority. Alternatively, one can read it as what obligations and permissions follow from those explicitly given by a normative authority. The former intuitively corresponds to explicit deontic modality, the latter to implicit deontic modality.

Those who, by default, read the given deontic operator in the explicit sense might deem it nearly trivial that there are persuasive examples showing the failure of substitution of provable equivalents in the deontic context. In comparison, those who are inclined to read the given deontic operator in the implicit sense might deem our examples unsatisfactory. Of course, we can substitute a formula ϕ for a formula ψ within the scope of a deontic modal in a system of deontic logic if the two are provably equivalent and the rule of substitution of provably equivalent formulas holds. But in other sense we cannot do so as soon as there are good reasons for saying that $O\phi$ is not equivalent

---

they are provably equivalent in the given system of classical modal logic, such as standard deontic logic (SDL). For the ease of explanation, as well as for generality, we will omit references to specific systems.

to $O\psi$. Of course, there are often conflicting intuitions with respect to the seriousness of various counterexamples; we do not assume that everyone will share our intuitions. Nonetheless, we take the absence of consensus on the question of hyperintensionality of deontic modalities as a crucial motivation for diving into the matter once more. The present paper aims to explore an unusual combination of two research areas: to examine the hyperintensionality of deontic modals in terms of a semantic theory of information, in particular, a procedural theory of information developed by Duží [7]. Doing so, we aim to sketch a systematic explanation of the hyperintensionality of deontic modals in informational terms.

2 Hyperintensionality

What is hyperintensionality? There are both positive and negative specifications of this notion. No consensus has been reached on how to specify hyperintensionality positively: different hyperintensional theories come with their own specifications. Hyperintensional frameworks include impossible world approaches, syntactic approaches and primitive hyperintensions, a version of truth-maker semantics, various structural and quasi-structural approaches, such as procedural semantics or syntax tree approaches, and more (see, for instance, [20, 10, 28, 14, 32, 9, 24]). In general, positive specifications are usually provided in terms of hyperintensions, which are fine-grained entities, as opposed to coarse-grained intensions and extensions. Beware of a possible confusion: as Jespersen and Duží [22] note, the term *intensional* is sometimes employed to describe what is in fact *hyperintensional*.

The consensus is much broader with respect to the *negative specification of hyperintensionality*, even though this consensus is not universal. One possible negative specification of hyperintensionality is in terms of *necessity*: a phenomenon X is hyperintensional if the substitution of necessary equivalents within the scope of X fails. One often encounters such a specification in analytic metaphysics (see, for instance, [27, p. 151]). Another such specification is in terms of *syn-*

onymy: a phenomenon X is hyperintensional if the substitution of synonyms within the scope of X fails. The latter specification is often employed in the most fine-grained contexts, such as direct quotation and poetry (see, for instance, [25, p. 195]).

We will presuppose the following broad understanding: a phenomenon X is hyperintensional if the substitution of provable equivalents within the scope of X fails. That is, X is hyperintensional if the substitution of a formula ϕ for a formula ψ that is provably equivalent to ϕ within the scope of X fails to guarantee the preservation of truth. Such failures have been identified in various domains, such as attitudes (as indicated above), meanings, definitions, inference, grounding, causality, and, more recently, truth in fiction, deontic modals, and responsibility.[2] While these failures are widespread, classical modal logics, which are habitually employed in the philosophical analysis, include the following rule, which supports the opposite intuition (that is, that such failures are impossible because the "contents" of formulas ϕ and ψ are indistinguishable from the perspective of a possible world semantics):

$$(RE) \quad \textit{If} \vdash \phi \equiv \psi, \textit{ then } \vdash \Box\phi \equiv \Box\psi$$

This rule allows us to substitute any provably equivalent formulas ϕ and ψ within the scope of a modal operator $\Box$ and thereby renders failures of such substitutions impossible, however problematic a given substitution may appear.

To begin with, the present paper will provide direct arguments for the hyperintensionality of deontic modals by arguing against the acceptability of (RE) in deontic logic. Recall that what one needs for a direct argument for hyperintensionality is to find two provably

---

[2]Apart from attitudes and deontic modals, various works discuss the hyperintensionality of the above phenomena. The hyperintensionality of meanings and inference is discussed (see, for instance, [32, 9]); on the hyperintensionality of definitions (see [4]); on the hyperintensionality of grounding, (see, for instance, [29]); on the hyperintensionality of causality and responsibility (see, for instance, [11, 19]); and finally, on the hyperintensionality of truth in fiction (see [2]). It is an open question which of these are representational in nature and which are non-representational (cf. Nolan 2014).

equivalent formulas ϕ and ψ such that it is not acceptable to say that formulas $O\phi$ and $O\psi$ are provably equivalent. Of course, $O\phi$ and $O\psi$ will still be provably equivalent in many systems of deontic logic. Nevertheless, as soon as one describes a plausible situation where $O\phi$ holds but $O\psi$ does not, this suffices for saying that they should not be provably equivalent (since there is a possible situation where they *are not* equivalent). Admittedly, however, this is not a criticism of a certain deontic logic where (RE) holds insofar as the logical system in question does not aim to cover situations such as the one described. For instance, several examples below employ logical truths and logical falsities. As soon as one assumes a system of deontic logic that is not conflict-tolerant (such as standard deontic logic, SDL), examples that assume normative conflicts cannot be used to put forth a conclusive criticism of these systems (even though, admittedly, normative conflicts are widespread in ordinary life, and thus their exclusion is already a theoretical vice). Similarly, if one denies that logical truths can be obligatory or permitted, examples that assume otherwise cannot be used to criticise such logics (note also that in this case it cannot be added that logical truths within the scope of deontic operators are widespread). In other words, we are aware of the limited nature of our examples as counterexamples to systems of deontic logic. Whether some example is a counterexample to a logical system is thus relative to its expressive power (can we express deontic operators with logical truths within their scope?) and its aims (does the logic aim to model normative conflicts nontrivially?). That being said, we believe that it is worth examining these examples in a general way.

3   The Failure of Substitution

The present section will focus on counterexamples to (RE) in the above sense. Here we would like to note that some of these examples may appear artificial or haphazard. However, those that are unnatural are nearly always those we deem a result of unacceptable substitutions — that is, if they look unnatural, it speaks in favour of our claims rather than against them.

To begin with, Faroldi [13] suggests the following counterexample to (RE): the proposition $O\phi$ expressed by the sentence *You ought to drive* is logically equivalent to the proposition $O(\phi \vee (\phi \wedge \psi))$ expressed by the sentence *You ought to drive or to drive and drink*, due to (RE). Despite this, the two obligations are (intuitively) different. While the former is a plausible obligation, the latter is apparently not. Now this is precisely the kind of scenario we are looking for: there might be a situation where you ought to drive (for instance, you are a chauffeur), but it is not the case that you ought to drive or to drive and drink. The latter suggests that one can choose between driving and driving and drinking, which is not the case in most countries (provided the drinking in question involves alcoholic beverages). Note that this free choice reading is invoked by natural languages rather than the logical connective of disjunction.[3]

A whole set of counterexamples to (RE) can be formulated in connection to tautologies and contradictions. Note that while tautologies within the scope of a deontic modal might appear futile, *SDL* has it that all classical tautologies are obligatory. Moreover, contradiction within the scope of a deontic modal might correspond to something that is genuinely obligatory. This happens when an agent faces a normative conflict of a certain kind (that is, when both ϕ and $\neg\phi$ are obligatory; we will return to the notion of normative conflicts below). Let us start with formulas containing contradictions within the scope of a deontic modal.

Formulas of the form $\phi \wedge \neg\phi$ are classically equivalent to formulas of the form $(\phi \wedge \neg\phi) \wedge \psi$. A contradiction can never be true, and in the same way, no conjunction that contains a contradiction as one of its conjuncts can ever be true. By (RE), $O(\phi \wedge \neg\phi)$ should be equivalent to $O((\phi \wedge \neg\phi) \wedge \psi)$. Before attempting to show that this is problematic, allow us to comment briefly on normative conflicts.

In its simplest form, a normative conflict is an unfortunate situation where both ϕ and $\neg\phi$ are obligatory. While it is common to

---

[3]As is well known, the free choice principle $O(\phi \vee \psi) \rightarrow (O\phi \wedge O\psi)$ is not a valid principle of SDL. Nevertheless, there are many logics that support free choice reasoning, as can be witnessed in the broad literature on free choice permission.

represent normative conflicts as $O\phi \wedge O\neg\phi$, we can also represent them in terms of $O(\phi \wedge \neg\phi)$ in the language of SDL.[4] A classic example of a normative conflict is as follows: Antigone ought to bury her brother Polyneices but also ought to leave him unburied. Both burying and not burying Polyneices are therefore obligatory for Antigone.

However, by the above inference, the obligation to bury and not to bury Polyneices is provably equivalent to the obligation to bury and not to bury Polyneices and to murder Creon, the ruler of the city of Thebes. Generally, if both ϕ and $\neg\phi$ are obligatory, then ϕ and $\neg\phi$ AND any other formula are all obligatory. This is counterintuitive, since facing a normative conflict does not entail also being obliged to do anything whatsoever.

In a similar vein, $(\phi \wedge \neg\phi) \vee \psi$ is equivalent to ψ. This is because both of these formulas are true if and only if ψ is true. By (RE), $O\psi$ is equivalent to $O((\phi \wedge \neg\phi) \vee \psi)$. But this means that the obligation to feed Thomas the cat is equivalent to the obligation to feed Thomas or to meet and not meet with a friend. This, again, is not very plausible.

Let us now turn to tautologies, such as the formula $\phi \vee \neg\phi$. The formula $\phi \vee \neg\phi$ is equivalent to $(\phi \vee \neg\phi) \vee \psi$. This is because a tautology or any disjunctive formula that contains a tautology as one of its disjuncts is always true. By (RE), the formula $O(\phi \vee \neg\phi)$ is equivalent to $O((\phi \vee \neg\phi) \vee \psi)$. So if one ought to drink or not drink, one ought to drink or not drink — or kill a politician (also not very plausible).

The formula ψ is equivalent to $(\phi \vee \neg\phi) \wedge \psi$. This is because both are true if and only if ψ is true. By (RE), the formula $O\psi$ is equivalent to $O((\phi \vee \neg\phi) \wedge \psi)$. So if one ought to pay taxes, one ought to pay taxes and take bribes from students or not take bribes from them. The two of these, once again, appear to be different obligations. The free choice reading suggests itself once again, and one is tempted to read the latter obligation as requiring that one pay taxes and choose between taking or not taking bribes from students.

---

[4]Normative conflict can be also understood more broadly, as involving a conflict between an obligation and a lack of it (this option, however, amounts to a contradiction).

We can also formulate a related example to that formulated by Faroldi: the formula ϕ is equivalent to the formula $\phi \wedge (\phi \vee \psi)$. This is because both formulas are true if and only if ϕ is true. So the formula $O\phi$ is equivalent to the formula $O(\phi \wedge (\phi \vee \psi))$. However, that would mean that the obligation to pay taxes is equivalent to the obligation to pay taxes and to pay taxes or dance naked in the streets. Here again, the free choice reading that is intuitively suggested by the natural language causes the most trouble.

Interesting examples can also be formulated in terms of paradoxes of material implication. For instance, it ought to be the case that if you drink, then if you drive, you drink, which should be equivalent to *it ought to be the case that you do not drink, or do not drink, or that you drink* (not to mention that, due to their tautological content, these are logically valid formulas).

In some cases, the free choice reading of a disjunctive modal formula was the main source of the worry. Other cases were connected to the notion of normative conflict (those containing contradictory propositions within the scope of a deontic modal). What is common to these examples is that the formulas ϕ and ψ are equivalent in classical propositional logic, but $O\phi$ for this or that reason appears to be a different, non-equivalent obligation than ψ when expressed in natural language.

4 Duží's Procedural Theory of Information

Let us now turn to Marie Duží's procedural theory of information (PTI), which employs the framework of Transparent intensional logic (TIL) in the context of semantic information (see, most importantly, [7]). We will use PTI in the upcoming section for a positive explanation of hyperintensional shift in deontic logic — a shift which is motivated by counterintuitive examples of the preceding section.

The original purpose in [7] has been to propose a solution to the well-known paradox of inference, but thanks to its explanatory power, it can expand way beyond the borders of this problem. Building on the work of Bar-Hillel and Carnap [3], who are recognized fathers of the classic theory of empirical semantic information, which leads to the paradox inference, Duží offers a solution to this problem based on distinguishing analytic information from empirical information within the TIL framework.

The empirical content of sentence S is defined as the set of possible worlds and times excluded by S. If P is the proposition denoted by a sentence S,[5] the empirical content of S, $EC(S)$, is the set of world-time pairs at which P does not take the truth-value T, because the proposition P either takes value F or is undefined at the given pair [7, p. 492]. In comparison, she defines analytic information conveyed by an expression E as the literal meaning of E. Meanings in TIL are constructions, which are formally defined abstract procedures (for a classical definition, see [32, pp. 56–65]). Any adequate analysis in TIL "contains constructions of all and only objects that receive explicit mention"; a definition of literal meanings adds "the demand that the extra-logical objects that receive mention by semantically simple subexpressions of E should be constructed by their Trivialisations" [7, p. 495].

Here we introduce necessary bits of terminology: Trivialisation is a simple construction ^0D that takes an object and returns the very same object. Trivialisation is crucial for capturing hyperintensionality in TIL: it allows us to mention a construction, which can be subsequently subject of an attitude. Variables, such as x, w, or t, are constructions that construct objects with respect to some valuation. Composition $[X\ Y_1 \ldots Y_n]$ is a construction that applies a function to some arguments and returns the value of this function on the given arguments (if there is such a value). Closure $[\lambda x_1 \ldots x_n Y]$ is a construction that construes a function by abstraction. Constructions have constituents: "Because procedures are inherently structured, they consist of one or multiple constituent subprocedures that are to be executed in order to arrive at the product (if any) produced by the respective procedure" [7, p. 481].

---

[5]In this study (following [7, pp. 492–493]) terms "empirical information" and "empirical information content" are used interchangeably.

The analytic content of an expression E is then defined as the set of constituents of the literal meaning of E [7, p. 498]. Consider the following construction which aims to literally analyse the sentence S "Antigone buries Polyneices" (for simplicity, we ignore the fictional context altogether, treating Antigone and Polyneices as individuals and burying as a binary relation between individuals):

$$\lambda w \lambda t [[[^0 Bury\, w]t]\, ^0 Antigone\, ^0 Polyneices]$$

In this example, the empirical content of S is the set of world-time pairs at which the proposition constructed by the above construction takes the value F (since partiality plays no role in this particular example, given the above simplification). The above construction is the analytic information conveyed, and the set of its constituents

$$\{\lambda w \lambda t [[[^0 Bury\, w]t]\, ^0 Antigone\, ^0 Polyneices],$$
$$[[[^0 Bury\, w]t]\, ^0 Antigone\, ^0 Polyneices],$$
$$[[^0 Bury\, w]t],\, ^0 Antigone,\, ^0 Polyneices,^0 Bury, w, t\}$$

is the corresponding analytic content.

The paradox of inference states that deductive inferences cannot be both valid and useful. If the conclusion of a deductive inference is not contained in the premises, it cannot be valid. On the other hand, if the conclusion is not different from the premises, it is useless. But the conclusion cannot be contained in the premises and also possess novelty; hence deductive inferences cannot be both valid and useful [5, p. 173]. Translated into informational terms, in the deductive inferences, as the empirical information (information about the state of the world) contained in the premises is the same as the empirical information contained in the conclusion, we do not gain any new empirical information in deductive reasoning. Hence deductive reasoning is useless. But is this really the case? Duží introduces the analytic information to show quite the opposite. Consider the following example, whereby one infers that Polyneices is dead via Modus Ponens from the premises that Antigone buries Polyneices and that if Antigone buries Polyneices, Polyneices is dead:

59

$$\lambda w \lambda t[[[^0 Bury\ w]t]\ ^0 Antigone\ ^0 Polyneices]$$
$$\lambda w \lambda t[^0 \supset [[[^0 Bury\ w]t]\ ^0 Antigone\ ^0 Polyneices]$$
$$[[[^0 Dead\ w]t]\ ^0 Polyneices]]$$
$$\therefore \lambda w \lambda t[[[^0 Dead\ w]t]\ ^0 Polyneices]$$

Even though the empirical information carried by the premises is the same as the empirical information in the conclusion (the conclusion does not exclude any different world-time pairs from those already excluded by the premises), their analytical information differs. In particular, the closure $\lambda w \lambda t[[[^0 Dead\ w]t]^0 Polyneices]$ is different from the literal meaning of both premises, so despite the appearances originally invoked by the paradox of inference, deductive reasoning yields new information — new analytic information. Other applications of analytic information vary. For instance, using the notion of procedural meaning, Duží explains the difference between the analytic and logical validity of sentences and arguments or formulates criteria to compare the information yield of analytically equivalent sentences. As a new direction of PTI research, our objective is to extend the use of PTI by the distinction between implicit and explicit deontic modalities in order to explain failures of substitution of provable equivalents within the scope of deontic modal operators. This is the task for the upcoming section.

5   A Sketch of a Proposal

We will employ a distinction between the analytic and the empirical semantic information content carried by sentences [3, 7] along with a distinction between implicit and explicit [6, 26, 8, 17] deontic operators. The two distinctions together appear to be crucial to a deeper understanding of the motivations for a hyperintensional shift in deontic logic. Explicit obligation (or permission) can be informally understood as an obligation (or permission) given (granted) explicitly by some normative authority. In comparison, implicit obligation (or permission) can be informally understood as an obligation (or permission) logically following from those explicitly given (granted)

by a normative authority. Explicit modalities have a higher priority over implicit modalities because they represent the input given
by the normative authority, or what the normative authority *really*
prescribes and allows. Unlike implicit modalities, which represent all
that logically follows from explicit obligations and permissions, explicit modalities are not prone to lead us in unintended directions,
which may contradict the design of the normative system.

Note that a procedural theory of information developed within
the framework of TIL has both of these distinctions in its conceptual
toolkit. We would also like to note that while the distinction between
implicit and explicit deontic operators already provides some explanatory work in the desired direction, we deem its combination with a
distinction between analytic and empirical semantic information content even more fruitful.

As an illustration, we will focus on one of our cases of a failure
of substitution discussed in Section 3 throughout the present section,
and provide an explanation of the case in terms of the above described
apparatus of the procedural theory of information (an explanation of
the other cases would be analogous). (*RE*)consider, for example, a
pair of deontic propositions $O(\phi \wedge \neg\phi)$ and $O((\phi \wedge \neg\phi) \wedge \psi)$, which
read as follows (if ϕ stands for a proposition that Antigone buries
Polyneices, and ψ stands for a proposition that Antigone murders
Creon): Antigone ought to bury and not to bury Polyneices; Antigone
ought to bury and not to bury Polyneices and to murder Creon. Our
claim was that intuitively, the two obligations are different, and in
turn they should not be treated as semantically indistinguishable by
deontic logic. However, they are intensionally indistinguishable. Distinguishing intensionally indistinguishable deontic formulas in turn
amounts to going hyperintensional in deontic logic.

Note that in (an) amodal context, the provable equivalence be-
tween $\phi \wedge \neg\phi$ and $(\phi \wedge \neg\phi) \wedge \psi$ does not seem to be problematic. Nor
there seem to be a problem in (b) alethic (or temporal) context. In
this example, neither of the above formulas can be realised throughout
the modal space, so both are temporally and alethically impossible,
so they have the same truth-conditions. In comparison, (c) deontic

context described above makes the equivalence problematic. So does (d) epistemic (or doxastic) context. Just consider the following: it makes a difference whether you believe in a contradiction, or in a contradiction and some particular additional proposition. Graham Priest might believe that there is a box that is empty but also isn't, but not believe in a conjunction that there is a box that is empty but also isn't, and that COVID-19 vaccines do not work. A belief in a contradiction in its own right does not tell anything about your views on the effectiveness of vaccines. In a similar vein, a knowledge database might have a contradictory information about Antigone burying and not burying her brother, but not have an additional information about Antigone murdering Creon.

By the rules of classical logic, formulas $\phi \wedge \neg\phi$ and $(\phi \wedge \neg\phi) \wedge \psi$ are equivalent. As we have seen in Section 4, it means they carry the same empirical information since they are true in the very same world-time pairs (namely, in no world-time pairs whatsoever). However, since their constructions clearly differ, they carry different analytic information (below we shift to a simplified infix notation without Trivialisation):

$$\lambda w \lambda t [[{}^0 Bury_{wt}\ {}^0 Antigone\ {}^0 Polyneices] \wedge$$
$$\neg [{}^0 Bury_{wt}\ {}^0 Antigone\ {}^0 Polyneices]]$$
$$\lambda w \lambda t [[[{}^0 Bury_{wt}\ {}^0 Antigone\ {}^0 Polyneices] \wedge$$
$$\neg [{}^0 Bury_{wt}\ {}^0 Antigone\ {}^0 Polyneices]]$$
$$\wedge [{}^0 Murder_{wt}\ {}^0 Antigone\ {}^0 Creon]]$$

When it comes to the above propositions embedded in the context of implicit deontic modals, the outcome is the same under a classical, conflict intolerant reading of implicit modals. In particular, the empirical information content is identical, but the analytic information differs. This is so because obligation is analysed in terms of a set of possible-world propositions that are obligatory at a given world and time (the same world-time pairs are thus excluded in each case, so the empirical information is the same). Nonetheless, there are different ways how to arrive at the same set, and the two constructions below depict two such ways (thereby the analytic information differs):

$$\lambda w \lambda t [\,{}^{0}O_{wt}[\lambda w' \lambda t'[[\,{}^{0}Bury_{w't'}\ {}^{0}Antigone\ {}^{0}Polyneices]$$
$$\wedge \neg[\,{}^{0}Bury_{w't'}\ {}^{0}Antigone\ {}^{0}Polyneices]]]]$$
$$\lambda w \lambda t [\,{}^{0}O_{wt}[\lambda\, w' \lambda t'[[[\,{}^{0}Bury_{w't'}\ {}^{0}Antigone\ {}^{0}Polyneices]$$
$$\wedge \neg[\,{}^{0}Bury_{w't'}\ {}^{0}Antigone\ {}^{0}Polyneices]]$$
$$\wedge[\,{}^{0}Murder_{w't'}\ {}^{0}Antigone\ {}^{0}Creon]]]]$$

If our observations finished here, we would be no closer to an adequate explanation of the hyperintensional shift in deontic logic than where we had ended in Section 3. On the one hand, there are plausible reasons to recognize equivalence between formulas in classical and alethic (or temporal) contexts. On the other hand, there are considerable intuitions that formulas in the deontic (or epistemic and doxastic) context are inequivalent. Moreover, formulas in all cases have the same empirical information, and due to their procedural structure, they all carry different analytic information. However, we believe that an informational perspective on explicit deontic modals can still enlighten us.

We propose that the mode of the analytic information distribution in the *explicit* reading of deontic modals in our Antigone example is different from the cases just presented. In this case too, the literal meaning of the two sentences consists of different constructions. However, unlike in the amodal and implicit modal contexts, in this context, the two combinations of analytic information define two distinct truth conditions to be taken into account when evaluating our two formulas in the set of assumed possible world-time pairs. As we have shown in Section 3, the situation where Antigone ought to bury her brother Polyneices but also ought to leave him unburied intuitively differs from one where Antigone ought to bury her brother Polyneices but also ought to leave him unburied and ought to murder Creon, the ruler of the city of Thebes. Can we provide some ground to show why it is so? Consider the familiar pair of deontic propositions in their 'explicit' reading:

$$\lambda w \lambda t [{}^0 O^*_{wt} \ {}^0 [\lambda w' \lambda t' [[{}^0 Bury_{w't'} \ {}^0 Antigone \ {}^0 Polyneices]$$
$$\wedge \neg [{}^0 Bury_{w't'} \ {}^0 Antigone \ {}^0 Polyneices]]]]$$
$$\lambda w \lambda t [{}^0 O^*_{wt} \ {}^0 [\lambda w' \lambda t' [[[{}^0 Bury_{w't'} \ {}^0 Antigone \ {}^0 Polyneices]$$
$$\wedge \neg [{}^0 Bury_{w't'} \ {}^0 Antigone \ {}^0 Polyneices]]$$
$$\wedge [{}^0 Murder_{w't'} \ {}^0 Antigone \ {}^0 Creon]]]]$$

The difference between the previous (implicit) analysis is as follows: obligation is analysed in terms of a set of constructions that are obligatory at a given world-time pair. As before, the analytic information is different, since the two constructions differ. However, the novelty is this: what is explicitly obligatory is not an eternal matter. The first of the two constructions depicts the "actual" scenario Antigone found herself in, and the corresponding sentence (with an explicit modality) is in turn true in such a scenario. A normative authority (which can be for the present purposes understood as a union of what the laws of the city prescribe, and what the moral code prescribes) gave rise to the famous dilemma. Note that while we are simplifying here by treating the two constructions as the relevant explicit obligations in the given scenario, nothing hinges on this choice, since analogous explanations would be available for more subtle obligations. However, the latter construction does not depict an actual scenario: the very same normative authority does not prescribe to murder Creon. The corresponding sentence (with an explicit modality) is thus false in such a scenario. But this means that the difference between the two is not of a merely analytic nature: world-time pairs excluded differ too, and so does the empirical information.

Deontic sentences corresponding to the above two cases thus inherently differ because they carry different analytic information, which forces us to evaluate these sentences as true in different world-time pairs. Consequently, unlike previous cases, in the case of explicit reading of deontic sentences, precisely due to different distribution of analytic information, these sentences also carry different empirical information. In this way, we use the described mechanics of the tension between analytical and empirical information in the case of explicit reading of deontic sentences as the explanatory vehicle for the hyper-intensional shift in deontic logic. We cannot substitute one provably equivalent formula for another within the scope of modal operator, because under explicit reading, corresponding deontic sentences carry both different analytic and different empirical information, i.e., they are true in different world-time pairs.

6 Concluding Remarks

All in all, (*RE*) and its counterparts in richer languages seem to be at the heart of numerous implausible consequences for deontic logic. These implausible results directly motivate the claim that the sub-stitution of provable equivalents within the scope of deontic modal operators fails. This in turn provides direct motivation for hyper-intensionality since going hyperintensional consists, in a nutshell, in abandoning (*RE*) along with its versions and suggesting an alternative logic and semantics.

The failures of substitution presented in the Section 3 are rooted in various interconnected reasons: one obligation sounds plausible but the other does not; one obligation entails that an agent faces a nor-mative conflict but the other does not; we tend to read disjunction occurring within the scope of a deontic formula in the *free choice* man-ner (as in a *free choice permission*), but the choice in question appears absurd; in many cases, it appeared that the satisfaction conditions of the two obligations or permissions differ.

We suggested that the hyperintensional shift in the context of deontic logic has its roots in the role played by the distribution of analytic and empirical information. While this study started from formulating direct arguments arising from natural reading of the con-troversial pairs of sentences, using the procedural theory of informa-tion along with a distinction between implicit and explicit deontic modalities as our core explanatory tools, we also paved the way for a more systematic explanation of the problem of hyperintensionality of deontic modals.

We have argued that unlike in amodal or implicit contexts, under explicit reading, corresponding deontic sentences carry not only different analytic information but also different empirical information. This means that they are true in different world-time pairs, thereby having different truth conditions, and as such should not be formally modelled as provably equivalent.

Acknowledgements

The work on this paper was supported by the Slovak Research and Development Agency under the contract no. APVV- 21-0405. We are also grateful to Lukáš Bielik for his comments on the paper.

References

[1] Anglberger, A. J. J., Faroldi, F. L. G., and Korbmacher, J. (2016): An Exact Truthmaker Semantics for Permission and Obligation. In: Olivier Roy — Allard Tamminga — Malte Willer (eds.): *Deontic Logic and Normative Systems: 13th International Conference, DEON 2016*, Bayreuth, Germany, July 18–21, 2016. Milton Keynes: College Publications, 16–31.

[2] Badura, C., Berto, F. (2019): Truth in fiction, impossible worlds, and belief revision. *Australasian Journal of Philosophy*, 97 (1), 178–193.

[3] Bar-Hillel, Y., and Carnap, R. (1953): Semantic information. *British Journal for the Philosophy of Science*, 4 (14), 147–157.

[4] Bielik, L., Gahér, F., and Zouhar, M. (2010): O definíciách a definovaní. *Filozofia*, 65 (8), 719–737.

[5] Cohen, M.R. and Nagel, E. (1934): *An Introduction to Logic and Scientific Method*, London: Routledge and Kegan Paul.

[6] Dretske, F. (1970): Epistemic Operators. *Journal of Philosophy*, 67 (24), 1007–1023.

[7] Duží, M. (2010): The Paradox of Inference and the Non-Triviality of Analytic Information. *Journal of Philosophical Logic*, 38 (5): 473–510.

[8] Duží, M., and Materna, P. (2001): Propositional Attitudes Revised. In: Childers, T. (ed.): *The LOGICA Yearbook 2000*. Praha: Filosofia, 163–173.

[9] Duží, M., Jespersen, B., and Materna, P. (2010): *Procedural Semantics for Hyperintensional Logic. Foundations and Applications of Transparent Intensional Logic.* Berlin: Springer series Logic, Epistemology, and the Unity of Science 17.

[10] Fagin, R., Halpern, J. Y., Moses, Y., and Vardi, M. Y. (1995): *Reasoning about Knowledge.* Cambridge — London: MIT Press.

[11] Faroldi, F. L. G. (2014): The Normative Structure of Responsibility. Milton Keynes: College Publications.

[12] Faroldi, F. L. G. (2019a): *Hyperintensionality and Normativity.* Cham: Springer. doi:10.1007/978-3-030-03487-0

[13] Faroldi, F. L. G. (2019b): Deontic Modals and Hyperintensionality. *Logic Journal of the IGPL,* 27(4), 387–410. doi:10.1093/jigpal/jzz011

[14] Fine, K. (2017): Truthmaker Semantics. In Hale, B., et al. (eds.): *A Companion to the Philosophy of Language.* Chichester: Wiley, 556–577. doi:10.1002/9781118972090.ch22

[15] Fitting, M., and Mendelsohn, R. L. (1998): *First-Order Modal Logic.* Dordrecht: Kluwer Academic Publishers.

[16] Floridi, L. (2011): *The Philosophy of Information.* Oxford: Oxford University Press.

[17] Glavaničová, D. (2015): K analýze deontických modalít v Transparentnej intenzionálnej logike. *Organon F,* 22 (2), 211 — 228.

[18] Glavaničová, D. (2019): Hyperintensionality of Deontic Modals: An Argument from Analogy. *Filozofia,* 74 (8), 652 — 662.

[19] Glavaničová, D., and Pascucci, M. (2019): Formal Analysis of Responsibility Attribution in a Multi-modal Framework. In: *PRIMA 2019: Principles and Practice of Multi-Agent Systems,* Springer Lecture Notes in Artificial Intelligence series.

[20] Jago, M. (2014): *The Impossible: An Essay on Hyperintensionality.* New York: Oxford University Press.

[21] Jarmużek, T., and Klonowski, M. (2020): On logic of strictly-deontic modalities. A semantic and tableau Approach. *Logic and Logical Philosophy,* 29 (3), 335–380.

[22] Jespersen, B., Duží, M. (2015): Introduction to the special issue on Hyperintensionality. *Synthese,* 192 (3), 525–534. doi:10.1007/s11229-015-0665-9

[23] Kamp, H. (1973): Free Choice Permission. *Proceedings of the Aristotelian Society,* 74, 57–74.

[24] King, J. C. (1995): Structured Propositions and Complex Predicates. *Noûs*, 29 (4), 516–535.

[25] Lepore, E. (2009): The heresy of paraphrase: when the medium really is the message. *Midwest Studies in Philosophy*, 33 (1), 177–197.

[26] Levesque, H. J. (1984): A Logic of Implicit and Explicit Belief. In: *Proceedings of the National Conference on Artificial Intelligence*. Cambridge: AAAI Press/MIT Press, 198–202.

[27] Nolan, D. (2014): Hyperintensional metaphysics. *Philosophical Studies*, 171 (1), 149–160.

[28] Pollard, C. (2008): Hyperintensions. *Journal of Logic and Computation*, 18 (2), 257–282.

[29] Schaffer, J. (2009): On what grounds what. In: Chalmers, D. J., et al. (eds.): *Metametaphysics*. New York: Oxford University Press, 347–383.

[30] Svoboda, V. (2016): Δ-TIL and Problems of Deontic Logic. *Organon F*, 23(4), 539–550.

[31] Svoboda, V. (202x): Are deontic modals hyperintensional? In This Volume.

[32] Tichý, P. (1988): *The Foundations of Frege's Logic*. Berlin, New York: de Gruyter.

TIL vs THL on Deduction

Miloš Kosterec

Institute of Philosophy SAS, v.v.i., Bratislava, Slovak Republic

`milos.kosterec@gmail.com`

Abstract

This chapter presents the investigation into the properties of the deduction systems of Transparent Intensional Logic (TIL) and Transparent Hyperintensional Logic (THL). The system of THL is presented to some depth with notes comparing it with TIL. These frameworks are demonstrated to be non-equivalent. Consequently, also their deduction systems are not fully equivalent. Furthermore, the strengths of the systems as well as the areas of application are discussed. The most important part of the chapter is the demonstration of invalidity of the deduction systems in both frameworks. The reasons for this are discussed and some remedies are suggested.

Introduction

The development of research within the formal investigations of the system of Transparent Intensional Logic (TIL) gained some momentum within the past decade. The fundamental work by Dužíet al. [1] presented an enhanced playground for the researchers willing to accept the transparent non-contextual nature of the contents of language terms. The work expanded on Tichý's [14] foundational work

This research was supported by the University of Oxford project 'New Horizons for Science and Religion in Central and Eastern Europe' funded by the John Templeton Foundation. The opinions expressed in the publication are those of the author(s) and do not necessarily reflect the view of the John Templeton Foundation along several different axes, thereby expanding the (conceptual) space of problems dealt within the system. I do not try to list all the various problems that TILians went on to solve. I will rather focus on one particular — that of presenting the deduction system *of* TIL. TIL is not a classical logic, either by look or by strength. It is so by design. However, as a logic, it used to be criticised as really a *non*-logic, as it seemingly did not present any system of deduction. That was, however, not a correct view then, as it is not now. Even before finalisation of *Foundations* Tichý presented a deduction system for a system not that much different from his final TIL.[1] The deduction system was applicable within the final system, even though it was not a part of the final presentation. It is no surprise that the development of these deduction system(s) among his followers over the notional basis of the *Foundations* employed this system presented as early as the 1980s (after some necessary augmentation) as a proper subpart. So not only there has been a deduction system within TIL since a long time ago; it underwent quite a development within the past decade or so.

TIL, as enhanced by Duží et al., is not the only system based on the basic notion of construction with its procedural semantics developed within the past years. Recently, a seemingly new system based on constructions emerged — that of Transparent Hyperintensional Logics (THL). THL has been, so far, mainly developed by Raclavský, Kuchyňka and Pezlar. THL was given a full presentation within the past two years, mainly by Raclavský [10, 11].[2] One of the most developed parts of THL is the deduction system, which is fully presented as a sequent calculus.

This chapter investigates and compares the two deduction systems developed within the field of transparent logics. My aim is not to compare the systems of TIL and THL per se, even though some results, and yes opinions, will be presented. I will aim at the comparison of the systems of deduction, in which I will aim at consideration of their:

- Relation: Are the systems equivalent or is one a subpart of the

---

[1] The reader is advised to consult Tichý [12, 13].

[2] Of course, Raclavský *et al.* [9] can be considered a main stepping stone towards the eventual formulation of the system of THL.

other?

- Areas of application: Where has been the main application of the systems so far? I will specially consider the area of logic of hyperintensional contexts.

- Metalogics — Soundness and Completeness: are the systems as presented sound or even complete?

The full consideration of both systems calls for a book, so I will not try to cover all the bases here. I will, however, present a stance on all the above features of both systems.

There are many ways to write a piece. One can, e.g., build a tension and wait with the presentation of the main results till the end. Here, I choose a different strategy. I will first present the results of my analysis and let the reader interested in understanding how I get there consult the rest of the paper. In a nutshell, deductions in TIL and in THL are not equivalent. One of the reasons is that the systems does not have equivalent foundational definitions. Another reason is that the systems, however sharing some of the deduction rules, differ in the other rules they include. The deduction system of TIL is stronger, so far, in its coverage of the logics of hyperintensional contexts. THL however, is quickly following up. Which leads me here, to the not-so-great part. Both systems are not sound (and therefore not complete). Deductions in TIL and THL both contain common invalid rules. Both systems also include invalid rules of their own. The main reason for the problems within both systems is that the influence of so-called Double Executions (or immersions) on freedom of variables as well as the rules of substitution is not covered well enough by either of the systems.

The structure of the chapter is as follows. First, I present the relevant systems of THL. I will not present the system of TIL (as I assume it was done several times over in this book). I will, however, present parts of the relevant definitions of TIL, when needed. This will enable me to demonstrate that TIL and THL are not equivalent systems. Then I present the basics of the deduction system of THL with several of its original rules. Then I will cover the strengths of the systems with focus on the logics of hyperintensional contexts. After that, I demonstrate the invalidity of various inference rules of THL as well as TIL, which will ground my claims of the systems not being sound. I follow with a discussion of the reasons for this unwelcoming state. The chapter ends with a conclusion.

1   THL Crash Course

I present the version of the system by Raclavský [11][3] accompanied with the relevant definitions from his [10]. One of the main differences between THL and TIL is in their presentation. TIL, following the non-classical view on classical logics of Tichý does not grant much time to the classic model-theoretic formalisation of the system. The presentation of THL tries to follow the classic style much more closely. THL first presents its language L_{TT^*} using BNF form:[4]

$$L_{TT^*}: \quad C ::= X|x|C_0(\bar{C}_m)|\lambda\tilde{x}_m.C_o|\lfloor\lfloor C_0\rfloor\rfloor_\tau$$

X are *constants*, x are *variables*, $C_0(\bar{C}_m)$ are *applications*, $\lambda\tilde{x}_m.C_0$ are *λ-abstracts*, $\ulcorner C_0 \urcorner$ are *acquisitions*, $\lfloor\lfloor C_0\rfloor\rfloor_\tau$ are *immersions*, τ stands for a type.[5] $E_1\ldots E_m$. $\tilde{E}_m$ stands for a string of entities $E_1\ldots E_m$. $\bar{E}$ stands for $E_1\ldots E_m$.

This language has two semantics. One is trivial — the terms shall stand for the constructions defined within the Tichý's system. The other semantics is called denotational. It is a semantics developed over the ramified typed system of objects developed by Tichý.[6] The basic notions of this semantics are as follows:

---

[3]THL is most exahustively presented by Raclavský [10]. The system there is developed within several gradual steps.

[4]The subscript TT^* hints on the fact system is developed over enhanced system of Tichý's type theory.

[5]These seem to be new labels for the corresponding ones in TIL: *applications are compositions, λ-abstracts are closures, acquisitions are trivialisations, immersions are Double Executions.*

[6]The reader can consult Raclavský [10, 37] for the full specification of the type system.

A *frame* F is $\{D_\tau\}_{\tau \in T}$, i.e. a family of sets indexed by τ from the set of types T. It consists of all interpreted TT^*'s types τ as domains D_t (for each τ). A *model* M is a couple $\langle F, I \rangle$, where I is the *interpretation function*. I maps constants of L_{TT^*} to objects of F. Each *valuation* v w.r.t F consists of infinite v-sequences sq^τ (for each type τ) of τ-objects, and so v supplies every variable (of every type) with a v-value. [11, 8]

The system of *models* over a *frame* is then used to present values of the terms of L_{TT^*}. These are specified by the *evaluation function*:

Expression '$[[C]]^{M,v}$' stands for 'on valuation v w.r.t. F, 'C' denotes in M'. Let X be an object or a construction, $C, \bar{D}_m$ constructions, c a variable for constructions, and $\mathsf{C}, \bar{\mathsf{C}}_m$ objects denoted by $C, \bar{C}_m$ in M on the valuation v w.r.t. F.[7] 'x/τ' stands for a statement that x is an object of type τ. Then,

$$[[x_i^\tau]]^{M,v} = \text{the } i\text{-th member } X_i \text{ of } sq^\tau, \text{ where } sq^\tau \text{ belongs to } v$$

$$[[C]]^{M,v} = I(C), \text{ where } C \text{ is a constant}$$

$$[[C(\bar{C}_m)]]^{M,v} = \mathsf{C}(\bar{\mathsf{C}}_m), \text{ if } [[C]]^{M,v} = \mathsf{C} \in \langle \bar{\tau}_m \rangle \to \tau, [[C_1]]^{M,v} = \mathsf{C}_1 \in \tau_1 \ldots [[C_m]]^{M,v} = \mathsf{C}_m \in \tau_m \text{ and } \exists x (x = \mathsf{C}(\bar{\mathsf{C}}_m))$$

$$= \_ \text{ (i.e. nothing) otherwise}$$

$$[[\lambda \tilde{x}_m.C]]^{M,v} = \text{the function } f \in \langle \bar{\tau}_m \rangle \to \tau, \text{ that takes each } [[C]]^{M,v'} \in \tau \text{ at the respective argument } \langle [[x_1]]^{M,v'}, \ldots, [[x_m]]^{M,v'} \rangle \text{ where } v' \text{ is like } v \text{ except for what it assigns to } \tilde{x}_m \text{ occurring in } \lambda \tilde{x}_m.C \text{ and for each } 1 \leq i \leq m, [[x_i^\tau]]^{M,v'} \in \tau_i$$

$$[[\ulcorner C \urcorner]]^{M,v} = C$$

$$[[\lfloor \lfloor C \rfloor \rfloor_\tau]] = X, \text{ if } X \text{ is the only } x/\tau \text{ such that } \exists c (x = [[c]]^{M,v} \wedge c = [[C]]^{M,v})$$

$$= \_ \text{ (i.e. nothing) otherwise}$$

This evaluation function is *partial*. Symbol '$\_$' captures the conditions under which a construction term is not assigned any value by this

---

[7]The italics play a role in the presentation of THL. C is a construction; C is the object denoted by the construction.

function.

Let's follow with the presentation of further notions of THL: *k-th order subconstruction, free variable* and *substitution function.*[8]

Def. *k*th- and $(k-1)$st-order subconstruction

Let $C_{(i)}$ and variables $x_1, x_2 \ldots x_n$ be kth-order constructions, for $1 \leq k$, and X be a $(k-j)$th-order non-construction or a $(k-j)$th-order construction, for $1 \leq j < k$.[9]

1. C is a *kth-order subconstruction of C.*

2. If X is a $(k-1)$st-order construction, then X is a $(k-1)st$-*order subconstruction of*

   (a) $C_{(i)}$ if $C_{(i)}$ is of form $\ulcorner X \urcorner, \lfloor \lfloor X \rfloor \rfloor$;

   (b) $C(\bar{C}_m)$ and also $\lambda \tilde{x}_m.C$ if $C_{(i)}$ is their kth-order subconstruction and is of form $\ulcorner X \urcorner, \lfloor \lfloor X \rfloor \rfloor$.

3. $C, C1, \ldots, C_m$ are the kth-order subconstruction of $C(\bar{C}_m)$.

4. C is a *kth-order subconstruction of* $\lambda \tilde{x}_m.C$.

5. Nothing is a kth-order or a $(k-1)$-st *order subconstruction* of C, unless it follows from 1–4.

Examples: Let $\ulcorner x \urcorner, \lfloor \lfloor x \rfloor \rfloor, \ulcorner \neg \urcorner (\ulcorner x \urcorner), \lambda x. \lfloor \lfloor x \rfloor \rfloor$ be constructions of order k.

- x is a $(k-1)$st order subconstruction of
  $\ulcorner x \urcorner, \lfloor \lfloor x \rfloor \rfloor, \ulcorner \neg \urcorner (\ulcorner x \urcorner), \lambda x. \lfloor \lfloor x \rfloor \rfloor$.

This notion of subconstruction is *richer* than the corresponding notion of subconstruction used within TIL. The notion of order of construction presents some information about the occurrence of the

---

[8]The concepts are from Raclavský [10, chapters 2.6.1, 2.6.2 and 3.4.2]; transcribed to the language used here. The single-executions are left out.

[9]kth-order construction is just a notational variant for construction of order k. kth order object is an object belonging to type of order k.

subconstruction within the main embedding construction. The reader will notice that the order of a subconstruction is lower than the order of its embedding construction only within the second point of the definition. kth order subconstruction of a kth order construction is also called its *direct subconstruction*.

The following is the notion of a free (occurrence) of a variable:

Def. free variable

Let x be an arbitrary variable and $C_{(i)}$ a construction.

1. The variable x is free in x.

2. If x is free in at least one of its occurrences in C, or in D while C is the of the form $\ulcorner D \urcorner$, then x is *free* in $\lfloor \lfloor C \rfloor \rfloor$.

3. If x is free in at least one of its occurrences in C, $C_1, \ldots$ or C_m, then x is *free* in $C(\bar{C}_m)$.

4. If x is free in at least one of its occurrences in C and is distinct from variables $x_1, \ldots, x_m$, then x is *free* in $\lambda \tilde{x}_m . C$.

5. Variable x is not *free* in a construction, unless it follows from 1–4.

Examples:

- x is free in $\ulcorner + \urcorner (\ulcorner 1 \urcorner, x), \lambda y . x, \lfloor \lfloor \ulcorner \ulcorner + \urcorner (\ulcorner 1 \urcorner, x) \urcorner \rfloor \rfloor$
- x is not free in $\ulcorner + \urcorner (\ulcorner 1 \urcorner, \ulcorner x \urcorner), \lambda x . x, \lfloor \lfloor \ulcorner \ulcorner x \urcorner \urcorner \rfloor \rfloor, y$

The above definitions cover enough ground for the presentation of the notion of the *substitution function*. '$C_{(D/x)}$' stands for 'the result of substitution of D for all free occurrences of the variable x in the construction C'.

Def. Substitution function Sub^k

Let C, D, x, B be kth-order constructions. Let '$FV(C)$' stand for the set of all free variables that are direct subconstructions of C. Then the value of the substitution function Sub^k, i.e. the result of substitution of a construction D for all free occurrences of variable x in the construction C (i.e. $\text{Sub}^k(D, x, C) = C_{(D/x)}$) is given by the following:

1. If the variable x is not free in C then $C_{(D/x)}$ is identical to C.

2. If the variable x is free in C then, if C is

   (a) x, $C_{(D/x)}$ is D.

   (b) $B(\bar{B}_m)$, $C_{(D/x)}$ is $B_{(D/x)}(\bar{B}_{m(D/x)})$.

   (c) $\lambda y.B$ and y does not belong to $FV(D)$, $C_{(D/x)}$ is $\lambda y.B_{(D/x)}$.

   (d) $\lambda y.B$ and y belongs to $FV(D)$, x belongs to $FV(B)$, z does not belong to $FV(D) \cup FV(B)$, then $C_{(D/x)}$ is $\lambda z.B_{(z/y)(D/x)}$.

   Let F, D, x be of order $(k-2)$,

   (f) $\lfloor\lfloor E \rfloor\rfloor$ and E is not of form $\ulcorner F \urcorner$, then $C_{(D/x)}$ is $\lfloor\lfloor E_{(D/x)} \rfloor\rfloor$.

   (g) $\lfloor\lfloor E \rfloor\rfloor$ and E is of form $\ulcorner F \urcorner$, then $C_{(D/x)}$ is $\lfloor\lfloor \ulcorner F_{(D/x)} \urcorner \rfloor\rfloor$.

Examples:

- $\ulcorner + \urcorner(\ulcorner 1 \urcorner, x)_{(\ulcorner 2 \urcorner/x)} = \ulcorner + \urcorner(\ulcorner 1 \urcorner, \ulcorner 2 \urcorner)$
- $\lambda y.x_{(\ulcorner 2 \urcorner/x)} = \lambda y.\ulcorner 2 \urcorner$
- $\lfloor\lfloor \ulcorner\ulcorner + \urcorner(\ulcorner 1 \urcorner, x)\urcorner \rfloor\rfloor_{(\ulcorner 2 \urcorner/x)} = \lfloor\lfloor \ulcorner\ulcorner + \urcorner(\ulcorner 1 \urcorner, \ulcorner 2 \urcorner)\urcorner \rfloor\rfloor$
- $y_{(\ulcorner 2 \urcorner/x)} = y$

This covers the presentation of THL.

2 TIL and THL are not equivalent

First, let me introduce some general remarks about the way the system of TIL is composed. The basis of the system, what I call its *primary definitions*, are those concerning the notions of *valuation; construction; constructing an object with respect to a valuation* and the *ramified type hierarchy*. These cover the relations among the objects within the ontology assumed by the system. In a nutshell, they are the *inductive* basis over which the objects of the system are specified. However, these solely are not enough to present an analysis of *logical* relations. The reason is simple. Even though the notion of *variable* is already present in the primary definitions, there are not many *logical concepts* it could model alone. There is no notion of free or bound variable inherent in the primary definitions. So it is of no wonder, that the next substantial part of the system, which I call *secondary definitions*, is covered usually by the notions of *subconstruction, free/bound occurrence of a variable, displayed/executed occurrence of a construction* (in case of TIL) and (*collisionless*) *substitution*. A lot of investigations of/into the logic of natural language is based on the use of these notions. I call them secondary, simply because they depend on, shall adhere and be coherent with the primary definitions. In other words, if any problem appears, which is based on the use of the secondary definitions, it shall not be solved by changes in primary definitions, but with the amendments of the secondary definitions first.[10]

But so far, I have not described TIL yet. The difference between TIL and the system of definitions it is built on, has not been given much focus in the past.[11] Even Tichý spent tens of pages and several chapters between introducing the notions covered by primary, (several of) secondary definitions and the notion of Transparent Intensional Logic.[12] In other words, the system of constructions is not yet TIL and TIL is not just the system of constructions. Only after the model of basic logical objects used for the analysis of language is presented we can speak about TIL. One uses TIL *after* s(he) models proposition as mappings from possible worlds-time pairs to truth

---

[10]I am fully aware that this is only a normative statement, though.

[11]Of course, except for the way the system was presented by [9].

[12]TIL is introduced in chapter twelve of *Foundations*, the presentation of the notions upon which it is built is finished after the sixth chapter.

values and intensions as functions from possible world-time pairs into further objects. In short, TIL is a system presented by *the models* of the standard logical objects built over the system of (typed) constructions. Therefore, one can present different models of the basic logical objects and relations and by that present a system different from TIL — that is at least the core of the difference assumed and presented by the proponents of THL. THL does not include the models that are usually assumed within the works of TIL.

This is, however, not why I claim that the systems of THL and TIL are not equivalent. I claim that these systems differ even in the systems they are built on. The systems of primary definitions are roughly[13] equivalent — the differences are rather in presentation than in the content. However, the secondary definitions of the system differ. To demonstrate my claim about the non-equivalency between TIL and THL, I will present one logically important difference — the notions of *free/bound* occurrence of variable differ in these systems. Consequently, the models of logical relations based on these notions are not equivalent as well. Let me elaborate.

The relevant piece of the definition of free/bound occurrence of variable in TIL is the point covering Double Executions:

> Let C be ^2X. Then any *occurrence* of ξ that is free, λ-*bound* in a constituent of C is, respectively, *free*, λ-*bound* in C. If an occurrence of ξ is ^0bound in a constituent ^0D of C and this occurrence of D is a constituent of $X'v$-constructed by X, then if the occurrence of ξ is free, λ-

---

[13]Roughly because the L_{TT^*} language contains a term for immersion: $\lfloor\lfloor C_o\rfloor\rfloor_\tau$, which contains a type term τ. The notion of Double Execution in TIL does not assume it is explicitly typed. This, however, opens up a possibility of various further *different* immersions within the base system of THL, for which there is only one Double Execution in TIL. This, however, leads to several problems I am not covering in this chapter. For example, consider a variable x that v-constructs constructions of order 1. Assume a valuation v_1 for which x v_1-constructs a construction of a truth value and assume a valuation v_2 for which x v_2-constructs a construction of an individual. This possibility is given by the way the notion of *construction of order n* is specified. For TIL, it is no problem to present this Double Execution: ^2x. How, however, shall the corresponding immersion be presented within THL? $\lfloor\lfloor x\rfloor\rfloor_o, \lfloor\lfloor x\rfloor\rfloor_\iota$ or both or neither of them?

bound in D it is *free*, λ-*bound* in C. Otherwise, any other occurrence of ξ in C is $^0 bound$ in C.

This part of the definition importantly depends on the notion of a *constitutent*. A constituent of a construction is its subconstruction, which is *executed* (not just displayed) during the procedure of constructing the resulting object by the whole construction. By example x in ^{20}x is considered free in TIL — and in THL too, for that matter. But consider the following example:

Let *Second* be the function, which is applied on tuple of arguments and returns the second of them as value, i.e., $Second(a, b) = b$. Consider now the following construction:

$$^2[^0 Second\ {}^0 y\ {}^0 x]$$

Is x free in its occurrence in this Double Execution? $^0 x$ is a constituent of $^2[^0 Second\ {}^0 y\ {}^0 x]$. $[^0 Second\ {}^0 y\ {}^0 x]$ v-constructs x by assumption. x is a constituent of x and x is free in x. Therefore, the occurrence of x shall be considered free in $^2[^0 Second\ {}^0 y\ {}^0 x]$. Consequently, it is open to binding by lambda operators etc. However, the occurrence of x in the corresponding construction term in THL:

$$\lfloor \lfloor \ulcorner Second \urcorner (\ulcorner y \urcorner, \ulcorner x \urcorner) \rfloor \rfloor_\tau$$

is not considered free. THL does not assume or present the notion (correspondent) to constituent of TIL. The relevant point of the definition of free variable does not enable us to consider x free in the above term. Therefore, the logical relations modelled by TIL are not coverable by THL for these kinds of cases. Therefore, TIL and THL differ even before the models of logical phenomena of (natural) languages are introduced.

3   Natural Deduction in THL

Derivation rules in ND_{TT*} present relations between *sequents* which are made of *matches*.[14] Let's present these notions in turn.

---

[14]I left out the notion of an *empty match* as it is not needed in the argumentation of the paper.

Def. Match

Let $\cong$ hold between two constructions, if they are assigned the same denotational value.

A *match* is a construction of form

1. $\ulcorner\cong\urcorner(\ulcorner C\urcorner, \ulcorner\ulcorner X\urcorner\urcorner)$ or
2. $\ulcorner\cong\urcorner(\ulcorner C\urcorner, \ulcorner x\urcorner)$

where the constructions $C, \ulcorner X\urcorner$ and x have an object of the same type as their denotational value. In what follows, I will use infix notation, e.g. '$\ulcorner\cong\urcorner(\ulcorner C, \ulcorner\urcorner x\urcorner)$' becomes '$C \cong x$')

Examples:

- $\ulcorner+\urcorner(\ulcorner 1\urcorner, \ulcorner 1\urcorner) \cong \ulcorner 2\urcorner, \ulcorner Sub\urcorner(\ulcorner\ulcorner 3\urcorner\urcorner, \ulcorner x\urcorner, \ulcorner\ulcorner+\urcorner(x, \ulcorner 1\urcorner)\urcorner) \cong \ulcorner+\urcorner(\ulcorner 3\urcorner, \ulcorner 1\urcorner)$

A match could be satisfied or unsatisfied with respect to a model and a valuation w.r.t. a frame:

Def. Satisfaction of a match

Let F be a frame, M a model.

A match is *satisfied* by a valuation v (w.r.t. F) in M iff $[[C]]^{M,v} = [[\ulcorner X\urcorner]]^{M,v}$ or $[[C]]^{M,v} = [[x]]^{M,v}$.

This relation is expressible within the system of THL. It is not just a part of the meta-system but can be grasped within the system. Sequents are built out of matches:

Def. (Valid) Sequent

A *sequent* is of the form $\Gamma \Rightarrow \Pi$, where Γ is a finite set of matches and $\Rightarrow$ stands for a (syntactic) consequence. Γ is *antecedent*. Π is a *succedent* of sequent. A sequent is *valid* in a model if every valuation that satisfies all members of Γ in the model also satisfies Π in the model.

The notion of a rule is then as follows:

Def. (Valid) Rule

Let $\bar{S}_m$ be a list of sequents. Let S be sequent. Then $\bar{S}_m \vdash S$ is a rule. A rule is *valid* in a model iff every valuation satisfying all members of $\bar{S}_m$ also satisfies S. The validity of a rule can be dependent on the satisfaction of some *conditions*.

These notions have their analogy with (at least one form of) the presentation of the deduction system within TIL. Let's now finally present some interesting rules of natural deduction of THL.[15] I will present those rules, which do not have a straightforward counterpart in deduction systems of TIL presented so far.

Rules for acquisition
The Rule of $\ulcorner . \urcorner$-identity

$$\ulcorner . \urcorner - ID \ \frac{\Gamma \Rightarrow \ulcorner X \urcorner \cong x \quad \Gamma \Rightarrow \ulcorner Y \urcorner \cong x \quad \Gamma \Rightarrow X \cong y}{\Gamma \Rightarrow Y \cong y}$$

Condition: x is not free in $X, Y, \Gamma, x/{*}k; y, X, Y/\tau, \tau$ is a type of order k

The Rule of Vacuous Sequent

$$\ulcorner . \urcorner - E. \ \frac{\Gamma \Rightarrow X \cong x \quad \Gamma \Rightarrow \ulcorner Y \urcorner \cong x}{\Gamma \Rightarrow M}$$

Condition: X is a subconstruction of Y. x is not free in $\Gamma, X, M.x, \ulcorner Y \urcorner, X/\tau$.

The Rule of $\ulcorner . \urcorner$-Instantiation

$$\ulcorner . \urcorner - INST \ \frac{\Gamma \ulcorner X \urcorner \cong x \Rightarrow M}{\Gamma \Rightarrow M}$$

Condition: x is not free in $M, \Gamma.x, \ulcorner X \urcorner /\tau$.

---

[15]The sequent style is much more used within THL so far. TIL deduction system included sequents no later than [3]. The proofs within TIL usually do not have the style of sequent calculi.

Rules for immersion

The Rule of $\lfloor\lfloor.\rfloor\rfloor$-Introduction

$$\lfloor\lfloor.\rfloor\rfloor - I \frac{\Gamma \Rightarrow X \cong x \quad \Gamma \Rightarrow \ulcorner Y \urcorner \cong x \quad \Gamma \Rightarrow Y \cong y}{\Gamma \Rightarrow \lfloor\lfloor X \rfloor\rfloor \cong y}$$

The Rule of $\lfloor\lfloor.\rfloor\rfloor$-Elimination

$$\lfloor\lfloor.\rfloor\rfloor - E \frac{\Gamma \Rightarrow X \cong x \quad \Gamma \Rightarrow \ulcorner Y \urcorner \cong x \quad \Gamma \Rightarrow \lfloor\lfloor X \rfloor\rfloor \cong y}{\Gamma \Rightarrow Y \cong y}$$

Condition $(\lfloor\lfloor.\rfloor\rfloor - I, \lfloor\lfloor.\rfloor\rfloor - E)$: x and y are not free in $X, Y, \Gamma.x, X/^*k.y/\tau, \tau$ is a type of order k.

The Rule of $\lfloor\lfloor.\rfloor\rfloor$-Instantiation

$$\lfloor\lfloor.\rfloor\rfloor - INST \frac{\Gamma \Rightarrow \lfloor\lfloor X \rfloor\rfloor \cong y \quad \Gamma, X \cong x \Rightarrow M}{\Gamma \Rightarrow M}$$

Condition: x and y are not free in $X, M, \Gamma.x, X/^*k.y/\tau, \tau$ is a type of order k.

Rules for Substitution

The Rule of Sub-Introduction

$$Sub - I \frac{\Gamma \Rightarrow D \cong x \quad \Gamma \Rightarrow C \cong a}{\Gamma \Rightarrow C_{(D/x)} \cong a}$$

The Rule of Sub-Elimination

$$Sub - E \frac{\Gamma \Rightarrow D \cong x \quad \Gamma \Rightarrow C_{(D/x)} \cong a}{\Gamma \Rightarrow C \cong a}$$

Condition (Sub-I, Sub-E): a is not free in $C, D, \Gamma.D, x/\tau$; $a, C/\tau_1.x$ is in C free for D.

This condition uses the notion of a variable being free in a construction for other construction. Following Raclavský [11, 113]: "A variable x is called not free in C iff (an occurrence of) x is either bound in C, or x has no occurrence in C at all." Together with Raclavský [10, 63]: "A variable x will be called free in C for D iff x is free in C and $D, x/\tau$".

This notion could, especially if used in the statement of condition of rules, preclude the application of the rule of these substitution rules on the cases of constructions *without* the free occurrence of a variable. This is rather interesting because the definition of substitution function seems to also cover these cases. I finish by the following rule containing the occurrence of the function Sub^k.

Explicit Substitution Rules
The Rule of Substitutivity of Identicals

$$SI \frac{\Gamma \to \lfloor \lfloor \ulcorner Sub^k \urcorner (\ulcorner D_1 \urcorner, \ulcorner x \urcorner, \ulcorner C \urcorner) \rfloor \rfloor \cong T \Gamma \Rightarrow \ulcorner = \urcorner (\ulcorner D_1 \urcorner, \ulcorner D_2 \urcorner) \cong T}{\Gamma \Rightarrow \lfloor \lfloor \ulcorner Sub^k \urcorner (\ulcorner D_1 \urcorner, \ulcorner x \urcorner, \ulcorner C \urcorner) \rfloor \rfloor \cong T}$$

Condition:
$D_1, D_2, x/\tau; C/o; x$ in. C is free for D.

4 Logics of Hyperintensional Contexts and Substitution

This is one of the main areas of the development of TIL in the past decade. Dužíand Jespersen presented the investigations concerning this area in at least three different articles (see [2, 4, 6]). I will present some of the rules provided by them in what follows. This section does not contain results from THL. Especially because even though Raclavský [10, especially chapter 5] contains a lot of *examples* of investigations into supposedly hyperintensional context created by the operator *believe*, the chapter does not contain a presentation of general deduction rules, or any proofs substantiating those.

One of the driving forces behind the investigations of hyperintensional contexts in TIL was the assumption that the basic logical rules do *hold* even in these contexts. Therefore, hyperintensional contexts are not taken as a place where a logical rule no longer applies, *if one applies them correctly*. In other words, the basic rules of substitution of identicals is *salva veritate* if one applies them for identical objects. Existential generalisation can be applied even in these contexts if one prepares them accordingly. Let us present some of the results.[16] First, lets present the rules of quantifying into the hyperintensional contexts:

The rule for quantifying into hyperintensional context

Let $Att \rightarrow_v (o\iota^* n)_{\tau\omega}$ be an arbitrary construction of a hyperintensional objectual attitude relation; $a \rightarrow_v \iota$ a construction of an individual. Let $C(X)$ be a construction of an attitude complement with a constituent $X/^* n \rightarrow_v \alpha; c/^*(n+1) \rightarrow_v {}^* n; y \rightarrow_v \alpha$. $C(X/y)$ is the result of the collisionless substitution of X for y in C. Then:

$$[Att_{wt}\ a^0 C(X/y)]$$
$$\text{------------------}$$
$$[^0\exists\lambda c[^0 Att_{wt}a[^0 Sub\ c^0 y^0 C(y)]]]$$

Notice that the conclusion only claims that there is a construction that is a constituent of the complement of the attitude of an agent. The conclusion *does not quantify over the object*, if any, constructed by the constituent, only over the *constituent* (i.e., a construction).

The rule for quantifying into hyperintensional context (case of Trivialization)

Let b/α. $C(^0 b/y)$ is the result of the collisionless substitution of $^0 b$ for y in C.

$$[Att_{wt}\ a^0 C(^0 b/y)]$$
$$\text{-----------------}$$
$$[^0\exists\lambda x[^0 Att_{wt}a[^0 Sub\ [^0 Tr x]^0 y^0 C(y)]]]$$

The rules for quantifying into hyperintensional contexts de re

Let $Att \rightarrow (o\iota^* n)_{\tau\omega}; a \rightarrow \iota; x, y/^* n \rightarrow_v \alpha; \exists/(o(o\alpha)); X/^* n \rightarrow \alpha;$ $C(y)/^* n$ has at least one free occurrence of y.

Active variant

$$[^0 Att_{wt}a[^0 Sub\ [^0 Tr X]^0 y^0 C(y)]]$$
$$\text{----------------}$$
$$[^0\exists\lambda x[^0 Att_{wt}a[^0 Sub\ [^0 Tr x]^0 y^0 C(x)]]]$$

---

[16]These were provided mainly by e.g. [1, 2, 4, 6].

Passive variant

$$[\lambda x[^0Att_{wt}a[^0Sub\ [^0Trx]^0y^0C(y)]X]$$
$$\text{-----------}$$
$$[^0\exists\lambda x[^0Att_{wt}a[^0Sub\ [^0Trx]^0y^0C(y)]]]$$

The rules for quantifying into hyperintensional contexts de dicto

Rule 1: Quantifying over a constituent of an attitude complement

Let $P(X/d)/^*n$ be a procedure with a constituent $X/^*n \to \alpha$ that has been substituted for the variable $d/^*n \to \alpha; c/^*n+1 \to {}^*n;\ {}^2c \to \alpha$.

$$[B_{wt}a^0P(X/d)]$$
$$\text{-----------}$$
$$[^0\exists\lambda c[B_{wt}a[^0Sub\ c^0d^0P(d)]]]$$

Rule 2: Quantifying over a Trivialized object

Let $P(^0b/y)/^*n$ be a procedure with a proper constituent $^0b/^*n \to \alpha$ that has been substituted for the variable $y/^*n \to \alpha; x/^*n \to \alpha$

$$[B_{wt}a^0P(^0b/y)]$$
$$\text{------------}$$
$$[^0\exists\lambda x[B_{wt}a[^0Sub[^0Tr^\alpha x]^0y^0P(y)]]]$$

The reader can see all these rules utilise the substitution function *Sub* which is defined by the notion of collisionless substitution. Therefore, to properly apply these rules one needs to properly understand the rules governing substitution.

One could worry whether there are any rules specified for such a substitution in TIL so far. We must apply the definitions of a free occurrence of a variable and that of a collisionless substitution in order to obtain the results. Furthermore, there are general norms stated within TIL that present the rules of correct application of substitution of identicals (*Leibniz's Law*) as follows. They are based on the notion of *collisionless replacement*:

'$C(D'/D)$' stands for the result of collisionless replacement of D by D' in C: construction C differs from $C(D'/D)$

85

only in replacing an occurrence of the subconstruction D of C by the construction D', and no occurrence of a free variable in D' is bound in C. [1, 273]

Let $\approx_v$ stand for the relation of v-congruency among constructions (they v-construct the same object, if any, with respect to valuation v. Let $\approx$ stand for the relation of equivalency. Let $=$ stand for the relation of identity (up to procedural isomorphism).[17] Then the rules for a correct replacement of construction within TIL are:[18]

The extensional rule of substitution. Let $D \rightarrow (\alpha\beta_1 \ldots \beta_m)$, $m \geq 1$, be a constituent of C and let D occur with extensional supposition in the non-generic context of C. Further, let D'_i, D_i be constituents of D, and let $C(D'_i/D_i)$ be the result of collisionless substitution of D'_i for D_i in C. Then:

$$^0D'_i \approx_v {}^0D_i$$
$$\text{------------}$$
$$^0C \approx_v {}^0C(D'_i/D_i)$$

In other words, the collisionless replacement of v-congruent constructions within a constituent occurring extensionally does not change the object v-constructed by the construction.[19]

---

[17]I am not discussing this notion in this chapter. The reader is advised to consult e.g., [5, 7]).

[18]The following is taken from [1, 274–275].

[19]One can be perplexed by the use of a *collisionless substitution* and not of the notion of a *collisionless replacement* here. The notion of a substitution is defined only for the substitution for free occurrences of variables. That makes the area of application of this rule unnecessarily limited. Following the ideas behind the formulation of rules for the substitution of identicals (here v-congruent constructions) it seems strange. I guess it's a typo. The use of general terms in $C(D'_i/D_i)$ points towards the idea that the notion of collisionless replacement is understood within this rule. This idea is heavily supported by the examples within [1, 275]. ... With a caveat that the example there does not respect the condition about the avoidance of collisions.

The intensional rule of substitution. Let $D \to (\beta_1 \dots \beta_m), m \geq 1$, be a constituent of C and let D occur with $(\beta_1 \dots \beta_m)$-intensional supposition or in a Υ-generic context of C for some Υ. Then

$$^0D' \approx {}^0D'$$
$$\text{------------}$$
$$^0C \approx {}^0C(D'/D)$$

Therefore, within these contexts, we can replace equivalent constructions without the change of object constructed with respect to any valuation.

The hyperintensional rule of substitution. Let the occurrence of D be mentioned* in C and let $=_{\text{isom}} /(o^*n^*n)$ be the relation of procedural isomorphism holding between constructions. Then, provided no collision of variables arises:

$$^0D =_{\text{isom}} {}^0D'$$
$$\text{------------}$$
$$^0C =_{\text{isom}} {}^0C(D/D')$$

Stating that by replacing mutually procedurally isomorphic constructions within hyperintensional contexts we keep the embedding construction (procedurally) identical.

These definitions themselves depend on the probably most complicated definitions within TIL (so far). I will omit the definitions of *extensional/intensional occurrence of a construction* as well as definitions concerning *generic contexts.*[20]

Now let's *assume* for the moment that the rules concerning substitution within both studied systems are correct. How do they compare? One can easily see that the TIL based substitution rules operate with the notion of *contexts* of three types (extensional, intensional, hyperintensional) as well as with the notion of intensional/extensional occurrence of constructions. The rules stated by THL do not have such a dependency. Furthermore, THL so far, lacks such delineation

---

[20]They are the subject matter of [1, chapter 2.6].

of contexts.[21] THL does not differenitate among intensional or extensional occurrences of constructions. Therefore, to understand the substitution rules within THL, one needs to accustom himself with far less definition background. We shall, nevertheless, not take this as a case in point to prefer the THL system of rules for substitution. The main reason is that THL rules concern *far fewer* cases than those formulated within TIL. See, TIL formulates the rules with the notion of *collisionless replacement* of constructions. THL rules are based on the notion of (collisionless) substitution for free occurrences of a variable. THL rules are therefore more general — they are not just for the variables. We can consider the collisionless substitution for a free occurrence of a variable a special case of collisionless replacement.[22] The rules of THL are therefore specified for a narrower space of cases than those of TIL. So, even though the system of THL seems simpler, it is also weaker (so far), at least in the point of covering the validity of the fundamental Leibniz's Law for the substitution of identicals.

5 Deduction systems are not sound

This section is a continuation of the analysis into the validity of rules within the deduction systems of THL from which some of the results have been already presented (see [8]). Those papers presented the counterexamples to the general validity of the Compensation Principle within TIL and within THL. Consequently, some of the basic rules of their deduction systems were demonstrated invalid (e.g., those concerned by the application of β-reduction by value). Therefore, we know already, that the deduction systems of TIL and THL are not sound in their present form.[23] This section presents further coun-

---

[21]Even though a rather classical negative delineation of hyperintensional contexts is assumed in THL (see [10, 11]).

[22]Just replace the term D in the definition by a variable and replace all of the free occurrences of a variable.

[23]Duží and Jespersen [6] changed an approach towards substitution. They use the notion of *correct substitution*, which applies to the constructions in their so called ^{20}-normal form (i.e., elimination of subconstructions of form ^{20}C by C within the main construction). This way, the definition of substitution in that article is

terexamples for other rules from the investigated deduction system — demonstrating that the problems are of a crucial nature.

5.1 Case of THL

I present the counterexamples for the validity of five basic deduction rules within THL. First, let's consider the two rules of substitution above (Sub-I) and (Sub-E).

Claim: The Sub-I rule is invalid.

Proof: By counterexample. Let f be a function. Assume the following case:

$$Sub-I\frac{\Gamma \Rightarrow \lfloor\lfloor y\rfloor\rfloor \cong z \quad \Gamma \Rightarrow y \cong x \quad \Gamma \Rightarrow x \cong D \quad \Gamma \Rightarrow \lambda x.z \cong \ulcorner f \urcorner}{\Gamma \Rightarrow \lambda x.z_{(\lfloor\lfloor y\rfloor\rfloor/z)} \cong \ulcorner f \urcorner}$$

Is a case of the application of Sub-I rule. Conditions are fulfilled: $\ulcorner f \urcorner$ is not free in $\lambda x.z$, $\lfloor\lfloor y\rfloor\rfloor$, Γ, z. z is free for $\lfloor\lfloor y\rfloor\rfloor$ in $\lambda x.z$. Correct typing is assumed. Let's now investigate whether the rule holds. Assume the premises are fulfilled. What kind of a function is f? It is a constant function from arguments of the type of objects *denotationally* assigned to x (i.e., v-constructed by x) into the value given by z. Variable z is free in $\lambda x.z$, f is a constant function. Let's now investigate the conclusion of the application of the rule. What is $\lambda x.z_{((\lfloor\lfloor y\rfloor\rfloor/z))}$? By the definition of Substitution function in THL

$$\lambda x.z_{(\lfloor\lfloor y\rfloor\rfloor/z)} = \lambda x.\lfloor\lfloor y\rfloor\rfloor.$$

The denotation value of $\lambda x.\lfloor\lfloor y\rfloor\rfloor$ is a function that takes each $[[\lfloor\lfloor y\rfloor\rfloor]]^{M,v'} \in \tau$ at the respective argument $[[x]]^{M,v'}$ where v' is like v except for what it assigns to x occurring in $\lambda x.\lfloor\lfloor y\rfloor\rfloor$. v' is like v except for what it assigns to x, therefore v and v' agree on the assignment to variable y. by assumption expressed in the premise, $[[y]]^{M,v'} = x$. Therefore, $[[\lfloor\lfloor y\rfloor\rfloor]]^{M,v'} = [[x]]^{M,v'}$. But then the function denoted by $\lambda x.\lfloor\lfloor y\rfloor\rfloor$ takes the value $[[x]]^{M,v'}$ at the respective

---

much closer to that within THL. Nevertheless, they are still not equivalent. Furthermore, the use of [20]-normal constructions does not block the counterexamples presented by [8].

argument $[[x]]^{M,v'}$. Therefore, it is an identity function. Because identity function is not a constant function, this is a counterexample for the rule of Sub-I. ∎

Claim: The Sub-E rule is invalid.
Proof: By counterexample.

$$Sub - E\frac{\Gamma \Rightarrow \lfloor\lfloor y\rfloor\rfloor \cong z \ \Gamma \Rightarrow y \cong x \ \Gamma \Rightarrow x \cong D \ \Gamma \Rightarrow \lambda x.z_{(\lfloor\lfloor y\rfloor\rfloor/z)} \cong \ulcorner f \urcorner}{\Gamma \Rightarrow \lambda x.z \cong \ulcorner f \urcorner}$$

The reasoning is analogous to the one above.[24]

The latest published investigations within THL (see [11]) introduced several further deduction rules into its deduction system. I will demonstrate them along with a counterexample to their validity.

Theorem: Conditionalised Existential Generalisation.
Let $D, d, x/\tau; C, \ulcorner T \urcorner / o; \exists^\tau / (\tau \to o) \to o; Sub/\langle {}^*n, {}^*n, {}^*n \rangle \to^* n.$
Then,

$$cond \ EG\frac{}{\Gamma, D \cong d, C_{D/x} \cong T \Rightarrow \exists^\tau \lambda x.C \cong T}$$

Claim: The cond EG rule is invalid.
Proof: By counterexample. Consider the following example:
Let's assume a correct assignment of types.

$$\frac{\Gamma \Rightarrow y \cong \ulcorner + \urcorner(x, \ulcorner 1 \urcorner), x \cong 3}{\Gamma, z \cong \lfloor\lfloor y\rfloor\rfloor, (\lambda x.\ulcorner \neq \urcorner(x, z))(z)_{((\lfloor\lfloor y\rfloor\rfloor/z))} \cong T \Rightarrow \exists^\tau \lambda z.(\lambda x.\ulcorner \neq \urcorner(x, z))(z) \cong T)}$$

Is a case of *cond EG rule*. *Cond EG* has *no* explicit premises. Therefore, it is applicable for *any* premises. $(\lambda x.\ulcorner + \urcorner(x, z))(z)_{((\lfloor\lfloor y\rfloor\rfloor/z))}$ stands for the result of the substitution of $\lfloor\lfloor y\rfloor\rfloor$ for free occurrences of z within the *application* of the function denoted by $\lambda x.\ulcorner + \urcorner(x, z)$ on the argument denoted by z. Now, assume a valuation under which the premises are satisfied and let's investigate the claim in the conclusion. First, the result of the substitution above is as follows:

$$(\lambda x.\ulcorner \neq \urcorner(x, z))(z)_{((\lfloor\lfloor y\rfloor\rfloor/z))} = (\lambda x.\ulcorner \neq \urcorner(x, z)_{((\lfloor\lfloor y\rfloor\rfloor/z))})(z)_{((\lfloor\lfloor y\rfloor\rfloor/z))} =$$
$$(\lambda x.\ulcorner \neq \urcorner(x, \lfloor\lfloor y\rfloor\rfloor))(\lfloor\lfloor y\rfloor\rfloor)$$

The denotational value of this application is as follows:
$[[(\lambda x.\ulcorner \neq \urcorner(x, \lfloor\lfloor y\rfloor\rfloor))(\lfloor\lfloor y\rfloor\rfloor)]]^{M,v}$ = the value of the function denoted

---

[24]The reader is invited to check the counterexample.

by $\lambda x.\ulcorner\neq\urcorner(x,\lfloor\lfloor y\rfloor\rfloor)$ on the argument given by $\lfloor\lfloor y\rfloor\rfloor$. So, what's the argument? By premises, $y = \ulcorner+\urcorner(x,\ulcorner1\urcorner), x = 3$. Therefore $[[x]]^{M,v} = 3$, and $[[\lfloor\lfloor y\rfloor\rfloor]]^{M,v} = [[[[y]]^{M,v}]]^{M,v} = [[\ulcorner+\urcorner(x,\ulcorner1\urcorner)]]^{M,v} = 4.^{25}$ So what's the function denoted by $\lambda x.\ulcorner\neq\urcorner(x,\lfloor\lfloor y\rfloor\rfloor)$ with respect to this model and valuation? It assigns any argument the value, which is given as a denotation of $\ulcorner\neq\urcorner(x,\lfloor\lfloor y\rfloor\rfloor)$ with respect to the valuation, which is as v but which assigns the argument 4 as a value to the variable x. In our case, the value of the function is given as a denotational value given by $[[\ulcorner\neq\urcorner(x,\lfloor\lfloor y\rfloor\rfloor)]]^{M,v'}$, where $[[x]]^{M,v'} = 4$. So, the value is given by the application of non-identity relation to the values given by $[[x]]^{M,v'} = 4$ and $[[\lfloor\lfloor y\rfloor\rfloor]]^{M,v'}$ which is, $[[\lfloor\lfloor y\rfloor\rfloor]]^{M,v'} = [[[[y]]^{M,v'}]]^{M,v} = [[\ulcorner+\urcorner(x,\ulcorner1\urcorner)]]^{M,v'} = 5.^{26}$ And because $4 \neq 5$, the value given by the function is Truth. Therefore, the antecedent of the conclusion of this application of the rule is satisfied. Let's investigate the succedent:

$$\exists^r\lambda z.((\lambda x.\ulcorner\neq\urcorner(x,z))(z)) \cong T.$$

There is no substitution mentioned within the term: $\exists^r\lambda z.((\lambda x.\ulcorner\neq\urcorner(x,z))(z))$. The consequent will be satisfied, if there is at least one model and a valuation, under which $\lambda z.((\lambda x.\ulcorner\neq\urcorner(x,z))(z))$ denotes non-empty set.^{27} So is there one?

$[[\lambda z.((\lambda x.\ulcorner\neq\urcorner(x,z))(z))]]^{M,v}$ denotes a function, which to an argument (say k) assigns the value given by the application $[[(\lambda x.\ulcorner\neq\urcorner(x,z))(z)]]^{M,v'}$, where v' is like v except that v' assigns k to z (i.e. $[[z]]^{M,v'} = k$). $[[(\lambda x.\ulcorner\neq\urcorner(x,z))(z)]]^{M,v'}$ is an application. Its denotational value is given as a value of the function given by $[[(\lambda x.\ulcorner\neq\urcorner(x,z))]]^{M,v'}$ on the argument given by $[[z]]^{M,v'} = k$. The function denoted by $[[(\lambda x.\ulcorner\neq\urcorner(x,z))]]^{M,v'}$ assigns to an argument k the value given by $[[\ulcorner\neq\urcorner(x,z)]]^{M,v''}$, where v'' is like v' except it assigns k to x. Therefore, $[[\ulcorner\neq\urcorner(x,z)]]^{M,v''}$ is the value of $\neq$ on arguments k and k, because $[[x]]^{M,v''} = k$ and $[[z]]^{M,v'} = k$. But then the value of the

---

^{25}Yeah, it stands for $x + 1$ when $x = 3$.

^{26}Remember, the valuation of x is 4 by the valuation v'.

^{27}This is just a shortcut — it denotes a characteristic function with the truth value Truth on at least one argument.

characteristic function denoted by $\lambda z.((\lambda x.^{\ulcorner}\neq^{\urcorner}(x,z))(z))$ on any argument is Falsity (F) assuming all objects are identical to themselves. We investigated the value of the function with respect to an arbitrary argument k. Therefore, the succedent of the conclusion is *not satisfied* under these assumptions. So, the whole conclusion is not valid. ■

Theorem: Instantiation and Exposure

Let $C/o; d, x/\tau; D/\tau_{(k,l)}; 0 \le k, l \le 1$. Then,

$$exp\text{-}INST \quad \frac{\Gamma \Rightarrow C_{(D_{(k,l)}/x)} \cong o \quad \Gamma, D_{(k,l)} \cong d \Rightarrow M}{\Gamma \Rightarrow M}$$

Claim: The exp — INST rule is invalid.

Proof: By counterexample.

Consider the following derivation:

$$\frac{\Gamma \Rightarrow (\lambda x.^{\ulcorner}T^{\urcorner}(^{\ulcorner}T^{\urcorner})).{}_{(D_{(k,l)}/x)} \cong T \quad \Gamma, D_{(0,0)} \cong 1 \Rightarrow {}^{\ulcorner}\div^{\urcorner}(^{\ulcorner}2^{\urcorner}, D) \cong 2}{\Gamma \Rightarrow {}^{\ulcorner}\div^{\urcorner}(^{\ulcorner}2^{\urcorner}, D) \cong 2}$$

Is a case of a derivation using the exp-INST rule. The premises are satisfied for all models and valuations.[28] See, $(\lambda x.^{\ulcorner}T^{\urcorner}(^{\ulcorner}T^{\urcorner})).{}_{(D_{(k,l)}/x)} = (\lambda x.^{\ulcorner}T^{\urcorner}(^{\ulcorner}T^{\urcorner}))$. Because the application does not contain any occurrence of the variable x. Furthermore, $[[(\lambda x.^{\ulcorner}T^{\urcorner}(^{\ulcorner}T^{\urcorner}))]]^{M,v} = T$. Simply because $\lambda x.^{\ulcorner}T^{\urcorner}$ denotes a constant function into T. The second premise is satisfied, because $2 \div 1 = 2$. But notice that $D_{(0,0)} \cong 1$ is *not a part of* Γ! Therefore, it is not guaranteed that *any* valuation satisfying Γ will also satisfy ${}^{\ulcorner}\div^{\urcorner}(^{\ulcorner}2^{\urcorner}, D) \cong 2$. Just consider a valuation under which D is 3.[29] Therefore, there are valuations that will satisfy premises but not conclusion. ■

Theorem: Existential Generalisation

Let $x, d/\tau; D/\tau_{(k,l)}; C, ^{\ulcorner}T^{\urcorner}/o; \exists^{\tau}/(\tau \to o) \to o; Sub/\langle {}^{*}n, {}^{*}n, {}^{*}n \rangle \to^{*} n; 0 \le k, l \le 1$. Then,

$$EG \quad \frac{}{\Gamma, C_{(D_{(k,l)}/x)} \cong T \Rightarrow \exists^{\tau} \lambda x C \cong T}$$

---

[28]Which assigns standard meanings to the letter T and $\div$.

[29]Notice that D is not specified as a construction with constant denotational value! It's value is specified only with respect to some of the valuations. Just pick some D which has a denotational value 1 for some valuations and different value for other valuations, e.g., pick D as a variable.

Claim: The EG rule is invalid.
Proof: By counterexample.
Consider the following derivation:

$$\overline{\Gamma, (\lambda x.^\ulcorner\neq^\urcorner(x,z))(z)_{(\lfloor\lfloor y\rfloor\rfloor/z)} \cong T \Rightarrow \exists^\tau \lambda z.(\lambda x.^\ulcorner\neq^\urcorner(x,z))(z)_{(\lfloor\lfloor y\rfloor\rfloor/z)} \cong T}$$

It is a case of application of EG rule. But above we established that the succedent is invalid for any valuations. And we know that there are at least some valuations, under which the antecedent is satisfied.[30] Therefore, this rule is invalid. ∎

This concludes the demonstration of the un-soundness of the deduction system within THL in its actual form.

5.2 Case of TIL

I present that even the most basic rule of substitution of identicals in *extensional contexts* does not hold in the way it is formulated in TIL. That is rather a surprising result.

Claim: The extensional rule of substitution is invalid.
Proof: By counterexample.

Consider a valuation v with respect to which xv-constructs 1, z v-constructs number 3, y v-constructs the Composition $[^0+x^02]$. The following Composition:

$$[\lambda x[^0\neq xz]^03]$$

v-constructs Falsity. Simply because $3 = 3$. Now, $\lambda x[^0\neq xz]$ is a constituent of $[\lambda x[^0\neq xz]^03]$ occurring with the extensional supposition in a non-generic context of $[\lambda x[^0\neq xz]^03]$, because $\lambda x[^0\neq xz]$ does v-construct a function to be applied within the Composition $[\lambda x[^0\neq xz]^03]$ and the genericity of the whole Composition is not raised by any Closure. Furthermore, ^2y is v-congruent with z. Both v-construct 3. z is a constituent of $\lambda x[^0\neq xz]$. So, it seems we can substitute z by ^2y within the whole Composition by following the rule.

---

[30] Just take one, which satisfies the premises of the counterexample for (cond EG) above.

We get:

$$^0z \approx_v^{02} y$$

$$\overline{}$$

$$^0[\lambda x[^0\neq\ xz]^03] \approx_v^0 [\lambda x[^0\neq xz]^03]_{(^2y/z)}$$

But $^0[\lambda x[^0\neq\ xz]^03]$ is not v-congruent with $^0[\lambda x[^0\neq\ xz]^03]_{(^2y/z)}$!
$[\lambda x[^0\neq xz]^03]_{(^2y/z)} = [\lambda x[^0\neq x^2y]^03]$. $[\lambda x[^0\neq x^2y]^03]$ v-constructs the
object which is the value of the function v-constructed by $\lambda x[^0\neq x^2y]$
on the argument 3. $\lambda x[^0\neq x^2y]$ v-constructs the function f which to an
argument 3 assigns the value (if any) $v(3/x)$-constructed by $[^0\neq x^2y]$.
$[^0\neq x^2y]v(3/x)$ presents an application of $\neq$ on arguments $v(3/x)$ constructed by x and (!) $v(3/x)$ constructed by $^2y.xv(3/x)$-constructs 3.
$yv(3/x)$-constructs $[^0+x^02]$ by assumption, $[^0+x^02]v(3/x)$-constructs
5, therefore $^2yv(3/x)$-constructs 5. And because $5 \neq 3$, the whole
Composition v-constructs Truth. Therefore, we have a case which in-
validates the validity of the extensional rule of substitution for TIL. ∎

This concludes the demonstration of the un-soundness of the deduction system within TIL in its current form.

6 Discussion

The reader probably noticed the root of the problems demonstrated in
the previous section. In one or other way, all the examples presumed
and applied the notions of Double Execution (immersion), Closure
(λ-abstract), with the use of the definition of collisionless (or correct)
substitution. The notion of substitution is based on the definition
of free (occurrence of) a variable. Some of these are parts of what
I have called primary definitions, some of the notions belong among
the secondary definitions. Assuming the precedence of the above-
mentioned primary definitions over the secondary definitions, one shall
be led to revisions within the secondary definitions, if such a way is
possible.

The situation is unwelcome since both deduction systems are
demonstrably unsound. The main reason behind the problem I see
is that the secondary definitions are based just on the structural ap-

proach to constructions. Freedom of a variable is specified recursively case by case following the kinds of constructions with respect to the freedom of the variable within the subconstructions of the main embedding construction. This approach has its limits precisely due to the character of Double Executions. See, if we assume the standard procedural approach, a Double Execution presents a procedure which *may* contain an execution of a procedure *not apparent* in the initial stages of the construction. Such a procedure need not be a subconstruction of the initial Double Execution. Consequently (and this was used in the counterexamples), even a variable can occur this way. Such a variable is, however, considered when looking for the result of the whole Double Execution. Therefore, it shall be counted among the constituents of the whole procedure. But it is not by the actual secondary definitions of the system(s). Moreover, the definition of constructing with respect to a valuation in the case of Closure uses what I call *valuation specifications*. The object produced by a Closure with respect to a valuation v — i.e., some function — is specified not only with the use of that particular valuation v but also with use of the further valuations $v(a/x_i)$, which have some of the values assigned to the variables specified explicitly. But the whole specification of the values of the function produced by the Closure is given by the Construction abstracted over. The values are defined by the notion of v-construction applied to this body of Closure *with respect to the valuation specifications*. Therefore, even an object constructed by a non-apparent variable (as in the cases above) is specified or influenced by the valuation specifications. In short if yv-constructs x, then this x should be considered bound by lambda within $\lambda x^2 y$. But the current states of the definitions within both systems do not cover these cases. Nevertheless, these cases are covered by the definition of Closure. Once you notice this omission, it is not that hard to produce counterexamples to the corresponding parts of the deduction systems.

The way out of this is possible along at least two routes. Either leave the definitions of the systems as they are. But then focus on proper specification of the areas of validity of the deduction rules of the system. Or change the relevant parts of the definitions so the deduction system would come out valid. Hopefully this can be arranged with the use of the analysis of the causes of the problems hinted above. One of the outcomes of the second approach would probably be the deepening of the procedurally based definitions. More of the properties of constructions would then be dependent on what they construct than it is now.

7 Conclusion

TIL and THL, in case of a deduction, already present strong systems. These differ not only by the form of presentation, but also by the areas of the application and the universes of valid arguments they are able to depict. The style of the deduction in THL is probably more concise. However, one could ask whether the stylish differences are enough to introduce the system, which is to a large part already present in the published literature, in this new way. This way of presentation, even if it has its merits, can make it harder for somebody who is not yet acquainted with TIL, to get connected with it. I know, the main differences between TIL and THL, by the proponents of THL, are the differences in models ascribed to the fundamental notions of philosophical logic and semantic analysis of (natural) languages. Even though the models are different, there are other parts of the systems that are common (and the deduction systems are a case in point). Also, both systems are demonstrably invalid, and I suppose the reasons for their invalidity are shared between the systems.

Acknowledgement

This research was supported by the University of Oxford project 'New Horizons for Science and Religion in Central and Eastern Europe' funded by the John Templeton Foundation. The opinions expressed in the publication are those of the author(s) and do not necessarily reflect the view of the John Templeton Foundation.

References

[1] Duží, M., Jespersen, B., Materna P. (2010): *Procedural Semantics for Hyperintensional Logic. Foundations and Applications of Transparent Intensional Logic*. First edition. Berlin: Springer, series Logic, Epistemology, and the Unity of Science, vol. 17, ISBN 978-90-481-8811-6.

[2] Duží, M., Jespersen B. (2012): Transparent quantification into hyperintensional contexts de re. *Logique et Analyse* 220, 513—554.

[3] Duží, M. (2013): Deduction in TIL: From simple to ramified hierarchy of types. *Organon* F, 20(supplementary issue 2), 5-36.

[4] Duží, M., Jespersen B. (2015): Transparent Quantification into Hyperintensional objectual attitudes. *Synthese*, 192(3), 635-677. doi: 10.1007/s11229-014- 0578-z

[5] Duží, M. (2019): If structured propositions are logical procedures then how are procedures individuated?. *Synthese special issue on the Unity of propositions*, 196(4), 1249-1283. doi: 10.1007/s11229-017-1595-5

[6] Duží, M., Jespersen B. (forthcoming): Transparent quantification into hyperpropositional attitudes de dicto.*Linguistics and Philosophy*.

[7] Jespersen, B. (2021): First among equals: co-hyperintensionality for structured propositions. *Synthese* 199(1-2), 1-15, https://doi.org/10.1007/s11229-020-02987-4

[8] Kosterec, M. (2021): Substitution inconsistencies in Transparent Intensional Logic. *Journal of Applied Non-Classical Logics*, doi: 10.1080/11663081.2021.1982553,

[9] Raclavský, J., Kuchyňka P., Pezlar I. (2015): *Transparentní intenzionální logika jako characteristica universalis a calculus ratiocinator [Transparent Intensional Logic as Characteristica Universalis and Calculus Ratiocinator]*. ISBN 978-80-210-7973-1, Brno: Masarykova univerzita (Munipress).

[10] Raclavský, J. (2020): *Belief Attitudes, Fine-Grained Hyperintensionality and Type-Theoretic Logic*. Studies in Logic 88. ISBN 978-1-84890-334-0, London: College Publications.

[11] Raclavský, J. (2021): The Rule of Existential Generalisation and Explicit Substitution. *Logic and Logical Philosophy*, 1–37, doi: 10.12775/LLP.2021.011.

[12] Tichý, P. (1982): Foundations of Partial Type Theory. *Reports on Mathematical Logic* 14, 57–72.

[13] Tichý, P. (1986): Indiscernibility of Identicals. Studia Logica 45(3), 257–273.

[14] Tichý, P. (1988): *The Foundations of Frege's Logic*. Berlin, Boston: De Gruyter. doi:10.1515/9783110849264

Evaluation of Automatically Constructed Word Meaning Explanations

Marie Stará, Pavel Rychlý and Aleš Horák
Faculty of Informatics Masaryk University
Botanická 68a, Brno 602 00
Czech Republic
{413827,pary,hales}@mail.muni.cz

Abstract

Preparing exact and comprehensive word meaning explanations is one of the key steps in the process of monolingual dictionary writing. In standard methodology, the explanations need an expert lexicographer who spends a substantial amount of time checking the consistency between the descriptive text and corpus evidence.

In the following text, we present a new tool that derives explanations automatically based on collective information from very large corpora, particularly on word sketches. We also propose a quantitative evaluation of the constructed explanations, concentrating on explanations of nouns. The methodology is to a certain extent language independent; however, the presented verification is limited to Czech and English.

We show that the presented approach allows to create explanations that contain data useful for understanding the word meaning in approximately 90% of cases. However, in many cases, the result requires post-editing to remove redundant information.

Keywords: explanations, word sketches, explanation construction

This work has been partly supported by the Ministry of Education of CR within the LINDAT-CLARIAH-CZ project LM2018101.

1 Introduction

When an expert lexicographer constructs a (monolingual) dictionary, one of the most challenging and time-consuming tasks is to create concise and comprehensive word meaning explanations, also referred to as (dictionary) definitions [5, 6, 17]. The standard approaches concentrate on selecting the shared vocabulary of terms used to describe the word and organize them in the order of the main word category followed by listing the characteristics which are specific to the word [22].

In this paper, we describe a new attempt to develop dictionary word explanations for Czech and English automatically, using statistical information aggregated from large text corpora. We work with the hypothesis that a meaning of a word can be deduced from its context [4]. Therefore, it is possible to abstract common collocations of a word and use them to explain the word meaning. Such an explanation helps the reader to understand a meaning of a new word unfamiliar beforehand.

In the following section, we discuss the related work and the uniqueness of the presented approach. In sections three and four, we discuss the method and evaluate the results. Section five concludes the text.

2 Related Work

The attempts to actually *create* explanations automatically have been rather scarce. Labropoulou *et al.* [15] generated dictionary definitions from a computational lexicon, i.e. a lexicon of formalized and explicitly encoded semantic information about words. The results were comprehensible and the generated definitions were consistent, the downside being the need for the ontological background in the form of the computational lexicon. That is why the authors were focusing only on selected concrete entities; our aim is to cover a broader part of the vocabulary.

The automated definition construction process has been mostly solved by text mining approaches. There were attempts to *find, mine,* or *extract* definitions. Early approaches, such as [13], used rule-based or pattern-based approaches to identify text passages containing the sought term and its explanation. Such an approach is usually limited to a selected domain and sources of texts to allow for acceptable precision and recall. The pattern-based approach was later adjusted for mining from very large corpora [14] which offered improved precision of 73–74% with Wikipedia corpora and 31–57% with large web corpora. Borg *et al.* [3] employed genetic programming techniques to generate the best definition templates and to learn to rank these templates by importance. The templates were then used to identify definitions in non-technical texts with high precision (up to 100%) but with about 50% recall. Later works [18, 10] solved the definition text search by annotating a corpus of definitions (from Wikipedia or from scientific papers) and then training a sequence labeling classifier to mark words as *term*, *definition* or *other*. This technique improved the F-score with Wikipedia benchmark corpus to 85%.

All these techniques concentrate on extracting (parts of) the definitions from existing human-made texts. Such an approach is useful for summarization of technical terminology and educative texts, but not for general notion explanations. Another possible problem with the extractive approaches lies in the authorship laws and their possible breach.

In the following text, we concentrate on extending our previous work published by Stará and Kovář [20, 19, 21]. The method and the involved tools are described in detail here and a quantitative evaluation of explanations is offered.

3   Method

The presented explanation creation method has been evaluated with nouns, adjectives and verbs in the Czech and English languages using Word Sketches [11, 12, 8] compiled with specific sketch grammars and the *csTenTen12*[1] and *enTenTen13*[2] corpora provided by the Sketch

---

[1] https://www.sketchengine.eu/cstenten-czech-corpus/
[2] https://www.sketchengine.eu/ententen-english-corpus/

```
=pronominal subjects of "%w"
        2:"PP" ADVERB{0,3} VERB_BE? ADVERB{0,2} 1:[tag="V.[^N]?"]
        2:"PP" ADVERB{0,5} VERB_BE ADVERB{0,2} 1:"V.N"

*DUAL
=modifiers of "%w"/nouns and verbs modified by "%w"
        2:"(JJ.*|N.*[^Z])" MODIFIER{0,3} NOUN{0,2} 1:NOUN NOT_NOUN
        2:"RB" 1:[tag="JJ.*"|tag="V.*"]
        1:VERB ADVERB{0,2} 2:"RB" [tag!="RB.*" & tag!="JJ.*"]
```

Figure 1: An example of sketch grammar relations of *pronominal subjects* and *modifiers of a word*

Engine corpus management system.[3]

A sketch grammar is a set of syntactic queries written in the corpus query language (CQL [9]) to identify inter-word relations based on their position, distance, part-of-speech tag and word form (see Figure 1 for an example). The grammar, i.e. the set of word relation rules, can be applied to large text corpus to create the word sketches. The word sketches show statistically meaningful collocations of a given word organized by the relation rules; an example is shown in Figure 2.

The definition grammars are partially based on existing grammars for the above-mentioned corpora. Specific modifications are focused mainly on identifying hypernyms/hyponyms and meronyms/holonyms for nouns, opposites and specific noun collocations for adjectives, and prepositional phrases for verbs. Apart from word sketches, we also employ the thesaurus tool [11, 12] to obtain synonyms.

These tools provide the grounding of the information that should be contained in the explanation. Following the explanation schemata by renowned dictionary creation guidelines [1, 16, 2, 7], we have compiled a list of definition/explanation types. In the overview below, we show the links between the types and the sketch grammar relations.

The standard way of how to explain a word meaning consists of two parts: the *genus proximum* and *differentia specifica*. In other

---

[3]Sketch Engine is a tool analysing text corpora to identify instantly what is typical in language and what is rare, unusual or emerging usage. See https://www.sketchengine.eu/ for details.

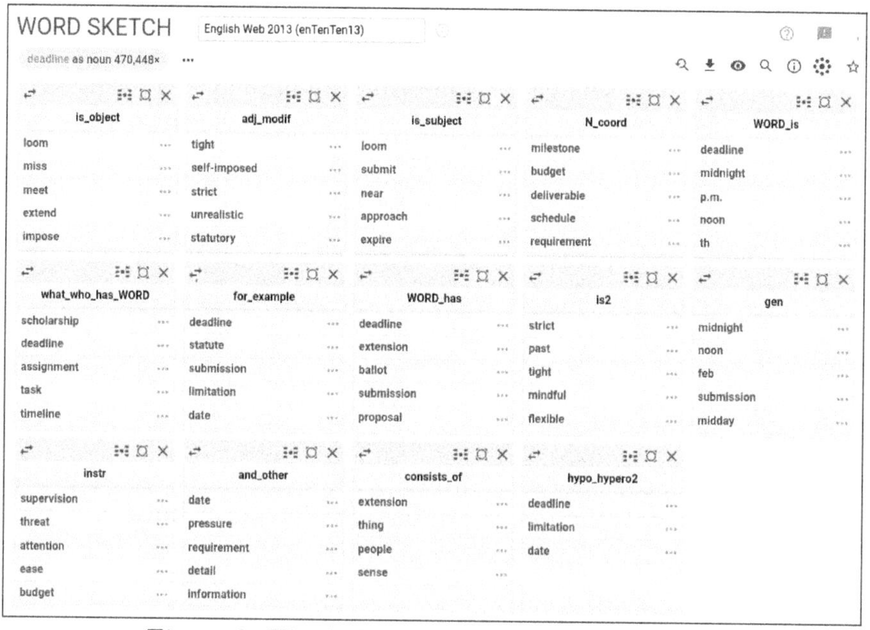

Figure 2: Word sketches for the word *deadline*

words, using a hypernym as the core determination and a set of features that distinguish the word from other related words. As long as the headword is a (concrete) noun, there is quite a clear path to a hypernym. Identification of the distinguishing features is, however, not so straightforward. They can have the form of a *verb* ("dog: an animal that barks"), a *noun* describing what the headword has got ("snake: an animal with forked tongue") or what the headword lacks ("snake: an animal with no legs"). In this regard, meronyms and holonyms are a specific case ("cutlery: spoon, fork, knife").

Another approach to explaining lies in using the *ostensive definition* or explaining by pointing. Pointing is quite helpful when explaining adjectives ("blue as the sky"). For adjective definitions, opposites can also bring clarification ("dead: not alive").

Using synonyms for explaining is not much encouraged; however, we find synonymy (or, in case of verbs, troponymy) generally useful. Here, we should remark that we consider a synonym quite loosely,

---

bone:

1. similar meaning as a/an **bone** can have (a/an) *tooth, joint, muscle, tissue, fracture, calcium, osteoporosis, skull, spine, injury, remain*

2. **bone** can be *bare, pubic, brittle*

3. for example (a/an) *femur, vertebra*

4. **bone** can have/contain (a/an) *marrow, skull, joint, tooth*

5. (a/an) *tissue, osteoporosis* can have/contain (a/an) **bone**

6. **bone** can *fragment, heal, fracture*

7. sth/sb can *break, strengthen, fracture* a/an **bone**

8. **bone** of (a/an) *contention, skull, spine*

9. **bone** with (a/an) *flesh, marrow, meat*

Figure 3: An example automatic explanation of the word *bone*

taking into account any words that have a similar meaning.

To describe the meaning of *verbs*, we mostly make use of valency, focusing especially on objects. We use valency with other parts-of-speech as well, mainly to find nouns and adjective modifiers.

To create the explanation as such, we use an automated script to combine all results together, with the main source being the word sketches and the thesaurus providing extra synonyms. The word sketches are sorted by their frequency score limited to the first three results. Some of the sketch relations are merged to one line in the explanation while removing duplicates. An example explanation for the word *bone* is presented in Figure 3. The lines are enumerated to make referencing easier.

The first line joins results from several relations – *cooperation, hypernymy*, and results from the *thesaurus*. The second line introduces significant *adjective modifiers*. The third line shows *examples*

104

Indicators in Czech Explanations				Indicators in English Explanations			
	N	J	V		N	J	V
synonym	92.96%	78.79%	90.48%	synonym	92.96%	72.73%	85.00%
J modifier	95.77%	-	-	Jmodifier	87.32%	-	-
subject	97.18%	-	90.48%	subject	69.01%	-	85.00%
object	94.37%	-	100.00%	object	80.28%	-	87.50%
hypernym	67.61%	-	-	hypernym	54.93%	-	-
hyponym	29.58%	-	-	hyponym	21.13%	-	-
meronym	54.93%	-	-	meronym	56.34%	-	-
holonym	46.48%	-	-	holonym	56.34%	-	-
A modifier	-	-	90.48%	Amodifier	-	75.76%	9.00%
(such) as	-	36.36	-	as	-	42.42%	-
troponym	-	-	45.24%	troponym	-	0.00%	52.50%
opposite	-	63.64%	-	opposite	-	45.45%	-
PP	-	-	97.62%	PP	-	-	-
infrequent	54.93%	57.58%	45.24%	infrequent	61.97%	18.18%	35.00%
data issues	81.69%	30.30%	95.86%	data issues	12.68%	36.36%	32.50%

N: noun, J: adjective, V: verb, A: adverb, PP: prepositional phrase

Table 1: Indicators in Czech and English Explanations

of the headword. The fourth and fifth lines present the *meronyms* and *holonyms*, respectively. The sixth and seventh lines list *verbs* that typically have the headword as a *subject* and *object*. The eighth and ninth line post *nouns* connected with the headword by *genitive* and *instrumental* case.

4 Evaluation

For both Czech and English, we manually evaluated a test set consisting of 71 nouns, 33 adjectives and 40 verbs.[4]

The evaluation proceeded in a quantitative way measuring the occurrence numbers of identified features. The resulting aggregated score should correspond to the decision about the usefulness of the explanation. The indicators of a (presumably) good explanation are the presence of (useful):

- synonyms: all parts-of-speech

- adjective modifiers: nouns

- adverbial modifiers: verbs, (adjectives)

---

[4]42 for Czech due to aspect variants.

- noun collocation: adjectives (*(such) as*), verbs (*is subject/object of*)

- verb collocation: nouns (*subject/object*)

- opposites: adjectives

- hypernyms/hyponyms: nouns

- meronyms/holonyms (part of): nouns

- troponyms: verbs

- prepositional phrases: verbs

On the other hand, certain features may also serve as negative indicators. Markers of possible problems with the explanation are the presence of:

- infrequent expressions

- errors caused by the corpus data (wrong lemma/tag; interchanging objects and subjects, meronyms and holonyms, etc.)

Table 1 lists the ratios of explanations that contain the said (positive or negative) indicator. The fact that the indicator is not present does not necessarily mean that the explanation is bad or insufficient: different words require different indicators, as discussed in Section 4.1.

Table 2 shows the total number of explanations that are sufficient as-is, denoted as *good* explanations. The explanations that contain *some* of the important data but are either incomplete (lack some of the necessary information) or contain too much junk data, or the data are misleading (e.g. opposites are presented as synonyms) are counted as *post-edit*; the last group of explanations, *bad*, consists of those that are completely insufficient or contain so many issues they would require rewriting, not just post-editing.

As the results were evaluated manually, we necessarily used our subjective view towards the evaluation based on our experience as a language and dictionary users. Nevertheless, we tried to minimize the

Czech	good	post-edit	bad
N	39.44%	52.11%	8.45%
J	42.42%	27.27%	30.30%
V	16.67%	78.57%	4.76%

English	good	post-edit	bad
N	39.44%	49.30%	11.27%
J	21.21%	60.61%	18.18%
V	23.81%	69.05%	2.38%

N: noun, J: adjective, V: verb

Table 2: Overall quality of Explanations

bias by following the above-mentioned indicators. We plan to engage more evaluators in the future to offer a broad objective assessment of the quality and intelligibility of the explanations.

4.1 Nouns

In this section, we present a more detailed evaluation of noun explanations, offering a comparison with existing dictionary definitions in the Macmillan Dictionary.[5]

A universally acknowledged truth says that a noun explanation should contain its hypernym. Even though this is generally true, as e.g. in Example 1 below, there are counterexamples, such as Examples 2 and 3 where the hypernym is too general or replaced by a synonym, respectively. When evaluating the testing dataset with the established dictionary, we see that a hypernym is present in 70.42% of the noun definitions, while 4.23% headwords are not defined in the dictionary.

Example 1. deer: *a large brown **animal** with long thin legs. The adult male **deer** is called a stag and may have*

---

[5]https://www.macmillandictionary.com/

*antlers growing from its head. The female **deer** is called a doe and a young **deer** is called a fawn.*[6]

Example 2. teacher: *someone whose job is to **teach***[7]

Example 3. stream: *a small narrow **river***[8]

Examples 1, 2, and 3 are all cases of a good definition, as they denote what does the headword mean. It is important to note that all these explanations use different strategies of what semantic relations to use.

In the second example, a verb describing the prototypical activity of the headword is necessary, while in the third example, only a synonym with a few modifiers is sufficient. Such an approach is not always applicable, as can be seen in Example 4. An explanation like this could be as well used to describe a *shrew*, *rat*, or *opossum*, even a *cat*. To avoid such confusion, we decided to prefer redundant data over data scarcity.

Example 4. mouse: *a small furry **animal** with a long tail*[9]

Example 5 shows an automatically created explanation that can be compared with the human-made one. The explanation contains hypernyms (*water, waterway*; possibly also *source, body*); synonyms (*river, tributary*; possibly also *lake, pond, channel*); and verb collocates for **stream** as a subject (*flow, meander*).

Example 5. stream:

- *similar meaning as a/an **stream** can have (a/an)* river, lake, pond, flow, channel, tributary, water, source, waterway, body

---

[6]https://www.macmillandictionary.com/dictionary/british/deer
[7]https://www.macmillandictionary.com/dictionary/british/teacher
[8]https://www.macmillandictionary.com/dictionary/british/stream_1
[9]https://www.macmillandictionary.com/dictionary/british/mouse_1

- **stream** *can be* steady, endless, constant
- *for example (a/an)* river, habitat
- **stream** *can have/contain (a/an)* trout, flow, water-fall
- *(a/an)* watershed, valley *can have/contain (a/an)* **stream**
- **stream** *can* flow, meander, replenish
- *sth/sb can* cross, never-end, flow *a/an* **stream**
- **stream** *of (a/an)* income, consciousness, revenue
- **stream** *with (a/an)* waterfall, trout

The results for nouns are encouraging, as a significant number of the explanations helps to understand the word meaning without the need of *excessive* post-editing. The other parts of speech seem to require a slightly different approach, mainly if we compare the explanations to existing dictionary definitions. As adjectives' main function is to modify, we need to change the paradigm and accept the fact that the results can be helpful even when they do not conform to the standard definitions. A similar problem occurs with verbs.

5 Conclusions

In this paper, we introduced a new tool for automatic construction of word meaning explanations for Czech and English, using large corpora, especially the word sketches technique. We have conceived a quantitative evaluation of explanations, focusing mainly on explanations of nouns.

The presented approach gathers enough data to construct explanations for 91.25% and 88.73% of nouns for Czech and English, respectively. As a majority of the results needs post-editing, the output in general is not yet in the state that could be presented to users as actual explanations. However, the status quo can be used as a basis for human-made explanations or definitions.

To further improve our work, the next tasks will be finding out which words need which specific approach, such as deciding which words do (not) require a hypernym, or for what words it is necessary to output a verb collocation. We believe such steps will further improve the results and reduce the need for post-editing.

References

[1] B. T. S. Atkins and Michael Rundell. *The Oxford guide to practical lexicography*. Oxford University Press, 2008.

[2] Renata Blatná and František Čermák. *Manuál lexikografie*. H&H, 1995.

[3] Claudia Borg, Mike Rosner, and Gordon Pace. Evolutionary algorithms for definition extraction. In *Proceedings of the 1st Workshop on Definition Extraction*, pages 26–32, Borovets, Bulgaria, September 2009. Association for Computational Linguistics.

[4] Kenneth Church. A pendulum swung too far. *Linguistic Issues in Language Technology*, 6(5):1–27, 2011.

[5] Peter Gilliver. The making of the Oxford English dictionary. *Lexikos*, 26(1):436–445, 2016.

[6] Jack C Gray. Creating the electronic new Oxford English dictionary. *Computers and the Humanities*, 20(1):45–49, 1986.

[7] Patrick Hanks. Do word meaning exist? In T. Fontenelle, editor, *Practical lexicography: a reader*, pages 125–134. Oxford University Pres, 2008.

[8] Aleš Horák and Pavel Rychlý. Discovering Grammatical Relations in Czech Sentences. In *RASLAN 2009*, pages 81–88, 2009.

[9] Miloš Jakubíček, Adam Kilgarriff, Diana McCarthy, and Pavel Rychlý. Fast syntactic searching in very large corpora for many languages. *PACLIC*, pages 741–47, 2010.

[10] Yiping Jin, Min-Yen Kan, Jun-Ping Ng, and Xiangnan He. Mining scientific terms and their definitions: A study of the ACL anthology. In *Proceedings of the 2013 Conference on Empirical Methods in Natural Language Processing*, pages 780–790, Seattle, Washington, USA, October 2013. Association for Computational Linguistics.

[11] Adam Kilgarriff, Vít Baisa, Jan Busta, Milos Jakubícek, Vojtech Kovár, Jan Michelfeit, Pavel Rychlý, and Vít Suchomel. The Sketch Engine: ten years on. *Lexicography*, 1:7–36, 2014.

[12] Adam Kilgarriff, Pavel Rychlý, Pavel Smrz, and David Tugwell. The Sketch Engine. In Geoffrey Williams and Sandra Vessier, editors, *Pro-*

ceedings of the 11th EURALEX International Congress, pages 105–115, Lorient, France, july 2004. Université de Bretagne-Sud, Faculté des lettres et des sciences humaines.

[13] Judith Klavans and Smaranda Muresan. Definder: Rule-based methods for the extraction of medical terminology and their associated definitions from on-line text. In *Proceedings of the AMIA Symposiumm*, 01 2000.

[14] Vojtěch Kovář, Monika Močiariková, and Pavel Rychlý. Finding Definitions in Large Corpora with Sketch Engine. In *Proceedings of the Tenth International Conference on Language Resources and Evaluation (LREC 2016)*, 2016.

[15] Penny Labropoulou, Elena Mantzari, Harris Papageorgiou, and Maria Gavrilidou. Automatic generation of dictionary definitions from a computational lexicon. In *Proceedings of the Second International Conference on Language Resources and Evaluation (LREC'00)*, Athens, Greece, May 2000. European Language Resources Association (ELRA).

[16] Sidney I. Landau. *Dictionaries: the art and craft of lexicography*. Cambridge University Press, 2nd edition, 2001.

[17] Margaret G McKeown. Creating effective definitions for young word learners. *Reading Research Quarterly*, pages 17–31, 1993.

[18] Roberto Navigli, Paola Velardi, and Juana Ruiz-Martínez. An annotated dataset for extracting definitions and hypernyms from the web. In *Proceedings of the International Conference on Language Resources and Evaluation*. European Language Resources Association, 01 2010.

[19] M. Stará. Automatically Created Noun Definitions for Czech. In Aleš Horák, Pavel Rychlý, and Adam Rambousek, editors, *Proceedings of the Twelfth Workshop on Recent Advances in Slavonic Natural Languages Processing, RASLAN 2018*, pages 63–68, Brno, 2018. Tribun EU.

[20] M. Stará. Automatically Created Noun Explanations for English. In Aleš Horák, Pavel Rychlý, and Adam Rambousek, editors, *Proceedings of the Thirteenth Workshop on Recent Advances in Slavonic Natural Language Processing, RASLAN 2019*, pages 83–87, Brno, 2019. Tribun EU.

[21] M. Stará and V. Kovář. Options for automatic creation of dictionary definitions from corpora. In Aleš Horák, Pavel Rychlý, and Adam Rambousek, editors, *Tenth Workshop on Recent Advances in Slavonic Natural Language Processing, RASLAN 2016*, pages 111–124, Brno, 2016. Tribun EU.

[22] B. Svensén. *A Handbook of Lexicography: The Theory and Practice of Dictionary-Making.* Cambridge University Press, Cambridge, 2009.

What Topic for *Off-Topic* in WK3?

Massimiliano Carrara
FISPPA, University of Padua, Italy
massimiliano.carrara@unipd.it

Filippo Mancini
FISPPA, University of Padua, Italy
filippo.mancini@unipd.it

Wei Zhu
FISPPA, University of Padua, Italy
wei.zhu@unipd.it

Abstract

Beall [1] proposes to read the middle-value of Weak Kleene logic as *off-topic*. This interpretation has recently drawn some attention: for instance, Francez [5] has pointed out that Beall's interpretation does not meet some important requirements to count as a truth value. Moreover, Beall is silent about what a topic (or a subject matter) is. But arguably, *what is a topic?* is a crucial question, and an answer is really important to fully understand his proposal. Thus, our goal here is to help to remedy this deficiency, and show how Beall's interpretation of Weak Kleene truth-values interacts with the notion of topic. To do that, we formalize his motivating ideas and draw some consequences from them.

We would like to thank Bjørn Jespersen and Pavel Materna for inviting us to contribute to this collected volume for Marie Duží. Massimiliano Carrara takes this opportunity to thank Marie for the change of working with her: it was a great opportunity for him to grow scientifically. This article was partially funded by the CARIPARO Excellence Project (CARR_ ECCE20_ 01): *Polarization of irrational collective beliefs in post-truth societies. How anti-scientific opinions resist expert advice, with an analysis of the antivaccination campaign (PolPost).*

1 Introduction

In the field of many-valued logics, Weak Kleene logic (WK3) is a greatly underdeveloped subject compared to its strong counterpart – i.e. K3. Despite the several attempts to provide a complete characterization for these system (e.g., [2], [6], and [8]), the problem of giving a philosophically sound interpretation for the semantic non-classical value, **u**,[1] still persists. Some different interpretations are now available, such as *nonsense, meaninglessness,* and *undefined.*[2] Among them, a new proposal by Beall [1] suggests to read **u** as *off-topic.* Thus, a proposition that obtains this value should be regarded as being off-topic.

Such an interpretation has recently drawn some attention and it has proved useful — e.g. it is adopted by Carrara and Zhu [3] to distinguish two kinds of computational errors. However, it still has some problem and lacks some explication. For example, Francez [5] has pointed out that **u** as *off-topic* does not satisfy the pre-theoretic understanding of a truth-value in a truth-functional logic.[3] If Francez's criticism is valid, a characterization of topic (or a subject matter) is needed to evaluate pro and cons of Beall's proposal. But Beall [1] is silent about what a topic is.[4] Thus, our goal here is to help to remedy this deficiency, and show how Beall's interpretation of WK3 truth-values could interact with the notion of topic.

This paper is divided into two sections. In §2 we introduce WK3. In §3 we elaborate on Beall's notion of topic and draw some consequences about how topics work according to his conception.

---

[1]**u** means *undefined* as in [8]. This value might also be referred to as the *third value,* the *middle value,* or 0.5.

[2]For a survey on new interpretations of **u** see [4].

[3]Specifically, Francez [5] claims that any notion that aspires to qualify as an interpretation of a truth-value has to satisfy certain requirements, and that Beall's interpretation of **u** as *off-topic* does not do that.

[4]"Topic" and "subject matter" are just synonyms, and throughout the paper we will use them interchangeably.

2 WK3

Let us briefly introduce WK3. Its language is the standard propositional language, L. Given a nonempty countable set $\mathsf{Var} = \{p, q, r, \dots\}$ of atomic propositions, the language of WK3 is defined by the following Backus–Naur Form:

$$\Phi_L ::= p \mid \neg\phi \mid \phi \vee \psi \mid \phi \wedge \psi \mid \phi \supset \psi$$

We use $\phi, \psi, \gamma, \delta \dots$ to denote arbitrary formulas, $p, q, r, \dots$ for atomic formulas, and $\Gamma, \Phi, \Psi, \Sigma, \dots$ for sets of formulas. Propositional variables are interpreted by a valuation function $V_a : \mathsf{Var} \longmapsto \{\mathbf{t}, \mathbf{u}, \mathbf{f}\}$ that assigns one out of three values to each $p \in \mathsf{Var}$. The valuation extends to arbitrary formulas according to the following definition:

Definition 2.1 (Valuation). A valuation $V : \Phi_L \longmapsto \{\mathbf{t}, \mathbf{u}, \mathbf{f}\}$ is the unique extension of a mapping $V_a : \mathsf{Var} \longmapsto \{\mathbf{t}, \mathbf{u}, \mathbf{f}\}$ that is induced by the tables from Table 1.

ϕ	$\neg\phi$
t	f
u	u
f	t

$\phi \vee \psi$	t	u	f
t	t	u	t
u	u	u	u
f	t	u	f

$\phi \wedge \psi$	t	u	f
t	t	u	f
u	u	u	u
f	f	u	f

$\phi \supset \psi$	t	u	f
t	t	u	f
u	u	u	u
f	t	u	t

Table 1: Weak tables for logical connectives in L

Table 1 provides the full *weak tables* from [8, §64], that obtain "by supplying [the third value] throughout the row and column headed by [the third value]".[5] Note that in WK3 negation works like in K3, but conjunction and disjunction work differently. Specifically, the interpretation of disjunction is not *max* and that of conjunction is not *min*.[6] The way **u** transmits is usually called *contamination* (or *infection*), since the value propagates from any $\phi \in \Phi_L$ to any construction $k(\phi, \psi)$, independently from the value of ψ (here, k is any complex formula made out of some occurrences of both ϕ and ψ and whatever combination of $\vee$, $\wedge$, $\supset$). To better capture the way **u** works in combination with the other truth-values, let us introduce the following definition:

Definition 2.2. For any $\phi \in \Phi_L$, *var* is a mapping from Φ_L to the power set of Var, which can be defined inductively as follows:

$$var(p) = \{p\},$$

$$var(\neg\phi) = var(\phi),$$

$$var(\phi \vee \psi) = var(\phi) \cup var(\psi),$$

$$var(\phi \wedge \psi) = var(\phi) \cup var(\psi),$$

$$var(\phi \supset \psi) = var(\phi) \cup var(\psi).$$

We can extend the above definition concerning *var* as follows:

---

[5]It is clear by Table 1 that $\wedge$ and $\supset$ could be defined from $\neg$ and $\vee$: it is easy to check from Table 1 that $\phi \wedge \psi = \neg(\neg\phi \vee \neg\psi)$ and $\phi \supset \psi = \neg\phi \vee \psi$. We prefer to introduce them all as primitives in order to have a complete overview with Table 1.

[6]It is for this reason that we find appropriate to use **u** rather than 0.5 for the non-classical value, since in WK3 it does not have the behavior of a "middle value".

Definition 2.3. *Let X be a set of sentences. var denotes a mapping from the power set of Φ_L to the power set of* Var. *That is, $var(X) = \bigcup\{var(\phi) \mid \phi \in X\}$.*

Then, the following fact expresses *contamination* very clearly:

Fact 2.1 (Contamination). For all formulas ϕ in L and any valuation V:

$$V(\phi) = \mathbf{u} \quad \textit{iff} \quad V_a(p) = \mathbf{u} \text{ for some } p \in var(\phi)$$

The left-to-right direction is shared by all the most common three-valued logics; the right-to-left direction is clear from Table 1, and it implies that ϕ takes value $\mathbf{u}$ if some $p \in var(\phi)$ has the value, and no matter what the value of q is for any $q \in var(\psi) \setminus \{p\}$.

The WK3 consequence relation is defined as preservation of the value $\mathbf{t}$, so that an important feature of WK3 is that Addition is invalid: $\phi \nvdash_{wk3} \phi \vee \psi$.

3 Beall's *Off-topic* Interpretation

According to Halldén [6] and Bochvar and Bergmann [2], the third value in WK3 is interpreted as *meaningless* or *nonsense*. However, such interpretations seem to suffer from some problems. For example, it is not at all obvious that we can make the conjunction or the disjunction of a meaningless sentence with one with a traditional truth-value. Observing this problem, Beall [1] proposes an alternative interpretation for $\mathbf{u}$: i.e. *off-topic*. Thus, *What is a topic?* is a crucial question for his proposal. For depending on how we answer this question we may have different consequences on such a reading. Unfortunately, Beall [1] is silent about that. But he gives some constraints we can use to explore how topics behave and how they relate to the WK3 truth-values.

Before presenting them, let us make some assumptions and define a simple notation to facilitate the discussion. We assume that topics can be represented by sets.[7] We use bold letters for topics, such as **s**, **t**, etc. ⊆ is the inclusion relation between topics, so that **s** ⊆ **t** expresses that **s** is included into (or is a subtopic of) **t**. Given that, we define a *degenerate* topic as one that is included in every topic.[8] Also, we define the overlap relation between topics as follows: **s** ∩ **t** iff there exists a non-degenerate topic **u** such that **u** ⊆ **s** and **u** ⊆ **t**. Further, it is assumed that every meaningful sentence α comes with a *least* subject matter, represented by $\tau(\alpha)$. $\tau(\alpha)$ is the unique topic which α is about, such that for every topic α is about, $\tau(\alpha)$ is included into it. Thus, we say that α is *exactly* about $\tau(\alpha)$.[9] But α can also be *partly* or *entirely* about other topics: α is entirely about **t** iff $\tau(\alpha)$ ⊆ **t**, whereas α is partly about **t** iff $\tau(\alpha)$ ∩ **t**.

3.1 Beall's Terminology and Motivating Ideas

Beall [1]'s new interpretation starts from setting a terminology concerning a *theory*, T. T is a set of sentences closed under a consequence relation, Cn. That is, $T = Cn(X) = \{\phi \mid X \vdash \phi\}$, where X is a given set of sentences and $\vdash$ is the consequence relation of the logic we are working with. As for WK3, theories are sets of sentences closed under WK3 logical consequence. Then, Beall puts forward the following motivating ideas for his proposal:

1. A theory is about all and only what its elements –

---

[7]This is a natural assumption. For note that topics are represented by sets in all the main approaches to subject matter as discussed in [7].

[8]The inclusion relation, ≤, is usually taken to be reflexive, so that every topic includes itself.

[9]Throughout this paper, when we talk about the topic of a sentence we mean its least topic. In case we want to refer to one of its topics that is not the least one, we will make it clear.

that is, the claims in the theory – are about.

2. Conjunctions, disjunctions and negations are about exactly whatever their respective subsentences are about:

   (a) Conjunction $\phi \wedge \psi$ is about exactly whatever ϕ and ψ are about.

   (b) Disjunction $\phi \vee \psi$ is about exactly whatever ϕ and ψ are about.

   (c) Negation $\neg \phi$ is about exactly whatever ϕ is about.

3. Theories in English are rarely about every topic expressible in English.

[1, p. 139]

As for the three WK3 semantic values, Beall proposes to "[...] read the value 1 not simply as *true* but rather as *true and on-topic*, and similarly 0 as *false and on-topic*. Finally, read the third value 0.5 as *off-topic*" [1, p. 140]. Thus, an arbitrary sentence ϕ is either true and on-topic, false and on-topic, or off-topic. And note that since both on-topic and off-topic sentences are arguably meaningful, the problem concerning the conjunction/disjunction of meaningless sentences vanishes.

3.2 Formalizing Beall's Ideas

We can arguably formalize Beall's motivating ideas as follows:[10]

Definition 3.1. *Let T be a WK3 theory and $\tau(T)$ be its topic. The following conditions show how the topics of WK3 sentences and $\tau(T)$*

---

[10]Note that we use the same notation, $\tau(\ldots)$, both for the topic of a sentence and the topic of a theory – i.e. a set of sentences.

are related.

1. $\tau(T) = \bigcup\{\tau(\phi) \mid \phi \in T\}$.

2. (a) $\tau(\phi \wedge \psi) = \tau(\phi) \cup \tau(\psi)$.

    (b) $\tau(\phi \vee \psi) = \tau(\phi) \cup \tau(\psi)$.

    (c) $\tau(\neg\phi) = \tau(\phi)$.

3. *for any T in English, there is some topic ζ expressible in English such that $\zeta \not\subseteq \tau(T)$.*

These formulations correspond to Beall's three motivating ideas concerning the off-topic interpretation. But the way we formally capture condition 3 requires a comment. Beall claims that "[t]heories in English are *rarely* about every topic expressible in English" (emphasis added). However, our formal translation ignores "rarely", and replaces it with "never". Nonetheless, as explained in §3.3, we do believe that there might be a theory that is about every topic. But since the existence of such a theory does not affect our considerations throughout the paper, we omit "rarely" in our formalization of Beall's condition 3.

For ease of understanding, some examples are presented below.

Example 3.1. *Let $\phi = q \vee r$ and $\psi = \neg p \wedge q$. Then, according to Def. 3.1 the following results follow:*

1. $\tau(\phi) = \tau(q) \cup \tau(r)$

2. $\tau(\psi) = \tau(p) \cup \tau(q)$

3. $\tau(\phi \vee \psi) = \tau(q) \cup \tau(r) \cup \tau(p)$

4. $\tau(\psi \vee \neg\psi) = \tau(p) \cup \tau(q)$

5. $\tau(\phi \wedge \neg\phi) = \tau(q) \cup \tau(r)$

These examples immediately follow from Def. 3.1. We note that what a classical tautology $\psi \vee \neg\psi$ is about is what its atomic components (p, q) are about, although neither p nor q is about **tautology**. Following this result, we can derive two further outcomes. First, a classical tautology might be off-topic of a theory about **tautology**. Suppose a theory T^*'s topic is about **tautology**. A claim like "The Moon is made of green cheese or the Moon is not made of green cheese" is off-topic, because this claim is about the **Moon** and **green cheese**, but not about **tautology**. Second, classical tautologies are not neutral to topics. In classical propositional logic, a tautology is true for all possible truth-value assignments to its atomic components. In weak Kleene logics, a tautology can be off-topic. That is, a tautology can be either true and on-topic or off-topic, but cannot be false and on-topic. Similar result holds for a contradiction: a contradiction can be either false and on-topic or off-topic, but cannot be true and on-topic.

From Def. 3.1 we can derive also the following results.

Corollary 3.1. *For any $\phi \in \Phi_L$, $\tau(\phi) = \bigcup\{\tau(p) \mid p \in var(\phi)\}$.*

Proof. We can prove it by induction.

1. Let ϕ be an atomic sentence p. Then $\tau(\phi) = \tau(p)$.

2. Let $\phi = \neg\psi$ and $\tau(\psi) = \bigcup\{\tau(p) \mid p \in var(\psi)\}$. Since $var(\phi) = var(\neg\psi) = var(\psi)$, we have $\tau(\neg\psi) = \tau(\psi) = \bigcup\{\tau(p) \mid p \in var(\phi)\}$.

3. Let $\phi = \gamma \wedge \delta$, $\tau(\gamma) = \bigcup\{\tau(p) \mid p \in var(\gamma)\}$, and $\tau(\delta) = \bigcup\{\tau(q) \mid q \in var(\delta)\}$. Since $var(\phi) = var(\gamma) \cup var(\delta)$, we can derive $\tau(\phi) = \tau(\gamma) \cup \tau(\delta) = \bigcup\{\tau(r) \mid r \in (var(\gamma) \cup var(\delta))\}$. That is, $\tau(\phi) = \bigcup\{\tau(r) \mid r \in var(\phi)\}$.

4. For $\phi = \gamma \vee \delta$, we can prove that $\tau(\phi) = \bigcup\{\tau(r) \mid r \in var(\phi)\}$ in the same way as above.

$\square$

Corollary 3.2. $\tau(T) = \bigcup\{\tau(p) \mid p \in var(T)\}$.

This corollary follows from the Def. 3.1 and Corollary 3.1. It shows that what a theory T is about boils down to the union of what the atomic components of each claims in T are about. Moreover, even if Beall does not mention what an arbitrary set (i.e. not necessarily a theory) is about, we buy the following very plausible definition:

Definition 3.2. *Let X be a set of sentences. Such a set is about all and only what its elements are about. That is, $\tau(X) = \bigcup\{\tau(\phi) \mid \phi \in X\}$.*

Thus, the following corollary follows from Def. 3.2:

Corollary 3.3. *Let X be a set of sentences. Then $\tau(X) = \bigcup\{\tau(p) \mid p \in var(X)\}$.*

By virtue of Corollary 3.3 and Def. 3.1, we can derive the following results:

Corollary 3.4. *For any $\phi, \psi \in \Phi_L$, $\tau(k(\phi, \psi)) = \tau(\{\phi, \psi\})$.*

Proof. By Def. 3.1, $\tau(k(\phi, \psi)) = \tau(\phi) \cup \tau(\psi)$. According to Def. 3.2, $\tau(\{\phi, \psi\}) = \tau(\phi) \cup \tau(\psi)$. Hence, $\tau(k(\phi, \psi)) = \tau(\{\phi, \psi\})$. $\qquad\square$

Corollary 3.5. *For any $\phi, \psi \in \Phi_L$, $\tau(k(\phi, \psi)) = \tau(var(\phi)) \cup \tau(var(\psi))$.*

Proof. According to Corollary 3.4, $\tau(k(\phi, \psi)) = \tau(\{\phi, \psi\})$. By virtue of Corollary 3.3, we can derive $\tau(k(\phi, \psi)) = \bigcup\{\tau(p) \mid p \in var(\{\phi, \psi\})\}$. According to Def. 2.2, $var(\{\phi, \psi\}) = var(\phi) \cup var(\psi)$. Hence, $\tau(k(\phi, \psi)) = \bigcup\{\tau(p) \mid p \in (var(\phi) \cup var(\psi))\} = \bigcup\{\tau(p) \mid p \in var(\phi)\} \cup \bigcup\{\tau(q) \mid q \in var(\psi)\}$. Since $var(\phi) = var(var(\phi))$ and $var(\psi) = var(var(\psi))$, we can derive $\tau(k(\phi, \psi)) = \tau(var(\phi)) \cup \tau(var(\psi))$ by Def. 3.2. $\qquad\square$

Corollary 3.6. *For any $\phi \in \Phi_L$ and WK3 theory T, $\tau(T) = \bigcup\{\tau(var(\phi)) \mid \phi \in T\}$.*

Proof. From Def. 3.2 and $var(\phi) = var(var(\phi))$, we can derive $\tau(\phi) = \tau(var(\phi))$. Since Def. 3.1 claims that $\tau(T) = \bigcup\{\tau(\phi) \mid \phi \in T\}$, we can derive $\tau(T) = \bigcup\{\tau(var(\phi)) \mid \phi \in T\}$ by substituting $\tau(\phi)$ with $\tau(var(\phi))$. ☐

Lemma 3.1. *For any $p \in \Phi_L$ and WK3 theory T, if $var(p) \subseteq var(T)$, then $\tau(p) \subseteq \tau(T)$.*

Proof. According to Def. 2.3, $var(T) = \bigcup\{var(\phi) \mid \phi \in T\}$, $var(var(\phi)) = \bigcup\{var(p) \mid p \in var(\phi)\}$. Then $var(T) = \bigcup\{var(p) \mid p \in var(T)\}$. If $var(p) \subset var(T)$, then $p \in var(T)$. Since $\tau(T) = \bigcup\{\tau(p) \mid p \in var(T)\}$, then $\tau(p) \subseteq \tau(T)$. ☐

However, the result does not hold in the opposite direction. That is, if $\tau(p) \subseteq \tau(T)$, it might not be the case that $var(p) \subseteq var(T)$. To understand this point, consider the following counterexample.

Example 3.2. *For any $r, q \in \Phi_L$ and WK3 theory T, let $var(r) \not\subseteq var(T)$, $var(q) \subseteq var(T)$, and $\tau(r) = \tau(q) \subseteq \tau(T)$. Therefore, even though $\tau(r) \subseteq \tau(T)$, $var(r) \not\subseteq var(T)$.*

This counterexample is possible because τ is not necessarily bijective. As a justification, consider the following line of reasoning. Suppose that if $\tau(r) \subseteq \tau(T)$, then $var(r) \subseteq var(T)$. In that case, we get that whatever sentence is in the theory, it is also on-topic; and that whatever sentence is not in the theory, it is also off-topic. But then, Beall's reading of the truth value 0 as false-and-on-topic is not available anymore. In other words, allowing for the bijection results in a conflict between Beall's conception of topic and his reading of the $WK3$ truth values.

To sum up: by Beall's ideas and the way WK3 works we get that (1) the topic of a sentence is completely determined by (is the union of the topics of) its propositional variables, and (2) the topic of a theory is completely determined by (is the union of the topics of) the propositional variables of its sentences. Moreover, we get also the following important result:

Theorem 3.1. *For any $\phi \in \Phi_L$ and WK3 theory T, if $var(\phi) \subseteq var(T)$, then $\tau(\phi) \subseteq \tau(T)$.*

Proof. We can prove it by induction.

1. If ϕ is an atomic sentence, this theorem holds for ϕ by virtue of Lemma 3.1.

2. If $\phi = \neg\psi$ and this theorem holds for $\neg\psi$. We can derive that this theorem holds for ϕ, because $var(\neg\psi) = var(\psi)$.

3. If $\phi = \gamma \vee \delta$, and this theorem holds for γ and δ. We can derive this theorem holds for ϕ, because $var(\gamma \vee \delta) = var(\gamma) \cup var(\delta)$.

4. We can prove this holds for $\phi = \gamma \wedge \delta$ in the same way.

$\square$

By virtue of Thm. 3.1, we can clarify Beall's on-topic/off-topic interpretation in the following way.

Corollary 3.7. *Let T be a WK3 theory and $\tau(T)$ be its topic. For any $\phi \in \Phi_L$,*

1. *ϕ is on-topic iff $\tau(\phi) \subseteq \tau(T)$. But note that this does not guarantee that $\phi \in T$. However, if $\phi \in T$, by Def. 3.1, it is definitively on-topic.*

2. *ϕ is off-topic iff $\tau(\phi) \nsubseteq \tau(T)$. This suffices to say that $\phi \notin T$.*

Finally, we can note that such an interpretation fits Beall's conditions as well as WK3 semantics. To see this, let's conjoin two propositional variables, p and q, to get $p \wedge q$. Suppose that both are on-topic, i.e. $\tau(p) \subseteq \tau(T)$ and $\tau(q) \subseteq \tau(T)$. According to 2(a), $\tau(p \wedge q) = \tau(p) \cup \tau(q)$. Thus, $\tau(p \wedge q) \subseteq \tau(T)$, that is $p \wedge q$ is on-topic, which is in line with WK3 semantics. Now, suppose that at least one of the conjuncts is off-topic, say q. Thus, $\tau(q) \nsubseteq \tau(T)$. Therefore, $\tau(p \wedge q) \nsubseteq \tau(T)$, which is also in line with WK3 semantics. Alternatively, we might also be tempted to consider the following different interpretation: for p to be on-topic means that $\tau(p) \cap \tau(T) \neq \varnothing$, whereas to be off-topic means that $\tau(p) \cap \tau(T) = \varnothing$. For instance, this is exactly what Hawke [7, p. 700] suggests: "[t]o say that a claim is *somewhat* on-topic is to say that its subject matter overlaps with the discourse topic".[11] However, such an interpretation is not compatible with Beall's constraints 1-3. For condition 2(a) clashes with WK3 semantics. To see this, suppose that $\tau(p) \cap \tau(T) \neq \varnothing$ but $\tau(q) \cap \tau(T) = \varnothing$. Thus, since $\tau(p \wedge q) = \tau(p) \cup \tau(q)$, it follows that $\tau(p \wedge q) \cap \tau(T) \neq \varnothing$ – i.e. $p \wedge q$ is on-topic. This contradicts WK3 semantics – namely, contamination. Moreover, our observations match [1, fn. 5]: "[a]n alternative account might explore 'partially off-topic', but I do not see this as delivering a natural interpretation of WK3". Here, Beall is suggesting to distinguish two notions: *off-topic* and *partially off-topic*. The latter might be legitimately taken to correspond to the alternative reading in terms of overlap between topics that we rejected – as indeed he does.

3.3   Some remarks about Beall's condition 3

As we anticipated in the previous section, some comments are required about Beall's condition 3: "[t]heories in English are rarely about every topic expressible in English". Now, we believe there are two possible ways to read such a condition: (3a) for Beall there is at least one theory the topic of which is a degenerate topic, so that it is included in every topic; (3b) for Beall there is at least one theory the topic

---

[11]Here, the discourse topic is what we call the topic of reference.

of which overlaps with every topic. Thus, (3a) reads the aboutness relation in Beall's condition 3 as *entire aboutness*, whereas (3b) reads it as *partial aboutness*. Now, we claim that (3a) should be rejected, based on the following reasons. According to the Corollary 3.6 above, the topic of a theory is the union of the topics of every atomic component of every sentence in that theory. Now, assume (3a) is the correct interpretation. Also, let us exclude the case $T = \varnothing$ due to vacuity. Thus, every theory has at least one propositional variable as a member. Therefore, the only way for a theory to have a degenerate topic is that there is only one propositional variable in the language, which is absolutely implausible. Because of that, we can dismiss (3a). Then, let us try (3b). There are two ways for a theory to have a topic which overlaps with every topic: either to have a degenerate topic, or to have a topic that includes every topic – i.e. a universal topic. Since the first option is rejected by the previous considerations, we are left with the second one. Let us try to elaborate a little on that and use some formalization. According to this reading, a universal theory T_U – i.e. a theory the topic of which is universal – can be defined in the following way:

Definition 3.3. *Let x be a variable ranging over sentences and set of sentences of the language. Thus, T_U is a universal theory iff $\forall x$ $\tau(x) \subseteq \tau(T_U)$. Also, we call the topic $\tau(T_U)$ of a universal theory, T_U, a universal topic.*

Then, we may wonder whether such a universal theory represents a coherent notion, as Beall seems to hold by using the word "rarely". We claim it is, based on the following considerations. We aim at showing that universality does not implies triviality – i.e. that a universal theory is not necessarily trivial.[12] To see that, consider that a universal topic must include the topic of every sentence of the language.

---

[12] Recall that a trivial theory is a theory that makes every sentence a theorem – i.e. everything is provable in a trivial theory. In other words, a trivial theory has any sentence expressible in the language as a member.

Thus, for any $\phi \in \Phi_L$, $\tau(\phi) \subseteq \tau(T_U)$. However, as we explained in Corollary 3.7, from that we cannot conclude that $\phi \in T_U$. Thus, we cannot infer that T_U is trivial. Therefore, a universal theory appears to be a coherent and available notion.[13]

4 Conclusion

In this paper we have formalized some of Beall's ideas about the notion of topic and drawn some facts from them, to see what kind of topic is the best one for his off/on-topic reading of the WK3 truth-values. The result is that, from Beall's perspective, for a claim ϕ to be on-topic means that $\tau(\phi) \subseteq \tau(T)$, where T is the WK3 theory at stake; whereas, for ϕ to be off-topic means that $\tau(\phi) \not\subseteq \tau(T)$. This is in line with WK3 semantics.

References

[1] J. Beall. Off-topic: A new interpretation of weak-Kleene logic. *The Australasian Journal of Logic*, 13(6):136–142, 2016.

[2] D. A. Bochvar and M. Bergmann. On a three-valued logical calculus and its application to the analysis of the paradoxes of the classical extended functional calculus. *History and Philosophy of Logic*, 2(1-2):87–112, 1981.

[3] M. Carrara and W. Zhu. Computational errors and suspension

---

[13]An interesting question we were asked by a referee is whether a universal theory is *ipso facto* a theory of everything. Arguably, the answer depends on how a theory of everything is defined. A good strategy might be to take a theory of everything as a theory stating all the truths about every object in the ontological domain. Given that, we cannot find a straightforward way to argue that being a universal theory implies being a theory of everything. But such an issue certainly requires a further and deeper investigation.

in a PWK epistemic agent. *Journal of Logic and Computation*, 31(7):1740–1757, 2021.

[4] R. Ciuni and M. Carrara. Semantical analysis of weak Kleene logics. *Journal of Applied Non-Classical Logics*, pages 1–36, 2019.

[5] N. Francez. On Beall's new interpretation of WK_3. *Journal of Logic, Language and Information*, pages 1–7, 2019.

[6] S. Halldén. *The Logic of Nonsense*. Uppsala Universitets Årsskrift, Uppsala, 1949.

[7] P. Hawke. Theories of aboutness. *Australasian Journal of Philosophy*, 96(4):697–723, 2018.

[8] S. C. Kleene. *Introduction to Metamathematics*. Bibliotheca Mathematica. North-Holland P. Noordhoff, Amsterdam Groningen, 1952.

TIL and Degrees of Belief

Lukáš Bielik*

Comenius University in Bratislava, Faculty of Arts, Department of Logic and Methodology of Sciences

lukas.bielik@uniba.sk

Abstract

This study explores the possibilities of using the theoretical apparatus of TIL to analyse a wider range of doxastic attitudes. Specifically, I propose a conservative extension of TIL's apparatus to the analysis of attitudes in which an agent adopts a degree of belief (i.e. credence) towards a particular proposition. Degrees of belief constitute a key part of Bayesian epistemology. Hence, if the presented TIL-based explication of doxastic attitudes with credences is correct, this shows that TIL can also be used to analyse those inferences in which agents use probabilities as formal counterparts of their degrees of belief.

1 Introduction

Transparent Intensional Logic (hereafter referred to as "TIL") is one of today's well-recognized and established logical systems. This system, whose foundations were laid by the Czech logician Pavel Tichý (especially in [38]),[1] and which was further developed in one form or another by others such as (in alphabetical order) Pavel Cmorej,

I want to thank Bjørn Jespersen for his excellent comments on an earlier version of this paper and Miloš Kosterec for many fruitful discussions on TIL. As always, any remaining shortcomings are my own.

*This work was supported by the project APVV-17-0057 *Analysis, Reconstruction and Evaluation of Arguments* and the project VEGA No. 1/0197/20 *Attitudes in Communication and Argumentation: Semantic and Pragmatic Aspects*.

[1]See also [37], and many other papers collected in [39].

Marie Duží, František Gahér, Daniela Glavaničová, Bjørn Jespersen, Miloš Kosterec, Petr Kuchyňka, Pavel Materna, Ivo Pezlar and Jiří Raclavský, is today—among others—already routinely and successfully used for the analysis of natural language.[2] TIL, as a ramified typed lambda calculus with a hyperintensional construction semantics, has now found its application in solving a wide range of problems. TIL has a theoretical apparatus with considerable expressive power, which enables the development of an independent theory of procedural semantics, the output of which is also a theory of concepts, conceptual systems (cf. [31], [32], [25], [8]) and propositions ([24], [26]). But the theory's applicative power manifests itself above all in its ability to solve those semantic problems that arise in inference when we use so-called conceptual and propositional attitudes (see [17]; [7]; [33]), or in contexts where certain expressions (usually certain descriptions) may occur with *de re* and *de dicto* supposition ([4]), or in the analysis of the grammatical tenses of sentences ([10]), in the analysis of questions ([12]), or in the analysis of the informativeness of analytic information (see [5]).[3]

It is the field of propositional attitudes to which I turn my attention in this paper. In particular, I will be concerned with extending previous analyses of doxastic attitudes from contexts where categorical propositional attitudes are used, represented by the attitude "Agent $\mathcal{A}$ believes that π" (where π is some proposition) to contexts where the subject takes a graded attitude towards the proposition, also known as a degree of belief or credence. The notion of degrees of belief is inherent in the so-called subjective interpretation of probability (cf. [22]), which in turn is part of Bayesian epistemology ("BE"), overlapping confirmation theory (cf. [3]; [23]) and Bayesian statis-

---

[2]A modern and comprehensive version of TIL can be found in [9], [11] and [6]. See also [34]. Specific research questions of TIL, both foundational or applied, have been the subject of countless studies. In Slovakia, they have been dealt with primarily by Pavel Cmorej, a friend of Pavel Tichý, in a number of studies, which are collected in [1] and [2].

[3]TIL is an open system which is still being developed. For a comparison of Tichý's version of TIL with Duží, Jespersen and Materna's [9], as well as for an alternative version of TIL, see [28]. See also [29].

tics.[4] In particular, my aim is to show in what sense the semantic apparatus of TIL is applicable to contexts where the agent's degrees of belief are expressible as real numbers, and where we use probability theory for epistemological purposes.

I will proceed as follows: in the following section 2, I will briefly introduce some basic elements of the semantic apparatus of TIL which underlie my analysis. I will then introduce those mathematical and philosophical principles that form the core of BE in its default form in section 3. In section 4, I will propose a TIL-based reconstruction of the formula known as Bayes' theorem. Finally, in section 5, I will propose a TIL-based analysis of sentences involving credences. The conclusion sums up the prospects and limits of this paper.

2 TIL in brief

The basic principles of TIL have been explained in several papers and monographs, and some modifications of this logical system can be discerned in the last couple of years of its development. The aim of this section is not to provide an introduction to TIL. However, if this paper is to be relatively self-contained, it will be useful for me to at least briefly characterize the concepts and principles of TIL on which I will rely in the subsequent sections.

TIL is a procedurally interpreted lambda-calculus with ramified type theory. Since TIL is defined recursively, it assumes the use of some fundamental objects over which the rest of the system is built. And since TIL finds its application primarily (although not exclusively) in natural language analysis, it turns out that a suitable system of basic objects—also called an *objectual base*—consists of four non-empty and mutually disjoint sets: the set of truth values True/-False, the set of individuals (the universe of discourse), the set of real numbers, and the set of logically possible worlds (o, ι, τ, ω). Each el-

---

[4] For a brief introduction to BE, see [14], [15]. A more comprehensive but still introductory account of the fundamental principles of BE is to be found in [40]. A tour de force of BE is represented by the work of Howson and Urbach (see [19]). But see also [13].

ement of such a base is a type of order 1 over the base. Moreover, if α and $\beta_1, \ldots, \beta_n$ are types of order 1 (over the base), then the set of all partial n-ary functions $((\alpha\beta_1 \ldots \beta_n))$ is also a type of order 1. Types of order 1 suffice to distinguish, at the semantic level, the two basic types of objects that natural language expressions can denote: intensions and extensions. Intensions are (partial) functions from possible worlds to any type, while extensions are functions defined on any type different from the possible worlds. Typical examples of intensions are properties of individuals—objects of type $(o\iota(\tau(\omega)))$ and propositions—objects of type $(o\tau(\omega))$. Typical examples of extensions are, for example, mathematical objects or functions, such as the multiplication function—an object of type $(\tau\tau\tau)$.

However, in addition to functions, TIL works with constructions of objects of various types at the semantic level. More precisely, constructions are a fundamental procedural element of the semantics of this system. Constructions represent certain idealized semantic procedures that (depending on the valuation function) either construct some other construction, construct a functional object, or construct nothing. TIL distinguishes two simple (variables, trivialization) and four compound kinds of constructions (composition, closure, execution, double execution).

Variables are a kind of construction that, depending on a valuation function v, construct an object of a certain type. *Trivialization* is a kind of construction that constructs an object of any type without any change. The pre-theoretical counterpart of trivialization is the procedure of the simple conceptual identification of a given object (construction or non-construction). In other words, if O is an object and the trivialization of O is denoted by ^0O, then ^0O constructs (identifies) the object O (without change of O). A trivialization is a construction that "supplies" other constructions with subconstructions as their constituents. Moreover, the (repeated) use of trivialization helps us to distinguish between the use of the construction and its being mentioned. (While ^0O constructs the object O, ^{00}O constructs the construction ^0O, and thus the construction ^0O is mentioned in the latter case, while in the former case it is used.)

Composition is a kind of construction that applies a certain function to a tuple of arguments. For example, if K is a construction that v-constructs a function f of type $(\alpha\beta_1 \ldots \beta_n)$, and $K_1, \ldots, K_n$ are constructions that v-construct arguments of type $\beta_1, \ldots, \beta_n$ (where f is defined on the given arguments), then the composition $[KK_1 \ldots K_n]$ v-constructs as a function value an object of type α. Otherwise, the composition does not construct anything; we may also say that it is improper. *Closure* is a type of construction that constructs a function by abstracting away the values of its arguments. This is a procedure that proceeds inversely to composition. Thus, if $x_1, \ldots, x_n$ are variables that v-construct objects of type $\beta_1, \ldots, \beta_n$, and a construction K constructs an object of type α, then $[\lambda x_1 \ldots x_n K]$ is a construction that v-constructs a function f of type $(\alpha\beta_1 \ldots \beta_n)$. *Execution* is a procedure that constructs something only when it "operates" on a construction. In this case, the execution of the construction K, denoted as ^1K, constructs whatever the construction K constructs. In the case of some more complex semantic phenomena (such as substitution), TIL also uses a kind of construction known as *double execution*. For example, if K is a construction that constructs a construction K^*, where K^* constructs an object of some type α, then the double execution of K, denoted as ^2K, constructs the object constructed by the construction K^*.

These kinds of constructions form the basis of the procedural semantics of TIL. It identifies the meaning of linguistic expressions with constructions, where what the constructions semantically convey are either other constructions, functions or elements of the objectual base.

We have already said that TIL is defined as a ramified theory of types. Thus, in addition to types of order 1, TIL defines constructions of order n, that is, constructions that v-construct objects (both constructions and non-constructions) of a type of order n. However, sets of constructions of order n, as well as every type of order n, are types of order $n + 1$. Moreover, sets of partial functions that consist of types of order $n + 1$ are also types of order $n + 1$.

This brief characterization is not meant to replace rigorous definitions of TIL. It does, however, approximate a number of key concepts and principles which I will rely on in the following sections.

3   What is Bayesian epistemology?

Traditional epistemology is concerned with problems of knowledge, justification and truth. The bearers of knowledge are primarily agents, or subjects, who form (on the basis of various sources) beliefs about themselves and the world. While for traditional epistemology the central concept is the belief of a certain subject, which under certain conditions can become knowledge, for Bayesian epistemology it is the notion of graded beliefs, or degrees of belief ([20]). Bayesianism is a philosophical approach that employs the apparatus of probability theory with several epistemological principles, in order to explicate or address selected problems in the theory of knowledge and scientific inference ([14], [15], [35], and [36]).

Bayesianism takes its name from the English cleric Thomas Bayes (1701–1761), who formulated a mathematical relationship—an equation—between the so-called prior probability, likelihood and probability of evidence, on the one hand, and its so-called posterior probability on the other. This equation lies at the heart of Bayesian epistemology. When, in the 1930s, the Soviet mathematician Andrei Kolmogorov laid the foundations of axiomatized probability theory ([27]), it turned out that Bayes' formula could be derived as a consequence (a theorem) of the axioms of the probability calculus. In fact, Bayes' theorem expresses the relation between conditional and unconditional probabilities.

However, probability theory is a mathematical theory. In order for a part of mathematical theory to be an apparatus of formal epistemology, it needs to be supplemented by some important epistemological principles. To cut a long story short, the essentials of Bayesian epistemology are associated with these three theses (cf. [14, 312]; [40]):

1. Human agents possess (not only full belief or disbelief, but also) degrees of belief (i.e. credences) that can be modelled by real numbers.

2. For any agent, her system of degrees of beliefs is rational if and only if it is consistent (coherent) with the theorems of probability theory.

3. The way rational agents change their degrees of belief over time is determined by the conditionalization rule (or its alternative).

Each of these principles deserves a brief characterization.

Concerning (1): When a subject considers a state of affairs (corresponding to a proposition), in addition to fully believing that the state of affairs will occur, or that it will not occur, she may also have a graded relation with respect to the object of her belief. For example, I may be 30% convinced that I will get to New Zealand in two years (from now), and I may be 95% convinced that I will visit Prague this year.[5]

Or: I'm 99.99% convinced that there's a black hole at the centre of our galaxy, but I'm only 0.01% convinced that somewhere in the universe there is another intelligent form of life. Bayesianism considers degrees of belief to be mental states that (rational) people use to adequately characterize their attitude towards certain (potential) states of the world. More precisely, a degree of belief is a kind of doxastic attitude that a certain subject takes towards a proposition. Degrees of belief can thus be characterized by means of real numbers. As such, they play the fundamental role in contemporary formal epistemology, i.e. in Bayesian epistemology. They represent an agent's intensity of belief (also called her credence) that some proposition in question is true.

Concerning (2): Bayesianism combines probability theory with some specific interpretation of what the term "probability" means in the theorems of probability calculus. Specifically, Bayesianism postulates that in the axioms and theorems of probability theory, the term "Pr" represents the degrees of belief of a rational agent (cf. [14], [22], and [40]). More precisely, in order for any agent's system of degrees of beliefs to be rational, it must be consistent with the principles of

---

[5]On the relation of full belief to partial belief (i.e. credences), see [18]. See also [20].

probability theory. That is to say, the agent's degrees of beliefs have to be coherent in the sense that they form a probability distribution. For example, one of the theorems tells us that $Pr(\neg A) = 1 - Pr(A)$, for any proposition A. For example, if I were 30% confident that I will get to New Zealand in two years, and 50% confident that I will not get to New Zealand in two years, then my system of degrees of belief (which would include both the first proposition and its negation) would not be rational.

Bayesian epistemology contains a number of arguments designed to show the plausibility of linking a rational system of degrees of beliefs to the theorems of probability theory (for details, see [21]). The intertwining of probability theory and degrees of belief makes Bayes' theorem (see below) a powerful tool in areas such as confirmation theory and decision theory.

Concerning (3): The theorems of probability theory are the theory of synchronic probabilities. In other words, if we apply probability theory to the domain of an agent's degrees of belief, then the theory allows the following. If a given agent has certain degrees of belief at a given time, then all her other degrees of belief are derivable as consequences of the theorems. However, probability theory says nothing about how and under what conditions a given agent should change her degrees of belief over time. The standard form of Bayesianism fills this gap with a rule known as conditionalization. This rule states that if, between times t and t', where $t < t'$, an agent learns that proposition A is true, then her degree of belief at time t' that proposition B is true is (or should be) equal to her degree of belief at time t that proposition B is true, assuming that A is true; formally, $Pr_{t'}(B) = Pr_t(B|A)$. A simple example will help us to understand this rule. Suppose we have to roll a regular die. Let B be the proposition that 4 was rolled on the die, and let A be the proposition that an even number was rolled. If we assume that the die is regular, then $Pr(B) = 1/6$. Moreover, if we consider how likely it is that 4 was rolled given that an even number was rolled, then the answer is $1/3$ (since 4 is one of three possible even outcomes). More precisely, $Pr(B|A) = 1/3$. Now suppose that at time t, when we have no further information about the die or the possible outcome of a single roll, we believe that $Pr_t(B) = 1/6$. However, suppose that the person rolling the die tells us at time t' that an even number was rolled. How should this information (i.e. our learning that A is true) be reflected in our degree of belief that B is true at time t'?

By the Bayesian rule of conditionalization, we have to set the value of our degree of belief at time t' that B is true as our degree of belief at the previous time t that B is true, assuming that A is true; thus $Pr_{t'}(B) = 1/3 = Pr_t(B|A)$. Our degree of belief at time t that B is true had a value of $1/6$, but at time t', after learning that A is true, its value has changed to $1/3$. The rule of conditionalization thus describes the conditions on how agents are to change their beliefs if they discover (observe) that some other—conditioning—proposition is true.

Although conditionalization is not the only option for regulating a system of credences, it is a starting point for reasoning about how to adjust such a system to changes over time without impairing its consistency.

But let us return to the formal apparatus employed by Bayesian epistemology. Traditionally, probability theory assigns probabilities to sets, or more precisely, to a $\sigma-$algebra generated over a set. However, Bayesian epistemology models the doxastic attitudes of rational agents. The objects of these attitudes are certain states of affairs represented ideally by sentences or propositions; typically, these are hypotheses and evidential sentences/propositions that represent the knowledge-base of a given agent (at a given time).

Therefore, it is customary to define the probability function Pr on the language of propositional logic, or on the language of propositions—that is, on the set of propositions $\mathcal{L}$, which is closed under negation and (countable) disjunction (or under other conjunctions definable through them). Thus, a function Pr is a function whose domain is the set of propositions $A, A_1, \ldots, A_n, B, \ldots \in \mathcal{L}$, and whose range is the real numbers r, satisfying the following conditions:

A1 $Pr(A) \geq 0$; for any A

A2 $Pr(\mathrm{T}) = 1$ for any tautology T

A3 $Pr(A \lor B) = Pr(A) + Pr(B)$, provided $\models \neg(A \land B)$

The above axioms characterize the basic properties of unconditional probability (also known as absolute probability). In addition,
however, probability theory also distinguishes the conditional probability function. The conditional probability function of (proposition)
A given (proposition) B—$Pr(A|B)$—is expressed by the relation represented by the following axiom:[6]

A4 Given that $Pr(B) \neq 0$: $Pr(A|B) = \frac{Pr(A \land B)}{Pr(B)}$

It is the axiom **A4** from which the key formula of Bayesian epistemology can be derived: **Bayes' theorem**. Bayes' theorem relates
so-called prior probabilities, likelihoods, and posterior probabilities:

BT Given that $Pr(B) \neq 0$: $Pr(A|B) = \frac{Pr(A) \times Pr(B|A)}{Pr(B)}$

If we substitute the hypothesis H and the evidential statement E
for the statements A and B (in that order), then Bayes' theorem can
be written as:

BT' Given that $Pr(E) \neq 0$: $Pr(H|E) = \frac{Pr(H) \times Pr(E|H)}{Pr(E)}$

$Pr(H|E)$ is the posterior probability, that is, the probability that
the hypothesis is true given that the evidential statement E is true.
$Pr(H)$ is the so-called prior probability, that is, the probability of the
hypothesis H before we consider assuming the truth of the evidential
statement E. The next term in the equation—$Pr(E|H)$—is known as
the likelihood. This is the probability that E is true if we assume that
H is true. Finally, $Pr(E)$ is the so-called expectability of evidence,
which is determined as follows.

---

[6]There are alternative axiomatizations of probability theory which take conditional probabilities as primitive and define unconditional probabilities as a limiting
case of conditional probabilities where we conditionalize on tautologies. See [16].

If $H_1 \vee \ldots \vee H_n$ is a partition of a set of hypotheses, and for each H_i (i = 1, ..., n), $Pr(H_i) > 0$, then:

$$Pr(E) = \sum_{n=1}^{n} Pr(H_i) \times Pr(E|H_i)$$

That is, $Pr(E)$ expresses the expectability of E irrespective of which particular hypothesis from the set of hypotheses is under consideration.

From Bayes' theorem and the aforementioned principles (1)–(3) listed at the beginning of this section, it is only a short step to building an epistemologically relevant theory of confirmation. Then, in order to use **BT'** to evaluate a hypothesis in terms of relevant evidence, it is necessary to define the conditions under which some hypothesis is confirmed or disconfirmed by some evidence. These conditions are captured by the following definitions (cf. [19], [23]):

C E (incrementally) confirms H iff $Pr(H|E) > Pr(H)$

D E (incrementally) disconfirms H iff $Pr(H|E) < Pr(H)$

The case where $Pr(H|E) = Pr(H)$ is the case of neutral or irrelevant evidence. **BT'** expresses the relation that holds between $Pr(H)$ and $Pr(H|E)$, among others, with the definitions **C** and **D** describing the conditions when the hypothesis under consideration is confirmed or disconfirmed. Bayesian epistemology thus has at its disposal tools capable of representing not only the synchronic evaluation of a particular hypothesis in relation to the evidence in question, but also the evolving and changing evaluation of hypotheses over time. Take a simple example: let our hypothesis H be the statement that person $\mathcal{P}$ suffers from disease X, and let E represent the positive result of a medical test that is standardly used to detect disease X. No medical test is 100% reliable. Suppose that the experts estimate the sensitivity of the test $(Pr(E|H))$ to be 0.97 (97%) and its specificity $(Pr(\neg E|\neg H))$ to be 0.94 (94%). From the specificity of the test, the false positive rate can also be derived (i.e. $Pr(E|\neg H) = 1 - Pr(\neg E|\neg H) = 0.06$. If the estimate of the prevalence of disease X in the population from which person $\mathcal{P}$ comes is at the level of 1%, then $Pr(H)$ can be estimated (before considering E) at the level of 0.01. And since the only alternative hypothesis to H that we consider in this context is the negation of H, $(\neg H)$, we can compute its probability on the basis of one of the theorems of probability theory: $Pr(\neg H) = 1 - Pr(H)$; and hence $Pr(\neg H) = 0.99$. Suppose that person $\mathcal{P}$ undergoes a medical test for disease X and the test result is positive. Given this evidence, what is the probability that person $\mathcal{P}$ has disease X?

Bayes' theorem gives us the answer. First of all, we need to compute $Pr(E)$. In this case we get:

$$Pr(E) = Pr(H) \times Pr(E|H) + Pr(\neg H) \times Pr(E|\neg H)$$
$$Pr(E) = 0.01 \times 0.97 + 0.99 \times 0.06 \approx 0.07$$

If we plug the above values into **BT'**, we get $Pr(H|E) = 0.14$. Since $Pr(H|E) = 0.14 > 0.01 = Pr(H)$, then (by **C**) E confirms H.

The potential of Bayesian epistemology for epistemic and doxastic considerations (especially in philosophy of science) is huge (see especially [36], [35] and [30]). This brief exposition assumes many things, glosses over several details, and simply does not address the vast majority of the other principles and their applications. The interested reader is referred to both the classics and the excellent pieces cited from the more recent literature.

4 Bayes' theorem and TIL

As such, Bayes' theorem is a mathematical statement that is a consequence of axioms **A1**–**A4**. It is a statement that says that the value of the probability function $Pr(H|E)$ on the left-hand side of the equation is equal to the result we get if we divide the product of the values of the probability functions $Pr(H)$ and $Pr(E|H)$ on the right-hand side by the value of the probability function $Pr(E)$, where the value of $Pr(E)$ must be greater than zero. In other words, $Pr(H|E)$ is a function of the probability functions $[Pr(H), Pr(E|H), Pr(E)]$.

How can BT' be represented in TIL? Each expression "$Pr(H|E)$", "$Pr(H)$", "$Pr(E|H)$" and "$Pr(E)$" expresses a certain (compound) construction c_1, c_2, c_3 or c_4 in TIL. Each of them constructs some real number, and BT' expresses that the relation $[^0= c_1 \ [^0\colon \ [^0\cdot \ c_2 \ c_3] \ c_4]]$ holds. Moreover, if c_4 v-constructs the number 0, then the entire construction $[^0= c_1 \ [^0\colon \ [^0\cdot \ c_2 \ c_3] \ c_4]]$ is v-improper. (This, of course, is not the only case where the resulting construction can be v-improper.)

So what types of objects do the constructions c_1, c_2, c_3 and c_4 (v-)construct? In section 2, we said that TIL mostly makes use of an objectual base with four basic types: a collection of *individuals* (type ι), a collection of two *truth-values*—TRUE and FALSE (type o), a set of *real numbers* (type τ), which—among other things—are used to represent time (or the temporal dynamics within a given possible world), and a set of logically *possible worlds* (type ω). The choice of another objectual base is also admissible, but most of the objects of semantic analysis for which TIL is used make do with these four basic types. It is also usually assumed that the basic types represent non-empty sets that are mutually disjoint. Thus, if we adopt the standard objectual base, then we have a type τ at our disposal, which represents real numbers, and hence we can say that the constructions c_1, c_2, c_3 and c_4 v-construct objects of type τ.

Do we have another option? If we got rid of the requirement of mutual disjointness of basic types of objects, then we could introduce a special type for real numbers in the interval $[0, 1]$, let's say δ, in which case it would be true that $\delta \subseteq \tau$. Thus, the constructions $c_1, \ldots, c_4$ would v-construct objects of type δ, while the construction $^0 =$ in the construction $[^0= c_1 \ [^0\colon \ [^0\cdot \ c_2 \ c_3] \ c_4]]$ would v-construct the truth value TRUE just in case the construction c_1 constructs the same number (object of type δ) as the construction $[^0\colon \ [^0\cdot \ c_2 \ c_3] \ c_4]$, while the constructions $^0\colon$ and $^0\cdot$ construct the functions of division and multiplication, respectively—both of type $(\delta\delta\delta)$.

Choosing a more specific type than τ is not necessary for the purpose of logical analysis of mathematical theorems in probability theory. However, the TIL user clearly has a wealth of tools at her disposal with which she may refine or delimit the object of her investigation. In the next section, I will assume this finer-grained division of types, and take the types $\iota, o, \tau, \delta, \omega$ to be the elements of the objectual base.

5   TIL, doxastic attitudes and degrees of belief

Bayes' theorem is a mathematical formula, but Bayesian epistemology uses it as a norm for the degrees of belief of rational agents. As such, then, the formula $\mathbf{BT'}$ expresses not only the relation between mathematical objects (functions or their values, or—from TIL's perspective—between constructions of mathematical objects), but also the (normatively required) relation between certain degrees of beliefs and other degrees of beliefs, which are expressible as intensions of a certain type in the TIL semantics. What kind of intensions are these?

In TIL, it is common to view empirical *quantities* or *magnitudes*, which are the objects of the relations of various scientific laws, as intensions of the type $(\tau(\tau\omega))$—that is, as functions that assign real numbers as values to possible worlds and times. While it is true that such a representation of magnitudes lacks the dimension of (metric) units, this is not an obstacle that prevents us from modelling the numerical and empirical side of magnitudes as objects of type $(\tau(\tau\omega))$. Moreover, in the context of the quantities that the social sciences work with, units of measurement (in the manner of physical units) do not occur.

If we view degrees of beliefs (of any rational agent) in general—i.e. by disregarding particular bearers—as quantities, then in TIL we can represent them (in simplified form) as functions that assign real numbers from the interval of values $[0, 1]$ to possible worlds and instants of time. These are thus objects of type $(\delta(\tau\omega))$. For Bayesian rational agents, Bayes' theorem is an expression of a rule that requires the degree of belief (of any agent) on the left-hand side to be equal to a function of the degrees of belief on the right-hand side. Bayes' theorem, transformed into an equation between corresponding degrees of belief, can then be represented in the following form:

$\mathbf{BT''}$ Given that $Cr(E) \neq 0$: $Cr(H|E) = \frac{Cr(H) \times Cr(E|H)}{Cr(E)}$

The expressions "$Cr(H|E)$", "$Cr(H)$", etc., thus represent a relation between different degrees of beliefs—that is, objects of the type $(\delta(\tau\omega))$. This is a relation between quantities, that is, functions whose values may change depending on the possible world or time. In a particular possible world and at a particular time, the value of $Cr(H)$ of one Bayesian agent may be equal to 0.05, while the value of $Cr(H)$ of another Bayesian agent may be equal to 0.47, or, for one agent, this value may vary over time. (The same is true for the other elements of **BT″**.)

However, we have admitted that modelling degrees of belief as magnitudes of type $(\delta(\tau\omega))$ is somewhat of an oversimplification. Indeed, if we wanted to analyse natural language sentences in which the credences of some particular person occur, such an analysis would be inappropriate. Let us first consider the common case of analysing doxastic attitudes, with which TIL works as a standard.

We already pointed out that TIL has been recognized as a logical system that is able to provide a theoretically fruitful analysis of different kinds of sentences expressing propositional attitudes. A category of *doxastic* or *epistemic* attitudes has been of special interest for those engaged in searching for criteria to distinguish between valid and invalid inferences containing sentences that express such attitudes. In general, a doxastic attitude is traditionally conceived as an attitude of a belief, disbelief or a suspension of judgement which an agent has with respect to some proposition or proposition-like entity.

Let us consider a simple sentence expressing a doxastic attitude such as:

S1 Theo believes that Judy is pregnant.

In TIL, we could provide the following as a candidate analysis of this sentence's meaning:

T1 $\lambda w\lambda t[^{0}Believes_{wt} \; ^{0}Theo \; \lambda w\lambda t \; [^{0}Pregnant_{wt} \; ^{0}Judy]]$

T1 represents the meaning of S1 as a construction which v-constructs a proposition about Theo's believing the *proposition* that Judy is pregnant. Alternatively, S1 could be analysed as a construction which v-constructs the proposition that Theo believes a *construction* which v-constructs the proposition that Judy is pregnant. In that case, the TIL analysis would amount to:

143

T1' $\lambda w \lambda t[^0 Believes^*_{wt} \; {}^0Theo \; {}^0[\lambda w \lambda t \; [^0 Pregnant_{wt} \; {}^0Judy]]]$

There are arguments showing that T1' avoids problems that T1 gives rise to. However, this is not the right place to consider the provisos of either analysis. For the sake of simplicity, from now on I will assume that the object of a doxastic attitude is a proposition.

Let us now move from standard categorical doxastic attitudes of belief to graded beliefs, and suppose that we want to analyse the sentence:

S2 Theo's credence that Judy is pregnant is 0.95.

S2 can be understood as saying that Theo is 95% confident that Judy is pregnant. In other words, S2 tells us what numerical value is equal to his degree of belief that the proposition that Judy is pregnant is true. How do we analyse S2 by means of TIL?

My proposal is the following. The meaning of "credence" can be identified with the construction ${}^0Cred$, which v-constructs a function of the following type: possible worlds and times are assigned a function that assigns a real number of type δ to functions from propositions to individuals. It is thus an object of type $(\delta(\iota(o(\tau\omega))(\tau\omega)))$, or in simplified notation: $(\delta\iota o_{\tau\omega})_{\tau\omega}$. To put it in non-technical lingo, the meaning of "credence" is a procedure that identifies the relationship between an individual, a proposition, a state of the world, and a number, where the number expresses the strength of the individual's belief in the truth of the proposition in question in a given state of the world. Thus, our analysis of S2 will look like this:

T2 $\lambda w \lambda t[^0 =_{(o\delta\delta)} \; [^0 Cred_{wt} \; {}^0Theo \; \lambda w \lambda t[^0 Pregnant_{wt} \; {}^0Judy]] \; {}^0 0.95]$

Thus, the construction corresponding to T2 constructs a proposition that is true if and only if the value of Theo's degree of belief that Judy is pregnant equals 0.95.

If our analysis of sentence S2 is correct, we can generalize it as follows. Let $\mathcal{P}$ be some person, π an arbitrary proposition, and r a real number from the interval $[0, 1]$. Then the degrees of belief of a person $\mathcal{P}$ in a proposition π can be expressed by the construction:

T $\lambda w \lambda t [^0{=}_{(o\delta\delta)}\ [^0Cred_{wt}\ ^0\mathcal{P}\ ^0\pi]\ ^0r]$

T is a construction-scheme that can be used to express the degrees of beliefs of arbitrary agents.

6 Conclusion

TIL has been developed in both theoretical and applied dimensions in recent years. The aim of my paper has been to show how the TIL apparatus can be extended from the standard cases of analysing categorical doxastic attitudes to cases of graded beliefs, that is, credences. In doing so, I considered the possibility of specifying the established objectual base as a new subtype that satisfies the needs of displaying probabilities and their corresponding credences as real numbers from the interval $[0, 1]$. If the proposed analysis is correct, then TIL can also be employed in the analysis of inferences that deal with probabilities (as degrees of belief).

Of course, some aspects of TIL, Bayesian epistemology, and their interconnections have not been addressed here. However, if the proposed treatment is correct in principle, then it can be seen as a precursor to further development, both theoretical and applied, of this logical system.

References

[1] Pavel Cmorej. *Na pomedzí logiky a filozofie.* Veda, Bratislava, 2001. [In Slovak; *At the Frontiers of Logic and Philosophy*].

[2] Pavel Cmorej. *Analytické filozofické skúmania.* Filozofický ústav SAV, Bratislava, 2009. [In Slovak; *Analytic Philosophical Investigations*].

[3] Vincenzo Crupi. Confirmation. In Edward N. Zalta, editor, *The Stanford Encyclopedia of Philosophy.* Metaphysics Research Lab, Stanford University, Spring 2021 edition, 2021. Available at: <https://plato.stanford.edu/archives/spr2021/entries/confirmation/>.

[4] Marie Duží. Intensional logic and the irreducible contrast between *de dicto* and *de re. ProFil*, 5(1):1–34, 2004. Available at: <http://profil.muni.cz/01_2004/duzi_de_dicto_de_re.pdf>.

[5] Marie Duží. The paradox of inference and the non-triviality of analytic information. *Journal of Philosophical Logic*, 39(5):473–510, 2010.

[6] Marie Duží. Deduction in TIL: From simple to ramified hierarchy of types. *Organon F*, 20(Suppl. issue 2):5–36, 2013.

[7] Marie Duží and Bjørn Jespersen. Transparent quantification into hyperintensional objectual attitudes. *Synthese*, 192(3):635–677, 2015.

[8] Marie Duží, Bjørn Jespersen, and Daniela Glavaničová. Impossible individuals as necessarily empty individual concepts. In A. Jordani and J. Malinowski, editors, *Logic in High Definition: Trends in Logical Semantics*, pages 177–202. Springer, Cham, 2021.

[9] Marie Duží, Bjørn Jespersen, and Pavel Materna. *Procedural Semantics for Hyperintensional Logic. Foundations and Applications of Transparent Intensional Logic*. Springer, Berlin, 2010.

[10] Marie Duží and Jakub Macek. Analysis of time references in natural language by means of transparent intensional logic. *Organon F*, 25(1):21–40, 2018.

[11] Marie Duží and Pavel Materna. *TIL jako procedurální logika. Průvodce zvídavého čtenáře Transparentní intenzionální logikou.* aleph, Bratislava, 2012.

[12] Marie Duží and Martina Číhalová. Questions, answers and presuppositions. *Computación y Sistemas*, 19(4):647–659, 2015.

[13] John Earman. *Bayes or Bust?* MIT Press, Cambridge (MA), 1992.

[14] Kenny Easwaran. Bayesianism I: Introduction and arguments in favor. *Philosophy Compass*, 6(5):312–320, 2011.

[15] Kenny Easwaran. Bayesianism II: Applications and criticisms. *Philosophy Compass*, 6(5):321–332, 2011.

[16] Kenny Easwaran. Conditional probabilities. In R. Pettigrew and J. Weisberg, editors, *The Open Handbook of Formal Epistemology*, pages 131–198. An Open Access Publication of The PhilPapers Foundation, 2019. Available at: <https://jonathanweisberg.org/pdf/open-handbook-of-formal-epistemology.pdf>.

[17] František Gahér. Anaphora and notional attitudes. *Logica et Methodologica*, 7:1–23, 2003.

[18] Konstantin Genin. Full and partial belief. In R. Pettigrew and J. Weisberg, editors, *The Open Handbook of Formal Epistemology*, pages 437–498. An Open Access Publication of The PhilPapers Foundation, 2019. Available at: <https://jonathanweisberg.org/pdf/open-handbook-of-formal-epistemology.pdf>.

[19] Colin Howson and Peter Urbach. *Scientific Reasoning. The Bayesian Approach*. Open Court, La Salle (IL), third edition, 2006.

[20] Franz Huber and Christoph Schmidt-Petri, editors. *Degrees of Belief*. Springer, Dordrecht, 2009.

[21] Alan Hájek. Arguments for–or against–probabilism? In F. Huber and Ch. Schmidt-Petri, editors, *Degrees of Belief*, pages 229–251. Springer, Dordrecht, 2009.

[22] Alan Hájek. Interpretations of Probability. In Edward N. Zalta, editor, *The Stanford Encyclopedia of Philosophy*. Metaphysics Research Lab, Stanford University, Fall 2019 edition, 2019. Available at: <https://plato.stanford.edu/archives/fall2019/entries/probability-interpret/>.

[23] Alan Hájek and James Joyce. Confirmation. In M. Curd and S. Psillos, editors, *The Routledge Companion to the Philosophy of Science*, pages 146–159. New York, Routledge, second edition, 2014.

[24] Bjørn Jespersen. Recent work on structured meaning and propositional unity. *Philosophy Compass*, 7(9):620–630, 2012.

[25] Bjørn Jespersen. Structured lexical concepts, property modifiers, and transparent intensional logic. *Philosophical Studies*, 172:321–345, 2015.

[26] Bjørn Jespersen. Anatomy of a proposition. *Synthese*, 196:1285–1324, 2019.

[27] Andrei N. Kolmogorov. *Foundations of the Theory of Probability*. Chelsea Publishing Company, New York, second edition, 1950.

[28] Miloš Kosterec. *Transparent Logics*. Book manuscript (2022), submitted to Springer.

[29] Miloš Kosterec. Substitution inconsistencies in transparent intensional logic. *Journal of Applied Non-Classical Logics*, 31(3-4):355–371, 2021.

[30] Hannes Leitgeb. *The Stability of Belief*. Oxford University Press, Oxford, 2017.

[31] Pavel Materna. *Concepts and Objects*. Acta Philosophica Fennica. Philosophical Society of Finland, Helsinki, 1998.

[32] Pavel Materna. *Conceptual Systems*. Logos, Berlin, 2004.

[33] Jiří Raclavský. *Belief Attitudes, Fine-Grained Hyperintensionality and Type-Theoretic Logic*. College Publications, London, 2020.

[34] Jiří Raclavský, Peter Kuchyňka, and Ivo Pezlar. *Transparentní intenzionální logika jako characteristica universalis a calculus ratiocinator*. Masarykova univerzita, Brno, 2015.

[35] Jonah Schupbach. *Bayesianism and Scientific Reasoning*. Cambridge University Press, Cambridge, 2022.

[36] Jan Sprenger and Stephan Hartmann. *Bayesian Philosophy of Science*. Oxford University Press, Oxford, 2019.

[37] Pavel Tichý. Constructions. *Philosophy of Science*, 53:514–534, 1985. Reprinted in Tichý (2004, pp. 599-621).

[38] Pavel Tichý. *The Foundations of Frege's Logic*. De Gruyter, Berlin and New York, 1988.

[39] Pavel Tichý. *Collected Papers in Logic and Philosophy*. Filosofia, Czech Academy of Sciences and University of Otago Press, Prague and Dunedin, 2004. Edited by V. Svoboda, B. Jespersen, and C. Cheyne.

[40] Michael G. Titelbaum. Precise credences. In R. Pettigrew and J. Weisberg, editors, *The Open Handbook of Formal Epistemology*, pages 1–55. The PhilPapers Foundation, 2019. Available at: <https://jonathanweisberg.org/pdf/open-handbook-of-formal-epistemology.pdf>.

Administrative Normal Form and Focusing for Lambda Calculi

David Binder
University of Tübingen
david.binder@uni-tuebingen.de

Thomas Piecha
University of Tübingen
thomas.piecha@uni-tuebingen.de

Abstract

The Curry-Howard correspondence between deductive systems and computational calculi is one of the great unifying ideas. It links purely logical investigations to practical problems in computer science, in particular the design and implementation of programming languages. Many aspects of this correspondence are widely known, such as the correspondence between natural deduction for intuitionistic logic and the simply typed λ-calculus. On the other hand, the importance of the sequent

It is our pleasure to contribute to this *Festschrift* in honor of Marie Duží. Our acquaintance with Marie dates back to her visit to Tübingen in January 2004, where she gave a talk on "Attitudes: Hyperintensional vs. intensional contexts". Our exchange of ideas has continued during the *Second Conference on Proof-Theoretic Semantics* (Tübingen, 8-10 March 2013), where she made "A plea for beta-conversion by value", at the *Third Conference on Proof-Theoretic Semantics* (Tübingen, 27-30 March 2019), where she presented "Natural Language Processing by Natural Deduction in Transparent Intensional Logic" [8], at LOGICA 2011 (Hejnice Monastery, 20-24 June 2011) and LOGICA 2016 (Hejnice Monastery, 20-24 June 2016), as well as during other conferences. This paper relates to Marie's work on call-by-value β-conversion, which she developed in particular in the context of Transparent Intensional Logic (cf. Duží [7], Duží and Jespersen [9, 10, 11], Duží and Kosterec [12] and Jespersen and Duží [18]).

calculus in proof-theoretic investigations is not yet reflected in the study of programming languages, where languages based on the λ-calculus dominate. One of the principal reasons for this is, we think, the lack of introductory material that could serve in helping to translate between logicians and programming language theorists.

Our small contribution in this respect is to introduce and expose the correspondence between two normal forms and their respective normalization procedures: administrative normal form and ANF-transformation on the one hand, and focused normal form and static focusing on the other. Though invented for different purposes, compiler optimizations in the case of the ANF-transformation and proof search in the case of focusing, they are structurally very similar. Both transformations bring proofs into a normal form where functions and constructors are only applied to values and where computations are sequentialized. In this paper we make this similarity explicit.

1 Introduction

In 1935, Gentzen [16] introduced the two most important logical calculi used in proof theory today: natural deduction and the sequent calculus. Natural deduction is used widely in both proof theory and the theory of computation and programming. Its success in the latter is due to the Curry-Howard correspondence (cf. [23]) between natural deduction proofs and programs, or propositions and types. The sequent calculus, on the other hand, did not yet have a comparable impact in the theory of programming languages. Especially in the case of the *classical* sequent calculus, this can be explained by the difficulty to reconcile those of its features that are essential for obtaining classical logic with a good computational interpretation. Such an interpretation was provided when the relationship between classical axioms and control operators was discovered by Griffin [17]. This discovery led to the development of several term systems for encoding sequent calculus proofs. One such system is the $\lambda\mu\tilde{\mu}$-calculus, introduced by Curien and Herbelin [4].

We will use the λ-calculus and the $\lambda\mu\tilde{\mu}$-calculus, which are related by a translation function from λ-terms Λ to $\lambda\mu\tilde{\mu}$-terms $\Lambda_{\mu\tilde{\mu}}$. For the λ-calculus we define the administrative normal form Λ^{ANF}, together with a transformation from Λ to Λ^{ANF}. In distinction to the usual presentation of the ANF-transformation, we divide this transformation into two parts by using an intermediate normal form Λ^{Q} between Λ and Λ^{ANF}. For the $\Lambda_{\mu\tilde{\mu}}$-calculus we define the so-called *focused normal form* $\Lambda^{\text{Q}}_{\mu\tilde{\mu}}$ (which corresponds to the subsyntax LKQ of [4]). The focusing transformation from $\Lambda_{\mu\tilde{\mu}}$ to $\Lambda^{\text{Q}}_{\mu\tilde{\mu}}$ is adapted from [6]. We define a new normal form $\Lambda^{\text{ANF}}_{\mu\tilde{\mu}}$ for $\lambda\mu\tilde{\mu}$-terms, which exactly mirrors the syntactic restrictions that characterize the administrative normal form Λ^{ANF} for λ-terms.

As our main result, depicted in Fig. 1, we show how the ANF-transformation on λ-terms corresponds to static focusing of $\lambda\mu\tilde{\mu}$-terms. The first part of the ANF-transformation corresponds precisely to the static focusing transformation. That is, it commutes with focusing via the translation function up to α-equivalence. The second part of the ANF-transformation can be simulated in the $\lambda\mu\tilde{\mu}$-calculus by μ-reductions.

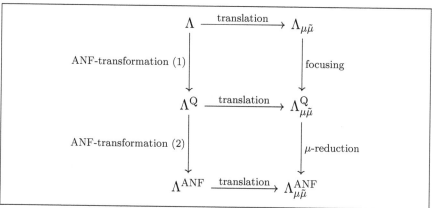

Figure 1: The relationship between the ANF-tranformation of λ-terms and focusing of $\lambda\mu\tilde{\mu}$-terms.

The paper is structured as follows. In Section 2 we present the main idea using an informal example. In Section 3 we formalize the syntax and type assignment rules for the λ-calculus and the $\lambda\mu\tilde{\mu}$-calculus, and in Section 4 we give the translation from the former to the latter. In Section 5 we provide the call-by-value operational semantics for both calculi. We introduce the ANF-transformation in Section 6 and static focusing in Section 7. The main result is presented in Section 8 and summarized in Section 9, which also contains an outlook to future work. The proofs of the main theorems can be found in Appendix A.

2   An informal example

We explain the main idea with an informal example. Consider the following program

$$\pi_2(\pi_1(1,4),3)$$

which consists of natural numbers 1, 4 and 3, pair constructors $(\sqcup, \sqcup)$ and projections $\pi_1 \sqcup$ and $\pi_2 \sqcup$ on the first and second element of a pair, respectively.

We expect this program to evaluate to the natural number 3. Using call-by-name we could immediately evaluate this program to its final value 3. However, using call-by-value we first have to evaluate the argument of π_2 to the value $(1,3)$ by evaluating $\pi_1(1,4)$ to 1.

There are different ways to formalize the evaluation of a term within a context. Here we choose the method of evaluation contexts (cf. Felleisen and Hieb [13] and Section 5.1 below). An *evaluation context* $E[-]$ is a term with a placeholder $\Box$, which is to be filled with the outermost redex to be evaluated next. We will use the symbol $\simeq$ throughout to express syntactic equality up to α-equivalence (i.e., up to the renaming of bound variables).

In our example, this allows us to evaluate the outermost redex $\pi_1(1,4)$ within the context $E[-] \simeq \pi_2(\Box, 3)$ as follows:

If $\pi_1(1,4) \triangleright 1$, then $\pi_2(\pi_1(1,4),3) \simeq E[\pi_1(1,4)] \triangleright E[1] \simeq \pi_2(1,3)$.

The translation (cf. Definition 4.1) of the program $\pi_2(\pi_1(1,4),3)$ into the $\lambda\mu\tilde{\mu}$-calculus (cf. Section 3.3) results in the program

$$\mu\alpha.\langle(\mu\beta.\langle(1,4) \mid \pi_1 \, \beta\rangle, 3) \mid \pi_2 \, \alpha\rangle.$$

We can recognize many familiar constructs from the initial program. We still have natural numbers 1, 4 and 3, the pair constructor $(\_, \_)$ and projections π_1 and π_2, but they are now organized and nested in a very different way with the help of two new constructs.

The first new construct is the *cut* $\langle \_ \mid \_ \rangle$ which is used to oppose a *proof* (or *proof term*) of a proposition with its *refutation* (or *refutation term*). In our example, we use the cut to oppose a proof $(1, 4)$ of the type $\mathbb{N} \wedge \mathbb{N}$ with a refutation $\pi_1\, \beta$ of the same type, where we assume that the refutation variable β stands for some unknown refutation of type $\mathbb{N}$. The reduction rules of the $\lambda\mu\tilde{\mu}$-calculus always replace a cut by another cut, and in the case of pairs the reduction rule allows to replace $\langle (1, 4) \mid \pi_1\, \beta \rangle$ by the new cut $\langle 1 \mid \beta \rangle$.

The second new construct is the μ-abstraction $\mu\alpha.\_$. We have more to say about this construct in Section 3.3, but for now it suffices to say that we use $\mu\alpha.\langle \_ \mid \_ \rangle$ to introduce a subcomputation (represented by the cut $\langle \_ \mid \_ \rangle$) returning to the output named by the variable α. For example, in order to represent the subcomputation $2 + 2$, we use the term $\mu\alpha.\langle 2 + 2 \mid \alpha \rangle$, which evaluates to $\mu\alpha.\langle 4 \mid \alpha \rangle$.

We cannot evaluate the program $\mu\alpha.\langle (\mu\beta.\langle (1, 4) \mid \pi_1\, \beta \rangle, 3) \mid \pi_2\, \alpha \rangle$ directly to its final value, since one can only evaluate cuts $\langle \_ \mid \_ \rangle$, whereas this program has the form of a μ-abstraction. This can be resolved by introducing a third construct, namely the *toplevel output* Top, which enables us to embed any μ-program in a cut whose second element is Top. Furthermore, a $\tilde{\mu}$-abstraction $\tilde{\mu}x.\langle \_ \mid \_ \rangle$ has to be used, which binds a value to the variable x in the subcomputation $\langle \_ \mid \_ \rangle$.

The example program then evaluates in the following way:

$$\langle \mu\alpha.\langle (\mu\beta.\langle (1, 4) \mid \pi_1\, \beta \rangle, 3) \mid \pi_2\, \alpha \rangle \mid \mathsf{Top} \rangle \tag{1}$$

$$\triangleright \langle (\mu\beta.\langle (1, 4) \mid \pi_1\, \beta \rangle, 3) \mid \pi_2\, \mathsf{Top} \rangle \tag{2}$$

$$\triangleright \langle \mu\beta.\langle (1, 4) \mid \pi_1\, \beta \rangle \mid \tilde{\mu}x.\langle (x, 3) \mid \pi_2\, \mathsf{Top} \rangle \rangle \tag{3}$$

$$\triangleright \langle (1, 4) \mid \pi_1(\tilde{\mu}x.\langle (x, 3) \mid \pi_2\, \mathsf{Top} \rangle) \rangle \tag{4}$$

$$\triangleright \langle 1 \mid \tilde{\mu}x.\langle (x, 3) \mid \pi_2\, \mathsf{Top} \rangle \rangle \tag{5}$$

$$\triangleright \langle (1, 3) \mid \pi_2\, \mathsf{Top} \rangle \tag{6}$$

$$\triangleright \langle 3 \mid \mathsf{Top} \rangle \tag{7}$$

Note that in step (5) we project from $(1,4)$ to 1 without being in an evaluation context. The evaluation within an evaluation context is instead simulated by steps (4) and (6). That is, steps (4) to (6) correspond to the single evaluation step

$$\pi_2(\pi_1(1,4),3) \triangleright \pi_2(1,3).$$

This sort of evaluation within a context, which is present in both the λ-calculus and the $\lambda\mu\tilde{\mu}$-calculus, poses no problem from a theoretical point of view. However, from a practical point of view, it is very inefficient to apply this kind of operational semantics since the search for a redex requires in general to traverse deeply into a term. Moreover, evaluations of this kind render the implementation of various compiler optimizations (cf. [22, 15]) more difficult. These difficulties can be avoided by using certain normal forms, for example, the so-called *administrative normal form* (*A-normal form* or *ANF*)[1] for the λ-calculus, and the *focused normal form* for the $\lambda\mu\tilde{\mu}$-calculus.

The ANF of the first example program

$$\pi_2(\pi_1(1,4),3)$$

is

$$\textbf{let } x = \pi_1(1,4) \textbf{ in } (\textbf{let } y = \pi_2(x,3) \textbf{ in } y), \tag{A}$$

whereas the focused normal form of the second program

$$\mu\alpha.\langle(\mu\beta.\langle(1,4) \mid \pi_1\ \beta\rangle,3) \mid \pi_2\ \alpha\rangle$$

is

$$\mu\alpha.\langle\mu\beta.\langle(1,4) \mid \pi_1\ \beta\rangle \mid \tilde{\mu}x.\langle(x,3) \mid \pi_2\ \alpha\rangle\rangle. \tag{F}$$

Comparing the ANF (A) with the focused normal form (F) makes the structural similarity between the two normal forms apparent: in both cases the subcomputation $\pi_1(1,4)$ (resp. $\langle(1,4) \mid \pi_1\ \beta\rangle$) was lifted out and then bound to the variable x in the subsequent computation

---

[1]While the "A" in "A-normal" originally had no special meaning, it was later given the meaning of "administrative normal form", due to the administrative redexes it introduces.

$\pi_2(x, 3)$ (resp. $\langle (x, 3) \mid \pi_2 \, \alpha \rangle$). The difference between (A) and (F) consists in the use of let-constructs in the λ-calculus on the one hand and the use of μ- and $\tilde{\mu}$-constructs in the $\lambda\mu\tilde{\mu}$-calculus on the other hand.

3 Syntax and type assignment

We present the syntax and type-assignment rules of the λ-calculus and of the $\lambda\mu\tilde{\mu}$-calculus. The syntax for types is the same in both calculi.

Definition 3.1 (Types). There are three kinds of *types* τ:

$$\tau ::= X \mid \tau \to \tau \mid \tau \wedge \tau.$$

That is, we have *atomic types* X, *implication types* $\tau \to \tau$ and *conjunction types* $\tau \wedge \tau$.

3.1 The λ-calculus

We use the standard simply typed λ-calculus with conjunction and a let-construct (cf., e.g., [20]). Since we only consider a call-by-value evaluation strategy, the values consist of variables, λ-abstractions and tuples of values.

Definition 3.2. The *syntax Λ of the λ-calculus* is defined as follows, where x are *term variables*:

1. *Terms:* $e ::= x \mid \lambda x.e \mid e \, e \mid (e, e) \mid \pi_1 \, e \mid \pi_2 \, e \mid \textbf{let } x = e \textbf{ in } e.$

2. *Values:* $v ::= \lambda x.e \mid (v, v) \mid x.$

A *judgement* is a sequent of the form $\Gamma \vdash e : \tau$, where Γ is a (possibly empty) set of declarations $\{x_1 : \tau_1, \ldots, x_n : \tau_n\}$.

Definition 3.3. The *type assignment rules of the λ-calculus* are:

$$\frac{}{\Gamma, x : \tau \vdash x : \tau} \text{ VAR} \qquad \frac{\Gamma \vdash e_1 : \sigma \quad \Gamma, x : \sigma \vdash e_2 : \tau}{\Gamma \vdash \textbf{let } x = e_1 \textbf{ in } e_2 : \tau} \text{ LET}$$

$$\frac{\Gamma, x : \sigma \vdash e : \tau}{\Gamma \vdash \lambda x.e : \sigma \to \tau} \text{ ABS} \qquad \frac{\Gamma \vdash e_1 : \sigma \to \tau \quad \Gamma \vdash e_2 : \sigma}{\Gamma \vdash e_1 \, e_2 : \tau} \text{ APP}$$

$$\frac{\Gamma \vdash e_1 : \sigma \quad \Gamma \vdash e_2 : \tau}{\Gamma \vdash (e_1, e_2) : \sigma \wedge \tau} \text{ PAIR} \qquad \frac{\Gamma \vdash e : \tau_1 \wedge \tau_2}{\Gamma \vdash \pi_i \, e : \tau_i} \text{ PROJ}$$

Note that rule PROJ comprises the two cases where either $i = 1$ or $i = 2$.

There are no structural rules since weakening and contraction are implicit. Note that the rule LET is derivable since any term $\textbf{let } x = e_1 \textbf{ in } e_2$ can be replaced by $(\lambda x.e_2)e_1$ without changing the type in the conclusion of a type assignment. However, let-bindings are used to make the evaluation order explicit; we will come back to this point in Section 7.

3.2 Towards the $\lambda\mu\tilde{\mu}$-calculus

The λ-calculus corresponds to natural deduction for the $\{\to, \wedge\}$-fragment of intuitionistic logic. The $\lambda\mu\tilde{\mu}$-calculus [4] was introduced as a system that corresponds to the classical sequent calculus, in which sequents have the form $\Gamma \vdash \Delta$ with (possibly empty) sets Γ, Δ of formulas on either side of the sequent symbol $\vdash$.

The usual interpretation of a valid classical sequent $\Gamma \vdash \Delta$ can be expressed as "If all the formulas in Γ are true, then at least one of the formulas in Δ is true." This interpretation has to be refined in order to understand the correspondence between the $\lambda\mu\tilde{\mu}$-calculus and the classical sequent calculus. The refinement consists in distinguishing three variants of the sequent $\Gamma \vdash \Delta$:

1. $\Gamma \vdash [\varphi], \Delta$

   "If all $\gamma \in \Gamma$ are true and all $\delta \in \Delta$ are false, then φ is true."

2. $\Gamma, [\varphi] \vdash \Delta$

   "If all $\gamma \in \Gamma$ are true and all $\delta \in \Delta$ are false, then φ is false."

3. $\Gamma \vdash \Delta$

"The assumption that all $\gamma \in \Gamma$ are true and all $\delta \in \Delta$ are false is contradictory."

The formula in square brackets $[\varphi]$ is called the *active* formula of the sequent. There can be at most one active formula in any sequent.

The $\lambda\mu\tilde{\mu}$-calculus has one syntactic category and one judgement form for each of these three interpretations:

1. The active formula φ in the succedent of a sequent $\Gamma \vdash [\varphi], \Delta$ is assigned to a *term e*, and the corresponding *judgement form* is

$$\Gamma \vdash e : \varphi \mid \Delta.$$

Here the symbol $\mid$ singles out a formula φ for which the proof e is currently constructed (cf. [4]).

2. The active formula φ in the antecedent of a sequent $\Gamma, [\varphi] \vdash \Delta$ is assigned to a *coterm s*, and the corresponding *judgement form* is

$$\Gamma \mid s : \varphi \vdash \Delta.$$

In this case, the symbol $\mid$ singles out a formula φ for which the refutation s is currently constructed.

3. A sequent $\Gamma \vdash \Delta$ with no active formula is interpreted by a *command c*, and the corresponding *judgement form* is

$$c : (\Gamma \vdash \Delta).$$

This judgement form can be read as follows: "If all $\gamma \in \Gamma$ are true and all $\delta \in \Delta$ are false, then c is a contradiction and a well-typed command."

3.3 The $\lambda\mu\tilde{\mu}$-calculus

We consider the syntax of the $\lambda\mu\tilde{\mu}$-calculus. We have to partition the set of λ-terms into the three syntactic categories of the $\lambda\mu\tilde{\mu}$-calculus, namely terms, coterms and commands. The basic idea is that the introduction forms $\lambda x.e$ and (e, e) (which correspond to the introduction rules in natural deduction) will remain terms of the $\lambda\mu\tilde{\mu}$-calculus. On the other hand, the elimination forms $\pi_i e$ and $e\,e$ (which correspond to the elimination rules in natural deduction) will become coterms. The terms of the $\lambda\mu\tilde{\mu}$-calculus therefore comprise the introduction forms $\lambda x.t$ and (t, t) of the λ-calculus, whereas the coterms comprise the elimination forms $\pi_i\,s$ and $t \cdot s$.

There are different ways to understand a coterm $t \cdot s$. First, since an implication $\varphi \to \tau$ is false if φ is true and τ is false, one can interpret $t \cdot s$ as a constructive refutation of an implication $\varphi \to \tau$, consisting of a proof t of φ and a refutation s of τ. Alternatively, in a computational context, $t \cdot s$ can be thought of as a stack frame in a call stack with argument t on top and s being the rest of the stack.

There is only one form of command in the $\lambda\mu\tilde{\mu}$-calculus: the *cut* $\langle t \mid s \rangle$, which combines a term with a coterm. The cut rule can be interpreted as a primitive way to construct a contradiction, namely by providing both a proof and a refutation of the same formula.

This leaves us with the two remaining constructs of μ- and $\tilde{\mu}$-abstraction, which, again, can be understood in two different ways. First, from a logical point of view, the μ-construct encodes a form of *reductio ad absurdum* at the level of judgements:

$$\frac{\begin{array}{c}[\varphi \text{ is false}]\\ \vdots \\ \text{contradiction}\end{array}}{\varphi \text{ is true}}\ (\mu)$$

This explains why the addition of μ-abstraction makes the logic classical. The $\tilde{\mu}$-construct, on the other hand, encodes the logical inference

$$\frac{\begin{array}{c}[\varphi \text{ is true}]\\ \vdots \\ \text{contradiction}\end{array}}{\varphi \text{ is false}}\ (\tilde{\mu})$$

Both inferences are on the level of judgements and do not involve logical constants; neither absurdity $\bot$ nor negation $\neg$ are used.

Second, from an operational point of view we see that $\tilde{\mu}$ behaves very similarly to the let-construct of the λ-calculus. In a command $\langle t \mid \tilde{\mu}x.c \rangle$, the $\tilde{\mu}$-abstraction is used to bind the term t in the remaining computation c. The μ-construct behaves similarly to control operators like call/cc or $\mathcal{C}$ (cf. [2, 17]).

Definition 3.4. The *syntax* $\Lambda_{\mu\tilde{\mu}}$ *of the* $\lambda\mu\tilde{\mu}$-*calculus is defined as* follows, where x are *term variables* and α are *coterm variables*:

1. *Terms:* $t ::= x \mid \lambda x.t \mid (t,t) \mid \mu\alpha.c$.

2. *Coterms:* $s ::= \alpha \mid t \cdot s \mid \pi_1 s \mid \pi_2 s \mid \tilde{\mu}x.c$.

3. *Commands:* $c ::= \langle t \mid s \rangle$.

4. *Values:* $w ::= \lambda x.t \mid (w,w) \mid x$.

In addition to term variable contexts $\Gamma \simeq \{x_1 : \tau_1, \ldots, x_n : \tau_n\}$, we now have to consider also coterm variable contexts $\Delta \simeq \{\alpha_1 : \tau_1, \ldots, \alpha_n : \tau_n\}$.

Definition 3.5. The *type-assignment rules of the* $\lambda\mu\tilde{\mu}$-*calculus* for the three judgement forms

1. *term typing:* $\Gamma \vdash t : \tau \mid \Delta$,

2. *coterm typing:* $\Gamma \mid s : \tau \vdash \Delta$, and

3. *command typing:* $c : (\Gamma \vdash \Delta)$

are the following:

<table>
<tr><td align="center">*Term typing*</td><td align="center">*Coterm typing*</td></tr>
</table>

$$\frac{}{\Gamma, x : \tau \vdash x : \tau \mid \Delta} \ \text{VAR}_x \qquad \frac{}{\Gamma \mid \alpha : \tau \vdash \alpha : \tau, \Delta} \ \text{VAR}_\alpha$$

$$\frac{\Gamma, x : \sigma \vdash t : \tau \mid \Delta}{\Gamma \vdash \lambda x.t : \sigma \to \tau \mid \Delta} \ \text{ABS} \qquad \frac{\Gamma \vdash t : \tau \mid \Delta \quad \Gamma \mid s : \sigma \vdash \Delta}{\Gamma \mid t \cdot s : \tau \to \sigma \vdash \Delta} \ \text{APP}$$

$$\frac{\Gamma \vdash t_1 : \tau_1 \mid \Delta \quad \Gamma \vdash t_2 : \tau_2 \mid \Delta}{\Gamma \vdash (t_1, t_2) : \tau_1 \wedge \tau_2 \mid \Delta} \text{ PAIR} \qquad \frac{\Gamma \mid s : \tau_i \vdash \Delta}{\Gamma \mid \pi_1 s : \tau_1 \wedge \tau_2 \vdash \Delta} \text{ PROJ}$$

$$\frac{c : (\Gamma \vdash \alpha : \tau, \Delta)}{\Gamma \vdash \mu\alpha.c : \tau \mid \Delta} \text{ MU} \qquad \frac{c : (\Gamma, x : \tau \vdash \Delta)}{\Gamma \mid \tilde{\mu}x.c \vdash \Delta} \text{ MU}_\sim$$

Command typing

$$\frac{\Gamma \vdash t : \tau \mid \Delta \quad \Gamma \mid s : \tau \vdash \Delta}{\langle t \mid s \rangle : (\Gamma \vdash \Delta)} \text{ CUT}$$

4   Translating λ-terms to $\lambda\mu\tilde{\mu}$-terms

We introduce a compositional translation from λ-terms to $\lambda\mu\tilde{\mu}$-terms and show that it preserves typeability.

Definition 4.1. The *translation* $[\![-]\!] : \Lambda \to \Lambda_{\mu\tilde{\mu}}$ is defined as follows:

$$[\![x]\!] :\cong x \tag{$\mathcal{T}_1$}$$
$$[\![\lambda x.e]\!] :\cong \lambda x.[\![e]\!] \tag{$\mathcal{T}_2$}$$
$$[\![(e_1, e_2)]\!] :\cong ([\![e_1]\!], [\![e_2]\!]) \tag{$\mathcal{T}_3$}$$
$$[\![e_1\, e_2]\!] :\cong \mu\alpha.\langle [\![e_1]\!] \mid [\![e_2]\!] \cdot \alpha \rangle \tag{$\mathcal{T}_4$}$$
$$[\![\pi_i\, e]\!] :\cong \mu\alpha.\langle [\![e]\!] \mid \pi_i\, \alpha \rangle \tag{$\mathcal{T}_5$}$$
$$[\![\textbf{let } x = e_1 \textbf{ in } e_2]\!] :\cong \mu\alpha.\langle [\![e_1]\!] \mid \tilde{\mu}x.\langle [\![e_2]\!] \mid \alpha \rangle \rangle. \tag{$\mathcal{T}_6$}$$

In the last three clauses, the coterm variable α has to be fresh.

Let e be any expression of the λ-calculus typeable with type τ in a context Γ. Then the translation $[\![e]\!]$ is a term of the $\lambda\mu\tilde{\mu}$-calculus that is typeable with the same type τ (in the same context Γ of term variables and with an empty context of coterm variables).

Theorem 4.2. *For all e, τ and Γ: if $\Gamma \vdash e : \tau$, then $\Gamma \vdash [\![e]\!] : \tau \mid \varnothing$.*

Proof. The proof is by induction on the derivation of $\Gamma \vdash e : \tau$ in the λ-calculus. The cases for variables, tuples and λ-abstractions are trivial; we will only discuss the following interesting cases.

The first case is for projections. Assume that the last rule in the typing derivation of e is PROJ. Then e has the form $\pi_i\, e$, whose translation is defined as $\mu\alpha.\langle [\![e]\!] \mid \pi_i\, \alpha\rangle$. We replace the λ-calulus derivation by the following $\lambda\mu\tilde{\mu}$-calculus derivation:

$$
\cfrac{
  \cfrac{\text{IH}}{\Gamma \vdash [\![e]\!] : \tau_1 \wedge \tau_2 \mid \varnothing} \qquad
  \cfrac{
    \cfrac{}{\varnothing \mid \alpha : \tau_i \vdash \alpha : \tau_i}\ \text{VAR}_\alpha
  }{\varnothing \mid \pi_i\, \alpha : \tau_1 \wedge \tau_2 \vdash \alpha : \tau_i}\ \text{PROJ}
}{
  \cfrac{\langle [\![e]\!] \mid \pi_i\, \alpha\rangle : (\Gamma \vdash \alpha : \tau_i)}{\Gamma \vdash \mu\alpha.\langle [\![e]\!] \mid \pi_i\, \alpha\rangle : \tau_i \mid \varnothing}\ \text{MU}
}\ \text{CUT}
$$

The second interesting case is for function applications. Assume that the last rule in the derivation of $\Gamma \vdash e : \tau$ is APP. Then e must have the form $e_1\, e_2$, whose translation is defined as $\mu\alpha.\langle [\![e_1]\!] \mid [\![e_2]\!] \cdot \alpha\rangle$. We replace the original derivation by:

$$
\cfrac{
  \cfrac{\text{IH}}{\Gamma \vdash [\![e_1]\!] : \sigma \to \tau \mid \varnothing} \qquad
  \cfrac{
    \cfrac{\text{IH}}{\Gamma \vdash [\![e_2]\!] : \sigma \mid \varnothing} \qquad
    \cfrac{}{\varnothing \mid \alpha : \tau \vdash \alpha : \tau}\ \text{VAR}_\alpha
  }{\varnothing \mid [\![e_2]\!] \cdot \alpha : \sigma \to \tau \vdash \alpha : \tau}\ \text{APP}
}{
  \cfrac{\langle [\![e_1]\!] \mid [\![e_2]\!] \cdot \alpha\rangle : (\Gamma \vdash \alpha : \tau)}{\Gamma \vdash \mu\alpha.\langle [\![e_1]\!] \mid [\![e_2]\!] \cdot \alpha\rangle : \tau \mid \varnothing}\ \text{MU}
}\ \text{CUT}
$$

The last case is for let-bindings. Assume that the last rule in the derivation of $\Gamma \vdash e : \tau$ is LET. Then e must have the form $\mathbf{let}\, x = e_1 \,\mathbf{in}\, e_2$, whose translation is defined as $\mu\alpha.\langle [\![e_1]\!] \mid \tilde{\mu}x.\langle [\![e_2]\!] \mid \alpha\rangle\rangle$. We replace the original derivation by:

$$
\cfrac{
  \cfrac{\text{IH}}{\Gamma \vdash [\![e_1]\!] : \sigma \mid \varnothing} \qquad
  \cfrac{
    \cfrac{
      \cfrac{\text{IH}}{\Gamma, x : \sigma \vdash [\![e_2]\!] : \tau \mid \varnothing} \qquad
      \cfrac{}{\varnothing \mid \alpha : \tau \vdash \alpha : \tau}\ \text{VAR}_\alpha
    }{
      \cfrac{\langle [\![e_2]\!] \mid \alpha\rangle : (\Gamma, x : \sigma \vdash \alpha : \tau)}{\Gamma \mid \tilde{\mu}x.\langle [\![e_2]\!] \mid \alpha\rangle \vdash \alpha : \tau}\ \text{MU}_\sim
    }\ \text{CUT}
  ]
}{
  \cfrac{\langle [\![e_1]\!] \mid \tilde{\mu}x.\langle [\![e_2]\!] \mid \alpha\rangle\rangle : (\Gamma \vdash \alpha : \tau)}{\Gamma \vdash \mu\alpha.\langle [\![e_1]\!] \mid \tilde{\mu}x.\langle [\![e_2]\!] \mid \alpha\rangle\rangle : \tau \mid \varnothing}\ \text{MU}
}\ \text{CUT} \qquad \square
$$

We will also need the following lemma about the translation of values:

Lemma 4.3 (Translation preserves values). *An expression e is a value of the λ-calculus if, and only if, $[\![e]\!]$ is a value of the $\lambda\mu\tilde{\mu}$-calculus.*

Proof. By inspection of the relevant cases. □

5 Call-by-value operational semantics

We introduce the evaluation rules for the λ-calculus and for the $\lambda\mu\tilde{\mu}$-calculus.

5.1 Evaluation in the λ-calculus

For the λ-calculus we first have to define how to reduce immediate redexes. We do this in Definition 5.1. One can note how all three rules implement the *call-by-value* strategy: a function application $(\lambda x.e_1)e_2$ can only be reduced if e_2 is a value; a projection $\pi_i(e_1, e_2)$ can only be reduced if both e_1 and e_2 are values; and a let-binding **let** $x = e_1$ **in** e_2 can only be reduced if e_1 is a value.

Definition 5.1. The *call-by-value evaluation rules for the λ-calculus* are:

$$(\lambda x.e)\, v \triangleright e[v/x] \qquad\qquad (\beta_\rightarrow)$$

$$\pi_i(v_1, v_2) \triangleright v_i \qquad\qquad (\beta_\wedge)$$

$$\textbf{let } x = v \textbf{ in } e \triangleright e[v/x]. \qquad\qquad (\beta_{\text{let}})$$

These rules are not sufficient, since none of the rules are applicable to the term $(\pi_1(v_1, v_2), v_3)$, for example. We therefore need to extend them to allow for the reduction of redexes within a term. Furthermore, since we want evaluations to be deterministic, we must extend Definition 5.1 in such a way that there is always exactly one redex within a term which can be evaluated next. For example, in a tuple (e_1, e_2) we must specify whether we want to evaluate e_1 or e_2 first (and similarly for function applications $e_1\, e_2$).

We specify deterministic evaluations within a context by using the concept of evaluation contexts, introduced in [13]. An evaluation context $E[-]$ is a term with one argument place marked by the symbol □, which indicates where we evaluate the next immediate redex. Deterministic evaluation is ensured by a *unique decomposition lemma*:

Every term e that is not a value can be uniquely decomposed into an evaluation context $E[-]$ and an immediate redex e' such that $e \simeq E[e']$.

Definition 5.2. The syntax of *evaluation contexts* $E[-]$ is defined as follows:

$$E[-] ::= \Box \mid E\,e \mid v\,E \mid (E, e) \mid (v, E) \mid \mathbf{let}\ x = E\ \mathbf{in}\ e \mid \pi_i\,E.$$

Evaluation contexts now allow us to properly define evaluation within a context:

$$e \triangleright e' \implies E[e] \triangleright E[e']. \tag{Congruence}$$

5.2 Evaluation in the $\lambda\mu\tilde{\mu}$-calculus

For the $\Lambda_{\mu\tilde{\mu}}$-calculus, we again first introduce the rules for evaluating immediate redexes. The choice of the *call-by-value* evaluation strategy is manifested in the following ways: first, a redex $\langle \lambda x.t \mid e \cdot s \rangle$ can only be reduced if the function argument e is a value; second, a redex $\langle (e_1, e_2) \mid \pi_i\,s \rangle$ can only be reduced if both e_1 and e_2 are values; third, the *critical pair* $\langle \mu\alpha.c_1 \mid \tilde{\mu}x.c_2 \rangle$, which could a priori be reduced to either $c_1[\tilde{\mu}x.c_2/\alpha]$ or $c_2[\mu\alpha.c_2/x]$, is resolved by requiring in the rule $(\tilde{\mu})$ that a redex $\langle e \mid \tilde{\mu}x.c \rangle$ can only be reduced if e is a value.

Definition 5.3. The *call-by-value evaluation rules for the $\lambda\mu\tilde{\mu}$-calculus* are:

$$\langle \lambda x.t \mid v \cdot s \rangle \triangleright \langle t[v/x] \mid s \rangle \tag{$\beta_\to$}$$

$$\langle (v_1, v_2) \mid \pi_i\,s \rangle \triangleright \langle v_i \mid s \rangle \tag{$\beta_\wedge$}$$

$$\langle \mu\alpha.c \mid s \rangle \triangleright c[s/\alpha] \tag{μ}$$

$$\langle v \mid \tilde{\mu}x.c \rangle \triangleright c[v/x]. \tag{$\tilde{\mu}$}$$

These rules are, again, not complete. For example, there is no rule applicable to the cut $\langle (\mu\alpha.c, v) \mid \pi_i\,s \rangle$, since the first element of the tuple is not yet a value. Instead of the evaluation contexts $E[-]$, we will add focusing contexts $F[-]$ and *dynamic focusing rules* ς. The focusing contexts play the role of the evaluation contexts for the λ-calculus, while the ς-rules correspond to the rule (Congruence).

Definition 5.4. The syntax of *focusing contexts* $F[-]$ is defined as follows:

$$F[-] ::= (\Box, t) \mid (w, \Box) \mid \Box \cdot s.$$

Definition 5.5. We extend the evaluation rules of Definition 5.3 by the following *dynamic focusing rules*:

$$\langle F[t] \mid s \rangle \triangleright \langle t \mid \tilde{\mu}x.\langle F[x] \mid s \rangle \rangle \quad \text{(if } t \text{ is not a value)} \tag{ς_1}$$
$$\langle v \mid F[t] \rangle \triangleright \langle t \mid \tilde{\mu}x.\langle v \mid F[x] \rangle \rangle \quad \text{(if } t \text{ is not a value).} \tag{ς_2}$$

6 The ANF-transformation

While the evaluation rules presented in Section 5 are sufficient for purely theoretical investigations into the reduction theory of the λ-calculus and the $\lambda\mu\tilde{\mu}$-calculus, they are less ideal for other purposes. In particular, they are not ideal for generating efficient code that can be run on a real computer. For example, consider the congruence rule in Definition 5.1. Its operational meaning implies that we have to *search* for the next redex in the term, and this redex can appear nested at an arbitrary depth within the term. If we implemented this search procedure naively for each reduction step, then the resulting program would be very inefficient indeed.

Various methods to efficiently evaluate terms of the λ-calculus have been proposed, for both the call-by-value and call-by-name evaluation orders. One of these methods is the compilation to *abstract machines*[2], like the SEK, SECD or Krivine machine, which provide much more efficient means of evaluating λ-terms. The evaluation of commands of the $\lambda\mu\tilde{\mu}$-calculus is, in fact, very similar to the evaluation of machine states of an abstract machine. Another class of methods for compiling terms of the ordinary λ-calculus is based on a translation into the so-called *continuation-passing style* (CPS), which was introduced in a seminal paper by Reynolds [21]. These translations have been studied

---

[2]For an introduction to the theory of abstract machines, cf. [14].

both in logic, where they correspond to double negation translations (cf. [23]), and in the theory of optimizing compilers (cf. [1]).

One important variation of these CPS translations is the so-called *ANF-transformation*, which was introduced by Sabry and Felleisen [22] and later elaborated by Flanagan et al. [15]. We first introduce the syntax of the administrative normal form in Definition 6.1. The ANF-transformation itself is introduced in Definitions 6.3 and 6.7.

Definition 6.1. The *syntax of the administrative normal form* Λ^{ANF} is defined as follows:

1. *Values:* $v ::= \lambda x.e \mid (v, v) \mid x.$

2. *Computations:* $c ::= v \mid v\,v \mid \pi_1\,v \mid \pi_2\,v.$

3. *Terms:* $e ::= c \mid \mathbf{let}\ x = c\ \mathbf{in}\ e.$

The administrative normal form has two characteristic properties. The first is reflected in the syntax of computations c: a projection π_i can only be applied to a value, and, similarly, a function application $v_1\,v_2$ can only be formed between two values. This excludes terms like $\pi_1(x, \pi_2(y, z))$ or $(\pi_1(f, g))(\pi_2(x, y))$. The second property is reflected in the syntax of terms e: a let-expression $\mathbf{let}\ x = c\ \mathbf{in}\ e$ can only bind the result of a computation c to a variable x. Let-expressions cannot bind other let-expressions, that is, expressions like $\mathbf{let}\ x = (\mathbf{let}\ y = e_1\ \mathbf{in}\ e_2)\ \mathbf{in}\ e_3$ are excluded by the second property.

Usual presentations of the ANF-transformation enforce both properties in a single transformation from Λ to Λ^{ANF}. Instead, we present the transformation to administrative normal form as a two-part transformation:

$$\Lambda \xrightarrow{\ \ \mathcal{A}\ \ } \Lambda^{\mathbb{Q}} \xrightarrow{\ \ \mathcal{L}\ \ } \Lambda^{\text{ANF}}.$$

The first part consists of a function $\mathcal{A} : \Lambda \to \Lambda^{\mathbb{Q}}$ that enforces only the first of the two characteristic properties described above. A second transformation $\mathcal{L} : \Lambda^{\mathbb{Q}} \to \Lambda^{\text{ANF}}$ then enforces the second property. By presenting the ANF-transformation in this way, we can make the relation to focusing clearer. In Section 8 we will show that the first part of this transformation corresponds to focusing, whereas the second part of the transformation can be simulated by μ-reductions in $\Lambda_{\mu\tilde{\mu}}$.

Definition 6.2. The *syntax of the intermediate normal form* Λ^Q is defined as follows:

1. *Values:* $v ::= \lambda x.e \mid (v,v) \mid x$.

2. *Terms:* $e ::= v \mid \text{let } x = e \text{ in } e \mid e\,v \mid \pi_1\,e \mid \pi_2\,e$.

Note that Definition 6.2 only guarantees that pairs (v,v) consist of values, and that functions are always applied to values in a function application $e\,v$. The two transformations $\mathcal{A}$ and $\mathcal{L}$ are introduced in turn.

6.1 From Λ to Λ^Q

Recall that the first property that we want to enforce is that pairs consist of syntactic values, and that in function applications the function argument is already a value. The transformation $\mathcal{A}$ defined next guarantees the first property by binding any non-value argument which would violate this property to a fresh variable in a let-binding. For example, the term $\pi_1(\pi_2(x,y))$ is transformed by generating a fresh variable z, and binding the computation $\pi_2(x,y)$ to z in the computation $\pi_1\,z$: $\mathcal{A}(\pi_1(\pi_2(x,y))) :\simeq \text{let } z = \pi_2(x,y) \text{ in } \pi_1\,z$.

Definition 6.3. The *transformation* $\mathcal{A}: \Lambda \to \Lambda^Q$ is defined as follows:

$$\mathcal{A}(x) :\simeq x \tag{$\mathcal{A}_1$}$$
$$\mathcal{A}(\lambda x.e) :\simeq \lambda x.\mathcal{A}(e) \tag{$\mathcal{A}_2$}$$
$$\mathcal{A}(\text{let } x = e_1 \text{ in } e_2) :\simeq \text{let } x = \mathcal{A}(e_1) \text{ in } \mathcal{A}(e_2) \tag{$\mathcal{A}_3$}$$
$$\mathcal{A}(\pi_i\,e) :\simeq \pi_i(\mathcal{A}(e)) \tag{$\mathcal{A}_4$}$$

$$\mathcal{A}((v_1,v_2)) :\simeq (\mathcal{A}(v_1),\mathcal{A}(v_2)) \tag{$\mathcal{A}_5$}$$
$$\mathcal{A}((v_1,e_2)) :\simeq \text{let } x = \mathcal{A}(e_2) \text{ in } (\mathcal{A}(v_1),x) \tag{$\mathcal{A}_6$}$$
$$\mathcal{A}((e_1,v_2)) :\simeq \text{let } x = \mathcal{A}(e_1) \text{ in } (x,v_2) \tag{$\mathcal{A}_7$}$$
$$\mathcal{A}((e_1,e_2)) :\simeq \text{let } x = \mathcal{A}(e_1) \text{ in } (\text{let } y = \mathcal{A}(e_2) \text{ in } (x,y)) \tag{$\mathcal{A}_8$}$$

$$\mathcal{A}(e_1\, v_2) :\simeq \mathcal{A}(e_1)\, \mathcal{A}(v_2) \tag{$\mathcal{A}_9$}$$

$$\mathcal{A}(e_1\, e_2) :\simeq \mathbf{let}\ x = \mathcal{A}(e_2)\ \mathbf{in}\ \mathcal{A}(e_1)\, x. \tag{$\mathcal{A}_{10}$}$$

Remark 6.4. Among the clauses of Definition 6.3, the clauses $(\mathcal{A}_5)$ to $(\mathcal{A}_7)$ are subsumed by $(\mathcal{A}_8)$. Similarly, the clause $(\mathcal{A}_9)$ is subsumed by $(\mathcal{A}_{10})$. This redundancy is an optimization which guarantees that the transformation behaves as the identity function on terms that are already in Λ^{Q}.

Example 6.5. The result of the transformation

$$\mathcal{A}(\pi_1(\pi_1(\pi_1(x_1, x_2), x_3), x_4))$$

is the term

$$\mathbf{let}\ z_1 = (\mathbf{let}\ z_2 = \pi_1(x_1, x_2)\ \mathbf{in}\ \pi_1(z_2, x_3))\ \mathbf{in}\ \pi_1(z_1, x_4),$$

where z_1 and z_2 are variables that are generated during the transformation. This example shows that the result of $\mathcal{A}$ is, in general, not yet in Λ^{ANF}.

6.2 From Λ^{Q} to Λ^{ANF}

The second property which we want to enforce is that in a let-construct $\mathbf{let}\ x = c\ \mathbf{in}\ e$ the computation bound to the variable x must be of a restricted form. This will be guaranteed by the transformation $\mathcal{L}$, given by Definition 6.7. In order to define this transformation, we need to define a meta-level operation @ that operates on continuations k and values v from Λ^{ANF}:

Definition 6.6. *Continuations* are defined as follows:

$$k ::= \mathrm{id} \mid \overline{\lambda v}.\mathbf{let}\ x = \pi_i\, \overline{v}\ \mathbf{in}\ e \mid \overline{\lambda v}.\mathbf{let}\ x = \overline{v}\, v\ \mathbf{in}\ e \mid \overline{\lambda v}.\mathbf{let}\ x = \overline{v}\ \mathbf{in}\ e,$$

where e and v range over expressions and values from Λ^{ANF}.

The *meta-level operation* @ takes a continuation k and a value v from Λ^{ANF} and returns an expression of Λ^{ANF}. It is evaluated as follows:

$$\mathrm{id} @ v :\simeq v \tag{$@_1$}$$

$$\overline{\lambda v}.\text{let } x = \pi_i \, \overline{v} \text{ in } e \, @ \, v :\simeq \text{let } x = \pi_i \, v \text{ in } e \qquad (@_2)$$

$$\overline{\lambda v}.\text{let } x = \overline{v} \, v_2 \text{ in } e \, @ \, v_1 :\simeq \text{let } x = v_1 \, v_2 \text{ in } e \qquad (@_3)$$

$$\overline{\lambda v}.\text{let } x = \overline{v} \text{ in } e \, @ \, v :\simeq \text{let } x = v \text{ in } e. \qquad (@_4)$$

Using this technical tool we can now define the transformation $\mathcal{L}$.

Definition 6.7. The *transformation* $\mathcal{L} : \Lambda^Q \to \Lambda^{\text{ANF}}$ is given as follows:

Values

$$\mathcal{L}(x) :\simeq x \qquad (\mathcal{L}_1)$$

$$\mathcal{L}(\lambda x.e) :\simeq \lambda x.\mathcal{L}_{\text{id}}(e) \qquad (\mathcal{L}_2)$$

$$\mathcal{L}((v_1, v_2)) :\simeq (\mathcal{L}(v_1), \mathcal{L}(v_2)) \qquad (\mathcal{L}_3)$$

Terms

$$\mathcal{L}(e) :\simeq \mathcal{L}_{\text{id}}(e) \qquad (\mathcal{L}_4)$$

$$\mathcal{L}_k(e_1 \, v_2) :\simeq \mathcal{L}_{\overline{\lambda v}.\text{let } x=\overline{v} \, \mathcal{L}(v_2) \text{ in } k \, @ \, x}(e_1) \qquad (\mathcal{L}_5)$$

$$\mathcal{L}_k(\pi_i \, e) :\simeq \mathcal{L}_{\overline{\lambda v}.\text{let } x=\pi_i \, \overline{v} \text{ in } k \, @ \, x}(e) \qquad (\mathcal{L}_6)$$

$$\mathcal{L}_k(v) :\simeq k \, @ \, \mathcal{L}(v) \qquad (\mathcal{L}_7)$$

$$\mathcal{L}_k(\text{let } x = e_1 \text{ in } e_2) :\simeq \mathcal{L}_{\overline{\lambda v}.\text{let } x=\overline{v} \text{ in } \mathcal{L}_k(e_2)}(e_1). \qquad (\mathcal{L}_8)$$

Example 6.8. As an example of the transformation $\mathcal{L}$, consider the term (from Example 6.5)

$$\text{let } z_1 = (\text{let } z_2 = \pi_1(x_1, x_2) \text{ in } \pi_1(z_2, x_3)) \text{ in } \pi_1(z_1, x_4),$$

which can be transformed into Λ^{ANF} as follows:

$$\mathcal{L}_{\text{id}}(\text{let } z_1 = (\text{let } z_2 = \pi_1(x_1, x_2) \text{ in } \pi_1(z_2, x_3)) \text{ in } \pi_1(z_1, x_4))$$

$$= \mathcal{L}_{\overline{\lambda v_1}.\text{let } z_1=\overline{v_1} \text{ in } \mathcal{L}_{\text{id}}(\pi_1(z_1,x_4))}(\text{let } z_2 = \pi_1(x_1, x_2) \text{ in } \pi_1(z_2, x_3))$$

$$= \mathcal{L}_{\overline{\lambda v_1}.\text{let } z_1=\overline{v_1} \text{ in } \pi_1(z_1,x_4)}(\text{let } z_2 = \pi_1(x_1, x_2) \text{ in } \pi_1(z_2, x_3))$$

$$= \mathcal{L}_{\overline{\lambda v_2}.\text{let } z_2=\overline{v_2} \text{ in } \mathcal{L}_{\overline{\lambda v_1}.\text{let } z_1=\overline{v_1} \text{ in } \pi_1(z_1,x_4)}(\pi_1(z_2,x_3))}(\pi_1(x_1, x_2))$$

$$= \mathcal{L}_{\overline{\lambda v_2.\text{let } z_2 = \overline{v_2} \text{ in (let } z_1 = \pi_1(z_2,x_3) \text{ in } \pi_1(z_1,x_4))}}(\pi_1(x_1,x_2))$$

$$= \text{let } z_2 = \pi_1(x_1,x_2) \text{ in (let } z_1 = \pi_1(z_2,x_3) \text{ in } \pi_1(z_1,x_4)).$$

The term was transformed into Λ^{ANF} by (in a certain way) moving the let-binding of z_2 to the outside of the let-binding of z_1.

7 The focusing transformation

In distinction to the dynamic focusing rules of Definition 5.5, we now consider only static focusing. We first introduce the focused subsyntax $\Lambda^{Q}_{\mu\tilde{\mu}}$ as a subset of $\Lambda_{\mu\tilde{\mu}}$ (Definition 3.4).[3]

Definition 7.1. The *focused subsyntax* $\Lambda^{Q}_{\mu\tilde{\mu}}$ for the call-by-value evaluation strategy is defined as follows:

1. *Terms:* $t ::= w \mid \mu\alpha.c$.

2. *Coterms:* $s ::= \alpha \mid w \cdot s \mid \pi_1 s \mid \pi_2 s \mid \tilde{\mu}x.c$.

3. *Commands:* $c ::= \langle t \mid s \rangle$.

4. *Values:* $w ::= \lambda x.t \mid (w,w) \mid x$.

The focused subsyntax $\Lambda^{Q}_{\mu\tilde{\mu}}$ differs in two respects from $\Lambda_{\mu\tilde{\mu}}$. First, terms t must now either be values w or abstractions $\mu\alpha.c$. This excludes terms like $(\mu\alpha.c, t)$ and $(t, \mu\alpha.c)$ from the subsyntax $\Lambda^{Q}_{\mu\tilde{\mu}}$, which are part of the syntax of terms of Definition 3.4. This corresponds precisely to the restriction that constructors can only be applied to values. Second, the syntax of coterms has been changed by requiring the function argument in a coterm $t \cdot s$ to be a value; that is, we require $w \cdot s$. This corresponds to the requirement that functions can syntactically only be applied to values.

Lemma 7.2. *For any term $e \in \Lambda^{Q}$, $[\![e]\!] \in \Lambda^{Q}_{\mu\tilde{\mu}}$.*

Proof. By induction on e.

---

[3] $\Lambda^{Q}_{\mu\tilde{\mu}}$ corresponds to the subsyntax LKQ defined in [4].

1. Case $e \simeq \mathbf{let}\, x = e_1 \,\mathbf{in}\, e_2$: the translation of e is $\mu\alpha.\langle\llbracket e_1 \rrbracket \mid \tilde{\mu}x.\langle\llbracket e_2 \rrbracket \mid \alpha\rangle\rangle$. Using the induction hypothesis for e_1 and e_2, this term is in the subsyntax $\Lambda^{\mathrm{Q}}_{\mu\tilde{\mu}}$.

2. If e is of the form $e_1 v_2$, then the translation of e is $\mu\alpha.\langle\llbracket e_1 \rrbracket \mid \llbracket v_2 \rrbracket \cdot \alpha\rangle$. By Lemma 4.3, $\llbracket v_2 \rrbracket$ is a value, and by the induction hypothesis both $\llbracket e_1 \rrbracket$ and $\llbracket v_2 \rrbracket$ are in the subsyntax $\Lambda^{\mathrm{Q}}_{\mu\tilde{\mu}}$, so the resulting term is in the subsyntax $\Lambda^{\mathrm{Q}}_{\mu\tilde{\mu}}$.

3. If e is of the form $\pi_i\, e_1$, then $\llbracket \pi_i\, e_1 \rrbracket$ is $\mu\alpha.\langle\llbracket e_1 \rrbracket \mid \pi_i\, \alpha\rangle$. Using the induction hypothesis for e_1, $\llbracket e_1 \rrbracket$ is in $\Lambda^{\mathrm{Q}}_{\mu\tilde{\mu}}$. Therefore $\llbracket \pi_i\, e_1 \rrbracket$ is also in $\Lambda^{\mathrm{Q}}_{\mu\tilde{\mu}}$.

4. If $e \simeq v$, then we have to distinguish the following cases:

   (a) If $v \simeq x$, then $\llbracket x \rrbracket \simeq x$, which is in $\Lambda^{\mathrm{Q}}_{\mu\tilde{\mu}}$.

   (b) If $v \simeq (v_1, v_2)$, then $\llbracket (v_1, v_2) \rrbracket \simeq (\llbracket v_1 \rrbracket, \llbracket v_2 \rrbracket)$. By Lemma 4.3, both $\llbracket v_i \rrbracket$ are values, and by the induction hypothesis they are in the subsyntax $\Lambda^{\mathrm{Q}}_{\mu\tilde{\mu}}$. Therefore $\llbracket v \rrbracket$ is also in $\Lambda^{\mathrm{Q}}_{\mu\tilde{\mu}}$.

   (c) If $v \simeq \lambda x.e$, then $\llbracket \lambda x.e \rrbracket \simeq \lambda x.\llbracket e \rrbracket$. By the induction hypothesis $\llbracket e \rrbracket$ is in $\Lambda^{\mathrm{Q}}_{\mu\tilde{\mu}}$, therefore $\llbracket v \rrbracket$ is also in $\Lambda^{\mathrm{Q}}_{\mu\tilde{\mu}}$. $\square$

The subsyntax does not restrict the set of derivable sequents, since any term, coterm or command in the unrestricted syntax can be translated into the focused subsyntax $\Lambda^{\mathrm{Q}}_{\mu\tilde{\mu}}$ by using the following *static focusing transformation*.

Definition 7.3. The *static focusing transformation* $\mathcal{F} : \Lambda_{\mu\tilde{\mu}} \to \Lambda^{\mathrm{Q}}_{\mu\tilde{\mu}}$ is defined as follows:

Terms

$$\mathcal{F}(x) :\simeq x \qquad (\mathcal{F}_1)$$
$$\mathcal{F}(\mu\alpha.c) :\simeq \mu\alpha.\mathcal{F}(c) \qquad (\mathcal{F}_2)$$
$$\mathcal{F}(\lambda x.e) :\simeq \lambda x.\mathcal{F}(e) \qquad (\mathcal{F}_3)$$
$$\mathcal{F}((w_1, w_2)) :\simeq (\mathcal{F}(w_1), \mathcal{F}(w_2)) \qquad (\mathcal{F}_4)$$
$$\mathcal{F}((w_1, t_2)) :\simeq \mu\alpha.\langle\mathcal{F}(t_2) \mid \tilde{\mu}x.\langle(\mathcal{F}(w_1), x) \mid \alpha\rangle\rangle \qquad (\mathcal{F}_5)$$

$$\mathcal{F}((t_1, w_2)) :\simeq \mu\alpha.\langle \mathcal{F}(t_1) \mid \tilde{\mu}x.\langle(x, \mathcal{F}(w_2)) \mid \alpha\rangle\rangle \qquad (\mathcal{F}_6)$$

$$\mathcal{F}((t_1, t_2)) :\simeq \mu\alpha.\langle \mathcal{F}(t_1) \mid \tilde{\mu}x.\langle\mu\beta.\langle\mathcal{F}(t_2) \mid \tilde{\mu}y.\langle(x, y) \mid \beta\rangle\rangle \mid \alpha\rangle\rangle \quad (\mathcal{F}_7)$$

Coterms

$$\mathcal{F}(\alpha) :\simeq \alpha \qquad (\mathcal{F}_8)$$

$$\mathcal{F}(\tilde{\mu}x.c) :\simeq \tilde{\mu}x.\mathcal{F}(c) \qquad (\mathcal{F}_9)$$

$$\mathcal{F}(\pi_i\, s) :\simeq \pi_i\, \mathcal{F}(s) \qquad (\mathcal{F}_{10})$$

$$\mathcal{F}(w \cdot s) :\simeq \mathcal{F}(w) \cdot \mathcal{F}(s) \qquad (\mathcal{F}_{11})$$

$$\mathcal{F}(t \cdot s) :\simeq \tilde{\mu}x.\langle \mathcal{F}(t) \mid \tilde{\mu}y.\langle x \mid y \cdot \mathcal{F}(s)\rangle\rangle \qquad (\mathcal{F}_{12})$$

Commands

$$\mathcal{F}(\langle t \mid s\rangle) :\simeq \langle \mathcal{F}(t) \mid \mathcal{F}(s)\rangle \qquad (\mathcal{F}_{13})$$

$$\mathcal{F}(\langle t_1 \mid t_2 \cdot s\rangle) :\simeq \langle \mathcal{F}(t_2) \mid \tilde{\mu}x.\langle\mu\alpha.\langle\mathcal{F}(t_1) \mid x \cdot \alpha\rangle \mid \mathcal{F}(s)\rangle\rangle. \qquad (\mathcal{F}_{14})$$

In general, when several clauses are applicable, the most specific clause should be applied. The clauses $(\mathcal{F}_4)$, $(\mathcal{F}_5)$ and $(\mathcal{F}_6)$ are subsumed by the more general clause $(\mathcal{F}_7)$, and $(\mathcal{F}_{11})$ is subsumed by the clause $(\mathcal{F}_{12})$. The presence of these additional clauses guarantees that $\mathcal{F}$ behaves as the identity function when it is applied to a term, coterm or command that is already in the subsyntax $\Lambda_{\mu\tilde{\mu}}^{Q}$. With these optimizations, our definition corresponds to the one given in [6, Fig. 18], with the exception of the clause $(\mathcal{F}_{14})$. The additional clause $(\mathcal{F}_{14})$ is necessary to guarantee that the functions $[\![-]\!]$, $\mathcal{A}$ and $\mathcal{F}$ commute up to α-equivalence, as shown by Theorem 8.1. Without the clause $(\mathcal{F}_{14})$, Theorem 8.1 has to be slightly weakened to Theorem 8.2.

Lemma 7.4 ($\mathcal{F}$ preserves typeability). *For all terms t, coterms s and commands c:*

1. *If $\Gamma \vdash t : \tau \mid \Delta$, then $\Gamma \vdash \mathcal{F}(t) : \tau \mid \Delta$.*

2. *If $\Gamma \mid s : \tau \vdash \Delta$, then $\Gamma \mid \mathcal{F}(s) : \tau \vdash \Delta$.*

3. *If $c : (\Gamma \vdash \Delta)$, then $\mathcal{F}(c) : (\Gamma \vdash \Delta)$.*

Proof. By simultaneous structural induction on t, s and c, respectively. □

8 The main result

As explained in Section 6, the ANF-transformation can be split into a purely local transformation $\mathcal{A}$ and a global transformation $\mathcal{L}$. We show what these two parts correspond to in the $\lambda\mu\tilde{\mu}$-calculus, and prove how the ANF-transformation on λ-terms relates to static focusing of $\lambda\mu\tilde{\mu}$-terms.

8.1 The correspondence between $\mathcal{A}$ and $\mathcal{F}$

Theorem 8.1 (Focusing reflects the ANF-transformation). *For all λ-terms e, we have $\mathcal{F}(\llbracket e \rrbracket) \eqsim \llbracket \mathcal{A}(e) \rrbracket$.*

Proof. See Appendix A. ☐

If we omit the focusing rule $(\mathcal{F}_{14})$ from Definition 7.3, then Theorem 8.1 no longer holds up to syntactic equality ($\eqsim$). Instead, the following weaker result (Theorem 8.2) holds for $\eta\mu$-equality $\equiv$, which includes η-equivalence

$$\tilde{\mu}x.\langle x \mid s \rangle \equiv_\eta s \qquad \text{(for } x \text{ not free in } s).$$

Theorem 8.2 (Focusing reflects the ANF-transformation; case $\equiv$). *For all λ-terms e, we have $\mathcal{F}(\llbracket e \rrbracket) \equiv \llbracket \mathcal{A}(e) \rrbracket$.*

Proof. See Appendix A. ☐

8.2 Simulating $\mathcal{L}$ in the $\lambda\mu\tilde{\mu}$-calculus

Our main contention in this section is that a special purpose transformation like $\mathcal{L}$ is not necessary in $\Lambda_{\mu\tilde{\mu}}$. In order to transform from $\Lambda_{\mu\tilde{\mu}}^{\mathrm{Q}}$ to $\Lambda_{\mu\tilde{\mu}}^{\mathrm{ANF}}$ we only have to apply μ-reductions and $\tilde{\mu}$-expansions. More concretely, the effect that $\mathcal{L}$ has on a term, namely to globally reorganize the ordering of let-bindings, can be simulated by simply reducing μ-redexes in the image of the translation. In order to illustrate this central point, let us come back to Examples 6.5 and 6.8. Recall that we showed in Example 6.8 that $\mathcal{L}$ has the effect of changing the order of the two let-bindings of z_1 and z_2:

$$\mathcal{L}(\text{let } z_1 = (\text{let } z_2 = \pi_1(x_1, x_2) \text{ in } \pi_1(z_2, x_3)) \text{ in } \pi_1(z_1, x_4))$$
$$\eqcirc \text{let } z_2 = \pi_1(x_1, x_2) \text{ in } (\text{let } z_1 = \pi_1(z_2, x_3) \text{ in } \pi_1(z_1, x_4)).$$

This can be simulated as follows:

$$[\![\text{let } z_1 = (\text{let } z_2 = \pi_1(x_1, x_2) \text{ in } \pi_1(z_2, x_3)) \text{ in } \pi_1(z_1, x_4)]\!]$$
$$\eqcirc \mu\alpha.\langle\mu\beta.\langle[\![\pi_1(x_1, x_2)]\!] \mid \tilde{\mu}z_2.\langle[\![\pi_1(z_2, x_3)]\!] \mid \beta\rangle\rangle \mid \tilde{\mu}z_1.\langle[\![\pi_1(z_1, x_4)]\!] \mid \alpha\rangle\rangle$$
$$\triangleright \mu\alpha.\langle[\![\pi_1(x_1, x_2)]\!] \mid \tilde{\mu}z_2.\langle[\![\pi_1(z_2, x_3)]\!] \mid \tilde{\mu}z_1.\langle[\![\pi_1(z_1, x_4)]\!] \mid \alpha\rangle\rangle\rangle$$
$$\triangleleft \mu\alpha.\langle[\![\pi_1(x_1, x_2)]\!] \mid \tilde{\mu}z_2.\langle\mu\beta.\langle[\![\pi_1(z_2, x_3)]\!] \mid \tilde{\mu}z_1.\langle[\![\pi_1(z_1, x_4)]\!] \mid \beta\rangle\rangle \mid \alpha\rangle\rangle$$
$$\eqcirc [\![\text{let } z_2 = \pi_1(x_1, x_2) \text{ in } (\text{let } z_1 = \pi_1(z_2, x_3) \text{ in } \pi_1(z_1, x_4))]\!].$$

Next, we define the subsyntax $\Lambda_{\mu\tilde{\mu}}^{\text{ANF}}$, which differs from $\Lambda_{\mu\tilde{\mu}}^{\text{Q}}$ (Definition 7.1) in two aspects. First, commands are now required to consist of a *value* and a coterm instead of a term and a coterm, i.e., they do not contain any μ-redexes. Second, the coterms for projections and function applications are required to give an explicit name to the value they bind in the coterm they contain, i.e., they are $\tilde{\mu}$-expanded.

Definition 8.3. The *focused subsyntax* $\Lambda_{\mu\tilde{\mu}}^{\text{ANF}}$ for the call-by-value strategy is defined as follows:

1. *Terms:* $t ::= w \mid \mu\alpha.c$.

2. *Coterms:* $s ::= \alpha \mid w \cdot \tilde{\mu}x.c \mid \pi_1 \tilde{\mu}x.c \mid \pi_2 \tilde{\mu}x.c \mid \tilde{\mu}x.c$

3. *Commands:* $c ::= \langle w \mid s \rangle$.

4. *Values:* $w ::= \lambda x.t \mid (w, w) \mid x$.

We have to refine Definition 4.1.

Definition 8.4. The *refined translation* $[\![-]\!]^* : \Lambda^{\mathrm{ANF}} \to \Lambda^{\mathrm{ANF}}_{\mu\tilde{\mu}}$ is defined as the first function in the following set of mutually defined recursive functions:

First function (on expressions)

$$*[\![e]\!]^* :\simeq \mu\alpha.[\![e]\!]^*_\alpha \qquad (\mathcal{T}^*_1)$$

Second function (on expressions)

$$*[\![\mathbf{let}\ x = c\ \mathbf{in}\ e]\!]^*_s :\simeq [\![c]\!]^*_{\tilde{\mu}x.[\![e]\!]^*_s} \qquad (\mathcal{T}^*_2)$$

$$[\![v_1\ v_2]\!]^*_s :\simeq \langle [\![v_1]\!]^* \mid [\![v_2]\!]^* \cdot s \rangle \qquad (\mathcal{T}^*_3)$$

$$[\![\pi_i\ v]\!]^*_s :\simeq \langle [\![v]\!]^* \mid \pi_i\ s \rangle \qquad (\mathcal{T}^*_4)$$

$$[\![v]\!]^*_s :\simeq \langle [\![v]\!]^* \mid s \rangle \qquad (\mathcal{T}^*_5)$$

Third function (on values)

$$*[\![x]\!]^* :\simeq x \qquad (\mathcal{T}^*_6)$$

$$[\![(v_1, v_2)]\!]^* :\simeq ([\![v_1]\!]^*, [\![v_2]\!]^*) \qquad (\mathcal{T}^*_7)$$

$$[\![\lambda x.e]\!]^* :\simeq \lambda x.[\![e]\!]^*. \qquad (\mathcal{T}^*_8)$$

Lemma 8.5. *For all terms* $e \in \Lambda^{\mathrm{ANF}}$, $[\![e]\!]^* \in \Lambda^{\mathrm{ANF}}_{\mu\tilde{\mu}}$.

Proof. By induction on terms e. $\square$

Theorem 8.6. *For all terms* $e \in \Lambda^{\mathrm{Q}}$, $[\![\mathcal{L}(e)]\!]^* \equiv_\mu [\![e]\!]$.

Proof. See Appendix A. $\square$

9 Summary and outlook

We can summarize our results in the following diagram, where both the lower and the upper part are commutative.

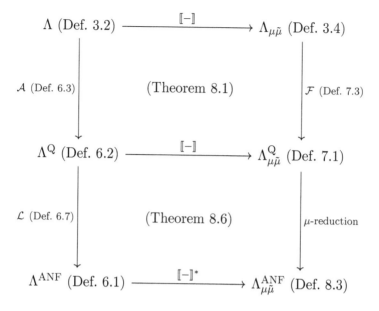

These results are embedded in a wider conceptual context. By the Curry-Howard correspondence, natural deduction for intuitionistic logic (more precisely, the $\{\rightarrow, \wedge\}$-fragment) corresponds to the λ-calculus on the one side, and the sequent calculus for classical logic corresponds to the $\lambda\mu\tilde{\mu}$-calculus on the other side. Our results thus establish a bridge between natural deduction for intuitionistic logic with its computational interpretation on the one side and the classical sequent calculus with its computational interpretation on the other side.

We would like to extend this work in two directions. The first concerns the asymmetry of the translation $[\![-]\!]$, which is only a mapping from Λ to $\Lambda_{\mu\tilde{\mu}}$, but not vice versa. In order to provide a translation in the opposite direction, from $\Lambda_{\mu\tilde{\mu}}$ to Λ, we will have to extend the λ-calculus with control operators. The seond extension concerns the treatment of evaluation orders other than call-by-value. While the treatment of call-by-name seems to be straightforward, the study of call-by-need (cf. [3, 19]) and its dual call-by-co-need (cf. [5]) in both the λ-calculus and the $\lambda\mu\tilde{\mu}$-calculus seems to be promising.

Acknowledgements

We would like to thank Jonathan Brachthäuser for discussions, and Philipp Schuster and Paul Downen for helpful remarks on an earlier version of this paper. This work was carried out within the DFG project "Constructive Semantics and the Completeness Problem", DFG grant PI 1174/1-1.

References

[1] Andrew W. Appel. *Compiling with Continuations*. Cambridge University Press, 1992.

[2] Zena M. Ariola and Hugo Herbelin. Minimal classical logic and control operators. In Jos C. M. Baeten, Jan Karel Lenstra, Joachim Parrow, and Gerhard J. Woeginger, editors, *Automata, Languages and Programming*, pages 871–885. Springer, 2003.

[3] Zena M. Ariola, John Maraist, Martin Odersky, Matthias Felleisen, and Philip Wadler. A call-by-need lambda calculus. In *Proceedings of the 22nd ACM SIGPLAN-SIGACT Symposium on Principles of Programming Languages*, POPL '95, pages 233–246, New York, NY, USA, 1995. Association for Computing Machinery. doi: 10.1145/199448.199507.

[4] Pierre-Louis Curien and Hugo Herbelin. The duality of computation. In *Proceedings of the Fifth ACM SIGPLAN International Conference on Functional Programming*, ICFP '00, pages 233–243, New York, NY, USA, 2000. Association for Computing Machinery.

[5] Paul Downen and Zena M. Ariola. Beyond polarity: Towards a multi-discipline intermediate language with sharing. In *27th EACSL Annual Conference on Computer Science Logic (CSL 2018)*. Schloss Dagstuhl – Leibniz-Zentrum für Informatik, 2018.

[6] Paul Downen and Zena M. Ariola. A tutorial on computational classical logic and the sequent calculus. *Journal of Functional Programming*, 28:e3, 2018. doi: 10.1017/S0956796818000023.

[7] Marie Duží. Structural isomorphism of meaning and synonymy. *Computación y Sistemas*, 18(3):439–453, 2014.

[8] Marie Duží. Natural Language Processing by Natural Deduction in Transparent Intensional Logic. In Thomas Piecha and Peter Schroeder-Heister, editors, *Proof-Theoretic Semantics: Assessment and Future Perspectives. Proceedings of the Third Tübingen Conference on Proof-Theoretic Semantics, 27-30 March 2019*, pages 41, 685–706. University of Tübingen, 2019. doi: 10.15496/publikation-35319.

[9] Marie Duží and Bjørn Jespersen. Procedural isomorphism, analytic information and β-conversion by value. *Logic Journal of the IGPL*, 21(2):291–308, 2012. doi: 10.1093/jigpal/jzs044.

[10] Marie Duží and Bjørn Jespersen. Transparent quantification into hyperpropositional contexts *de re*. *Logique et Analyse*, 55(220): 513–554, 2012.

[11] Marie Duží and Bjørn Jespersen. Transparent quantification into hyperintensional objectual attitudes. *Synthese*, 192:635–677, 2015.

[12] Marie Duží and Miloš Kosterec. A valid rule of β-conversion for the logic of partial functions. *Organon F*, 24(1):10–36, 2017.

[13] Matthias Felleisen and Robert Hieb. The revised report on the syntactic theories of sequential control and state. *Theoretical Computer Science*, 103(2):235–271, 1992.

[14] Matthias Felleisen, Robert Bruce Findler, and Matthew Flatt. *Semantics Engineering with PLT Redex*. MIT Press, 2009.

[15] Cormac Flanagan, Amr Sabry, Bruce F. Duba, and Matthias Felleisen. The essence of compiling with continuations. In *Proceedings of the ACM SIGPLAN 1993 Conference on Programming Language Design and Implementation*, PLDI '93, pages 237–247, New York, NY, USA, 1993. Association for Computing Machinery.

177

[16] Gerhard Gentzen. Untersuchungen über das logische Schließen. *Mathematische Zeitschrift*, 39:176–210, 405–431, 1935.

[17] Timothy G. Griffin. A formulae-as-type notion of control. In *Proceedings of the 17th ACM SIGPLAN-SIGACT Symposium on Principles of Programming Languages*, POPL '90, pages 47–58, New York, NY, USA, 1989. Association for Computing Machinery. doi: 10.1145/96709.96714.

[18] Bjørn Jespersen and Marie Duží. Transparent quantification into hyperpropositional attitudes de dicto. *Linguistics and Philosophy*, 45:1119–1164, 2022. doi: 10.1007/s10988-021-09344-9.

[19] John Launchbury. A natural semantics for lazy evaluation. In *Proceedings of the 20th ACM SIGPLAN-SIGACT Symposium on Principles of Programming Languages*, POPL '93, pages 144–154, New York, NY, USA, 1993. Association for Computing Machinery. doi: 10.1145/158511.158618.

[20] Benjamin C. Pierce. *Types and Programming Languages*. MIT Press, 2002.

[21] John C. Reynolds. Definitional interpreters for higher-order programming languages. In *Proceedings of the ACM Annual Conference – Volume 2*, ACM '72, pages 717–740, New York, NY, USA, 1972. Association for Computing Machinery. ISBN 9781450374927. doi: 10.1145/800194.805852.

[22] Amr Sabry and Matthias Felleisen. Reasoning about programs in continuation-passing style. In *Proceedings of the 1992 ACM Conference on LISP and Functional Programming*, LFP '92, pages 288–298, New York, NY, USA, 1992. Association for Computing Machinery.

[23] Morten Heine Sørensen and Paweł Urzyczyn. *Lectures on the Curry-Howard Isomorphism*, volume 149 of *Studies in Logic and the Foundations of Mathematics*. Elsevier, 2006.

A Proofs of the main theorems

Theorem 8.1 (Focusing reflects the ANF-transformation). *For all* λ-*terms* e, *we have* $\mathcal{F}(\llbracket e \rrbracket) \simeq \llbracket \mathcal{A}(e) \rrbracket$.

Proof. By induction on the structure of e.

1. Case $e \simeq x$: $\mathcal{F}(\llbracket x \rrbracket) \simeq x \simeq \llbracket \mathcal{A}(x) \rrbracket$.

2. Case $e \simeq \lambda x.e_1$:

$$\mathcal{F}(\llbracket \lambda x.e_1 \rrbracket) \simeq \lambda x.\mathcal{F}(\llbracket e_1 \rrbracket) \overset{\text{IH}}{\simeq} \lambda x.\llbracket \mathcal{A}(e_1) \rrbracket \simeq \llbracket \mathcal{A}(\lambda x.e_1) \rrbracket.$$

3. Case $e \simeq \pi_i \, e_1$:

$$
\begin{aligned}
\mathcal{F}(\llbracket \pi_i \, e_1 \rrbracket) &\simeq \mathcal{F}(\mu\alpha.\langle \llbracket e_1 \rrbracket \mid \pi_i \, \alpha \rangle) && (\mathcal{T}_5) \\
&\simeq \mu\alpha.\langle \mathcal{F}(\llbracket e_1 \rrbracket) \mid \pi_i \, \alpha \rangle && (\mathcal{F}_2, \mathcal{F}_{13}, \mathcal{F}_{10}, \mathcal{F}_8) \\
&\simeq \mu\alpha.\langle \llbracket \mathcal{A}(e_1) \rrbracket \mid \pi_i \, \alpha \rangle && (\text{IH}) \\
&\simeq \llbracket \pi_i \, \mathcal{A}(e_1) \rrbracket && (\mathcal{T}_5) \\
&\simeq \llbracket \mathcal{A}(\pi_i \, e_1) \rrbracket. && (\mathcal{A}_4)
\end{aligned}
$$

4. In case $e \simeq e_1 \, e_2$, we have to distinguish two subcases:

(i) Subcase $e \simeq e_1 \, v_2$:

$$
\begin{aligned}
\mathcal{F}(\llbracket e_1 \, v_2 \rrbracket) &\simeq \mathcal{F}(\mu\alpha.\langle \llbracket e_1 \rrbracket \mid \llbracket v_2 \rrbracket \cdot \alpha \rangle) && (\mathcal{T}_4) \\
&\simeq \mu\alpha.\langle \mathcal{F}(\llbracket e_1 \rrbracket) \mid \mathcal{F}(\llbracket v_2 \rrbracket) \cdot \alpha \rangle && (\mathcal{F}_2, \mathcal{F}_{13}, \mathcal{F}_{11}, \mathcal{F}_8) \\
&\simeq \mu\alpha.\langle \llbracket \mathcal{A}(e_1) \rrbracket \mid \llbracket \mathcal{A}(v_2) \rrbracket \cdot \alpha \rangle && (\text{IH}) \\
&\simeq \llbracket \mathcal{A}(e_1) \, \mathcal{A}(v_2) \rrbracket && (\mathcal{T}_4) \\
&\simeq \llbracket \mathcal{A}(e_1 \, v_2) \rrbracket. && (\mathcal{A}_9)
\end{aligned}
$$

(ii) Subcase $e \simeq e_1 \, e_2$:

$$
\begin{aligned}
\mathcal{F}(\llbracket (e_1 \, e_2) \rrbracket) &\simeq \mathcal{F}(\mu\alpha.\langle \llbracket e_1 \rrbracket \mid \llbracket e_2 \rrbracket \cdot \alpha \rangle) && (\mathcal{T}_4) \\
&\simeq \mu\alpha.\langle \mathcal{F}(\llbracket e_2 \rrbracket) \mid \tilde{\mu}x.\langle \mu\beta.\langle \mathcal{F}(\llbracket e_1 \rrbracket) \mid x \cdot \beta \rangle \mid \alpha \rangle \rangle \\
& && (\mathcal{F}_{14}, \mathcal{F}_2, \mathcal{F}_8)
\end{aligned}
$$

$$\simeq \mu\alpha.\langle[\![\mathcal{A}(e_2)]\!] \mid \tilde{\mu}x.\langle\mu\beta.\langle[\![\mathcal{A}(e_1)]\!] \mid x \cdot \beta\rangle \mid \alpha\rangle\rangle \quad \text{(IH)}$$
$$\simeq [\![\textbf{let } x = \mathcal{A}(e_2) \textbf{ in } \mathcal{A}(e_1)\ x]\!] \quad (\mathcal{T}_4, \mathcal{T}_6)$$
$$\simeq [\![\mathcal{A}(e_1\ e_2)]\!]. \quad (\mathcal{A}_{10})$$

5. In case $e \simeq (e_1, e_2)$, we have to distinguish four subcases:

(i) Subcase $e \simeq (v_1, v_2)$:

$$\mathcal{F}([\![(v_1, v_2)]\!]) \simeq \mathcal{F}([\![v_1]\!], [\![v_2]\!]) \quad (\mathcal{T}_3)$$
$$\simeq (\mathcal{F}([\![v_1]\!]), \mathcal{F}([\![v_2]\!])) \quad (\mathcal{F}_4)$$
$$\simeq ([\![\mathcal{A}(v_1)]\!], [\![\mathcal{A}(v_2)]\!]) \quad \text{(IH)}$$
$$\simeq [\![(\mathcal{A}(v_1), \mathcal{A}(v_2))]\!] \quad (\mathcal{T}_3)$$
$$\simeq [\![\mathcal{A}((v_1, v_2))]\!]. \quad (\mathcal{A}_5)$$

(ii) Subcase $e \simeq (v_1, e_2)$:

$$\mathcal{F}([\![(v_1, e_2)]\!]) \simeq \mathcal{F}(([\![v_1]\!], [\![e_2]\!])) \quad (\mathcal{T}_3)$$
$$\simeq \mu\alpha.\langle\mathcal{F}([\![e_2]\!]) \mid \tilde{\mu}x.\langle(\mathcal{F}([\![v_1]\!]), x) \mid \alpha\rangle\rangle \quad (\mathcal{F}_5)$$
$$\simeq \mu\alpha.\langle[\![\mathcal{A}(e_2)]\!] \mid \tilde{\mu}x.\langle(([\![\mathcal{A}(v_1)]\!], x) \mid \alpha\rangle\rangle \quad \text{(IH)}$$
$$\simeq [\![\textbf{let } x = \mathcal{A}(e_2) \textbf{ in } (\mathcal{A}(v_1), x)]\!] \quad (\mathcal{T}_3, \mathcal{T}_6)$$
$$\simeq [\![\mathcal{A}((v_1, e_2))]\!]. \quad (\mathcal{A}_6)$$

(iii) Subcase $e \simeq (e_1, v_2)$:

$$\mathcal{F}([\![(e_1, v_2)]\!]) \simeq \mathcal{F}(([\![e_1]\!], [\![v_2]\!])) \quad (\mathcal{T}_3)$$
$$\simeq \mu\alpha.\langle\mathcal{F}([\![e_1]\!]) \mid \tilde{\mu}x.\langle(x, \mathcal{F}([\![v_2]\!])) \mid \alpha\rangle\rangle \quad (\mathcal{F}_6)$$
$$\simeq \mu\alpha.\langle[\![\mathcal{A}(e_1)]\!] \mid \tilde{\mu}x.\langle(x, [\![\mathcal{A}(v_2)]\!]) \mid \alpha\rangle\rangle \quad \text{(IH)}$$
$$\simeq [\![\textbf{let } x = \mathcal{A}(e_1) \textbf{ in } (x, \mathcal{A}(v_2))]\!] \quad (\mathcal{T}_3, \mathcal{T}_6)$$
$$\simeq [\![\mathcal{A}((e_1, v_2))]\!]. \quad (\mathcal{A}_7)$$

(iv) Subcase $e \simeq (e_1, e_2)$:

$$\mathcal{F}([\![(e_1, e_2)]\!])$$
$$\simeq \mathcal{F}(([\![e_1]\!], [\![e_2]\!])) \quad (\mathcal{T}_3)$$
$$\simeq \mu\alpha.\langle\mathcal{F}([\![e_1]\!]) \mid \tilde{\mu}x.\langle\mu\beta.\langle\mathcal{F}([\![e_2]\!]) \mid \tilde{\mu}y.\langle(x, y) \mid \beta\rangle\rangle \mid \alpha\rangle\rangle \quad (\mathcal{F}_7)$$

$$\eqsim \mu\alpha.\langle[\![\mathcal{A}(e_1)]\!] \mid \tilde{\mu}x.\langle\mu\beta.\langle[\![\mathcal{A}(e_2)]\!] \mid \tilde{\mu}y.\langle(x,y) \mid \beta\rangle\rangle \mid \alpha\rangle\rangle \quad (\text{IH})$$
$$\eqsim [\![\text{let } x = \mathcal{A}(e_1) \text{ in } (\text{let } y = \mathcal{A}(e_2) \text{ in } (x,y))]\!] \qquad (\mathcal{T}_3, \mathcal{T}_6)$$
$$\eqsim [\![\mathcal{A}((e_1, e_2))]\!]. \qquad (\mathcal{A}_8)$$

6. In case $e \eqsim \text{let } x = e_1 \text{ in } e_2$ we have:

$$\mathcal{F}([\![\text{let } x = e_1 \text{ in } e_2]\!]) \eqsim \mathcal{F}(\mu\alpha.\langle[\![e_1]\!] \mid \tilde{\mu}x.\langle[\![e_2]\!] \mid \alpha\rangle\rangle) \qquad (\mathcal{T}_6)$$
$$\eqsim \mu\alpha.\langle\mathcal{F}([\![e_1]\!]) \mid \tilde{\mu}x.\langle\mathcal{F}([\![e_2]\!]) \mid \alpha\rangle\rangle$$
$$(\mathcal{F}_2, \mathcal{F}_{13}, \mathcal{F}_9, \mathcal{F}_8)$$
$$\eqsim \mu\alpha.\langle[\![\mathcal{A}(e_1)]\!] \mid \tilde{\mu}x.\langle[\![\mathcal{A}(e_2)]\!] \mid \alpha\rangle\rangle \qquad (\text{IH})$$
$$\eqsim [\![\text{let } x = \mathcal{A}(e_1) \text{ in } \mathcal{A}(e_2)]\!] \qquad (\mathcal{T}_6)$$
$$\eqsim [\![\mathcal{A}(\text{let } x = e_1 \text{ in } e_2)]\!]. \qquad (\mathcal{A}_3)$$

$\square$

Theorem 8.2 (Focusing reflects the ANF-transformation; case $\equiv$). *For all λ-terms e, we have $\mathcal{F}([\![e]\!]) \equiv [\![\mathcal{A}(e)]\!]$.*

Proof. We only have to modify subcase 4(ii) in the proof of Theorem 8.1. The modified proof is as follows (evaluated redexes are underlined).

4. In case $e \eqsim e_1 \, e_2$, we have to distinguish two subcases:

(i) Subcase $e \eqsim e_1 \, v_2$: identical to the proof of Theorem 8.1.
(ii) Subcase $e \eqsim e_1 \, e_2$:

$$\mathcal{F}([\![(e_1 \, e_2)]\!]) \eqsim \mathcal{F}(\mu\alpha.\langle[\![e_1]\!] \mid [\![e_2]\!] \cdot \alpha\rangle) \qquad (\mathcal{T}_4)$$
$$\eqsim \mu\alpha.\langle\underline{\mathcal{F}([\![e_1]\!])} \mid \tilde{\mu}y.\langle\mathcal{F}([\![e_2]\!]) \mid \tilde{\mu}x.\langle y \mid x \cdot \alpha\rangle\rangle\rangle$$
$$(\mathcal{F}_2, \mathcal{F}_{13}, \mathcal{F}_9, \mathcal{F}_1, \mathcal{F}_8, \mathcal{F}_{11})$$
$$\triangleright \mu\alpha.\langle\mathcal{F}([\![e_2]\!]) \mid \tilde{\mu}x.\langle\mathcal{F}([\![e_1]\!]) \mid x \cdot \alpha\rangle\rangle$$
$$\overset{\text{IH}}{\equiv} \mu\alpha.\langle[\![\mathcal{A}(e_2)]\!] \mid \tilde{\mu}x.\langle[\![\mathcal{A}(e_1)]\!] \mid x \cdot \alpha\rangle\rangle$$
$$\triangleleft \mu\alpha.\langle[\![\mathcal{A}(e_2)]\!] \mid \tilde{\mu}x.\langle\mu\beta.\langle[\![\mathcal{A}(e_1)]\!] \mid x \cdot \beta\rangle \mid \alpha\rangle\rangle$$
$$\eqsim [\![\text{let } x = \mathcal{A}(e_2) \text{ in } \mathcal{A}(e_1) \, x]\!] \qquad (\mathcal{T}_4 + \mathcal{T}_6)$$
$$\eqsim [\![\mathcal{A}(e_1 \, e_2)]\!]. \qquad (\mathcal{A}_{10})$$

$\square$

In order to prove that for all $e \in \Lambda^Q$, $[\![\mathcal{L}(e)]\!]^* =_\mu [\![e]\!]$ (Theorem 8.6), we introduce a transformation $\mathcal{M}$ that simulates the effect of applying $\mathcal{L}$ on a term from Λ^Q on its translation in $\Lambda^Q_{\mu\tilde{\mu}}$.

Definition A.1. The *μ-normalization operation* $\mathcal{M} : \Lambda^Q_{\mu\tilde{\mu}} \to \Lambda^{ANF}_{\mu\tilde{\mu}}$ is defined by the following clauses:

$$Values$$

$$\mathcal{M}(x) :\doteq x \qquad (\mathcal{M}_1)$$

$$\mathcal{M}(\lambda x.e) :\doteq \lambda x.\mathcal{M}(e) \qquad (\mathcal{M}_2)$$

$$\mathcal{M}((w_1, w_2)) :\doteq (\mathcal{M}(w_1), \mathcal{M}(w_2)) \qquad (\mathcal{M}_3)$$

$$Terms$$

$$\mathcal{M}(e) :\doteq \mu\alpha.\mathcal{M}_\alpha(e) \qquad (\mathcal{M}_4)$$

$$\mathcal{M}_s(w) :\doteq \langle \mathcal{M}(w) \mid s \rangle \qquad (\mathcal{M}_5)$$

$$\mathcal{M}_s(\mu\alpha.c) :\doteq \mathcal{M}(c[s/\alpha]) \qquad (\mathcal{M}_6)$$

$$Coterms$$

$$\mathcal{M}(\alpha) :\doteq \alpha \qquad (\mathcal{M}_7)$$

$$\mathcal{M}(\pi_i\, s) :\doteq \pi_i\, \tilde{\mu}x.\langle x \mid s \rangle \qquad (\mathcal{M}_8)$$

$$\mathcal{M}(w \cdot s) :\doteq w \cdot \tilde{\mu}x.\langle x \mid s \rangle \qquad (\mathcal{M}_9)$$

$$\mathcal{M}(\tilde{\mu}x.\langle e \mid s \rangle) :\doteq \tilde{\mu}x.\mathcal{M}_s(e) \qquad (\mathcal{M}_{10})$$

$$Computations$$

$$\mathcal{M}(\langle e \mid s \rangle) :\doteq \mathcal{M}_{\mathcal{M}(s)}(e). \qquad (\mathcal{M}_{11})$$

Definition A.2. We define the operation $\_;\_$ which takes a continuation k (cf. Definition 6.6) and a coterm s, and returns a coterm $k; s$:

$$\text{id}; s :\doteq s \qquad (\mathcal{C}_1)$$

$$\overline{\lambda v}.\text{let } x = \pi_i\, \overline{v} \text{ in } e; s :\doteq \pi_i\, \tilde{\mu}x.[\![e]\!]^*_s \qquad (\mathcal{C}_2)$$

$$\overline{\lambda v}.\text{let } x = \overline{v}\, v' \text{ in } e; s :\doteq v' \cdot \tilde{\mu}x.[\![e]\!]^*_s \qquad (\mathcal{C}_3)$$

$$\overline{\lambda v}.\text{let } x = \overline{v} \text{ in } e; s :\doteq \tilde{\mu}x.[\![v]\!]^*_s. \qquad (\mathcal{C}_4)$$

Lemma A.3. *We have $[\![k @ v]\!]_s^* \simeq \langle [\![v]\!]^* \mid k; s \rangle$.*

Proof. By case analysis on k:

1. Case $k \simeq \mathrm{id}$:

$$[\![\mathrm{id} @ v]\!]_s^* \simeq [\![v]\!]_s^* \tag{$@_1$}$$
$$\simeq \langle [\![v]\!]^* \mid s \rangle \tag{$\mathcal{T}_5^*$}$$
$$\simeq \langle [\![v]\!]^* \mid \mathrm{id}; s \rangle. \tag{$\mathcal{C}_1$}$$

2. Case $k \simeq \overline{\lambda v}.\mathbf{let}\ x = \pi_i\ \overline{v}\ \mathbf{in}\ e$:

$$[\![\overline{\lambda v}.\mathbf{let}\ x = \pi_i\ \overline{v}\ \mathbf{in}\ e @ v]\!]_s^* \simeq [\![\mathbf{let}\ x = \pi_i\ v\ \mathbf{in}\ e]\!]_s^* \tag{$@_2$}$$
$$\simeq [\![\pi_i\ v]\!]_{\tilde{\mu}x.[\![e]\!]_s^*}^* \tag{$\mathcal{T}_2^*$}$$
$$\simeq \langle [\![v]\!]^* \mid \pi_i\ \tilde{\mu}x.[\![e]\!]_s^* \rangle \tag{$\mathcal{T}_4^*$}$$
$$\simeq \langle [\![v]\!]^* \mid \overline{\lambda v}.\mathbf{let}\ x = \pi_i\ \overline{v}\ \mathbf{in}\ e; s \rangle. \tag{$\mathcal{C}_2$}$$

3. Case $k \simeq \overline{\lambda v}.\mathbf{let}\ x = \overline{v}\ v'\ \mathbf{in}\ e$:

$$[\![\overline{\lambda v}.\mathbf{let}\ x = \overline{v}\ v'\ \mathbf{in}\ e @ v]\!]_s^* \simeq [\![\mathbf{let}\ x = v\ v'\ \mathbf{in}\ e]\!]_s^* \tag{$@_3$}$$
$$\simeq [\![v\ v']\!]_{\tilde{\mu}x.[\![e]\!]_s^*}^* \tag{$\mathcal{T}_2^*$}$$
$$\simeq \langle [\![v]\!]^* \mid [\![v']\!]^* \cdot \tilde{\mu}x.[\![e]\!]_s^* \rangle \tag{$\mathcal{T}_3^*$}$$
$$\simeq \langle [\![v]\!]^* \mid \overline{\lambda v}.\mathbf{let}\ x = \overline{v}\ v'\ \mathbf{in}\ e; s \rangle. \tag{$\mathcal{C}_3$}$$

4. Case $k \simeq \overline{\lambda v}.\mathbf{let}\ x = \overline{v}\ \mathbf{in}\ e$:

$$[\![\overline{\lambda v}.\mathbf{let}\ x = \overline{v}\ \mathbf{in}\ e @ v]\!]_s^* \simeq [\![\mathbf{let}\ x = v\ \mathbf{in}\ e]\!]_s^* \tag{$@_4$}$$
$$\simeq [\![v]\!]_{\tilde{\mu}x.[\![e]\!]_s^*}^* \tag{$\mathcal{T}_2^*$}$$
$$\simeq \langle [\![v]\!]^* \mid \tilde{\mu}x.[\![e]\!]_s^* \rangle \tag{$\mathcal{T}_5^*$}$$
$$\simeq \langle [\![v]\!]^* \mid \overline{\lambda v}.\mathbf{let}\ x = \overline{v}\ \mathbf{in}\ e; s \rangle. \tag{$\mathcal{C}_4$}$$

$\square$

The μ-normalization operation $\mathcal{M}$ corresponds precisely to the transformation $\mathcal{L}$:

Lemma A.4. *The following statements hold:*

1. *For all values $v \in \Lambda^Q$: $[\![\mathcal{L}(v)]\!]^* \simeq \mathcal{M}([\![v]\!])$.*

2. *For all expressions $e \in \Lambda^Q$: $[\![\mathcal{L}(e)]\!]^* \simeq \mathcal{M}([\![e]\!])$.*

3. *For all $e \in \Lambda^Q$, continuations k and coterms s:*

$$[\![\mathcal{L}_k(e)]\!]_s^* \simeq \mathcal{M}_{k;s}([\![e]\!]).$$

Proof. We prove these three statements by simultaneous induction. For the first statement, Lemma A.4(1), we use induction on v:

1. Case $v \simeq x$:

$$[\![\mathcal{L}(x)]\!]^* \simeq [\![x]\!]^* \qquad (\mathcal{L}_1)$$
$$\simeq x \qquad (\mathcal{T}_6^*)$$
$$\simeq \mathcal{M}(x) \qquad (\mathcal{M}_1)$$
$$\simeq \mathcal{M}([\![x]\!]). \qquad (\mathcal{T}_1)$$

2. Case $v \simeq (v_1, v_2)$:

$$[\![\mathcal{L}((v_1, v_2))]\!]^* \simeq [\![(\mathcal{L}(v_1), \mathcal{L}(v_2))]\!]^* \qquad (\mathcal{L}_3)$$
$$\simeq ([\![\mathcal{L}(v_1)]\!]^*, [\![\mathcal{L}(v_2)]\!]^*) \qquad (\mathcal{T}_7^*)$$
$$\simeq (\mathcal{M}([\![v_1]\!]), \mathcal{M}([\![v_2]\!])) \quad \text{(IH for Lemma A.4(1))}$$
$$\simeq \mathcal{M}(([\![v_1]\!], [\![v_2]\!])) \qquad (\mathcal{M}_3)$$
$$\simeq \mathcal{M}([\![(v_1, v_2)]\!]). \qquad (\mathcal{T}_3)$$

3. Case $v \simeq \lambda x.e$:

$$[\![\mathcal{L}(\lambda x.e)]\!]^* \simeq [\![\lambda x.\mathcal{L}(e)]\!]^* \qquad (\mathcal{L}_2)$$
$$\simeq \lambda x.[\![\mathcal{L}(e)]\!]^* \qquad (\mathcal{T}_8^*)$$
$$\simeq \lambda x.\mathcal{M}([\![e]\!]) \quad \text{(IH for Lemma A.4(2))}$$
$$\simeq \mathcal{M}(\lambda x.[\![e]\!]) \qquad (\mathcal{M}_2)$$
$$\simeq \mathcal{M}([\![\lambda x.e]\!]). \qquad (\mathcal{T}_2)$$

For Lemma A.4(2), we show the following:

$$
\begin{aligned}
\llbracket \mathcal{L}(e) \rrbracket^* &\eqsim \mu\alpha.\llbracket \mathcal{L}(e) \rrbracket^*_\alpha && (\mathcal{T}_1^*) \\
&\eqsim \mu\alpha.\llbracket \mathcal{L}_{\mathrm{id}}(e) \rrbracket^*_\alpha && (\mathcal{L}_4) \\
&\eqsim \mu\alpha.\mathcal{M}_{\mathrm{id};\alpha}(\llbracket e \rrbracket) && \text{(IH for Lemma A.4(3))} \\
&\eqsim \mu\alpha.\mathcal{M}_\alpha(\llbracket e \rrbracket) && (\mathcal{C}_1) \\
&\eqsim \mathcal{M}(\llbracket e \rrbracket). && (\mathcal{M}_4)
\end{aligned}
$$

For Lemma A.4(3), we perform induction on e:

1. Case $e \eqsim v$:

$$
\begin{aligned}
\llbracket \mathcal{L}_k(v) \rrbracket^*_s &\eqsim \llbracket k @ \mathcal{L}(v) \rrbracket^*_s && (\mathcal{L}_7) \\
&\eqsim \langle \llbracket \mathcal{L}(v) \rrbracket^* \mid k; s \rangle && \text{(Lemma A.3)} \\
&\eqsim \langle \mathcal{M}(\llbracket v \rrbracket) \mid k; s \rangle && \text{(IH for 1)} \\
&\eqsim \mathcal{M}_{k;s}(\llbracket v \rrbracket). && (\mathcal{M}_5)
\end{aligned}
$$

2. Case $e \eqsim \mathbf{let}\ x = e_1\ \mathbf{in}\ e_2$:

$$
\begin{aligned}
\llbracket \mathcal{L}_k(\mathbf{let}\ x = e_1\ \mathbf{in}\ e_2) \rrbracket^*_s &\eqsim \llbracket \mathcal{L}_{\overline{\lambda v.\mathbf{let}\ x=\overline{v}\ \mathbf{in}\ \mathcal{L}_k(e_2)}}(e_1) \rrbracket^*_s && (\mathcal{L}_8) \\
&\eqsim \mathcal{M}_{\overline{\lambda v.\mathbf{let}\ x=\overline{v}\ \mathbf{in}\ \mathcal{L}_k(e_2)};s}(\llbracket e_1 \rrbracket) && \text{(IH)} \\
&\eqsim \mathcal{M}_{\tilde{\mu}x.\llbracket \mathcal{L}_k(e_2) \rrbracket^*_s}(\llbracket e_1 \rrbracket) && (\mathcal{C}_4) \\
&\eqsim \mathcal{M}_{\tilde{\mu}x.\mathcal{M}_{k;s}(\llbracket e_2 \rrbracket)}(\llbracket e_1 \rrbracket) && \text{(IH)} \\
&\eqsim \mathcal{M}_{\mathcal{M}(\tilde{\mu}x.\langle \llbracket e_2 \rrbracket \mid k; s\rangle)}(\llbracket e_1 \rrbracket) && (\mathcal{M}_{10}) \\
&\eqsim \mathcal{M}(\langle \llbracket e_1 \rrbracket \mid \tilde{\mu}x.\langle \llbracket e_2 \rrbracket \mid k; s\rangle\rangle) && (\mathcal{M}_{11}) \\
&\eqsim \mathcal{M}_{k;s}(\mu\alpha.\langle \llbracket e_1 \rrbracket \mid \tilde{\mu}x.\langle \llbracket e_2 \rrbracket \mid \alpha\rangle\rangle) && (\mathcal{M}_6) \\
&\eqsim \mathcal{M}_{k;s}(\llbracket \mathbf{let}\ x = e_1\ \mathbf{in}\ e_2 \rrbracket). && (\mathcal{T}_6)
\end{aligned}
$$

3. Case $e \eqsim e\, v$:

$$
\begin{aligned}
\llbracket \mathcal{L}_k(e\, v) \rrbracket^*_s &\eqsim \llbracket \mathcal{L}_{\overline{\lambda v'.\mathbf{let}\ x=v'\ v\ \mathbf{in}\ k @ x}}(e) \rrbracket^*_s && (\mathcal{L}_5) \\
&\eqsim \mathcal{M}_{\overline{\lambda v'.\mathbf{let}\ x=v'\ v\ \mathbf{in}\ k @ x};s}(\llbracket e \rrbracket) && \text{(IH)} \\
&\eqsim \mathcal{M}_{[v]\cdot\tilde{\mu}x.\llbracket k @ x \rrbracket^*_s}(\llbracket e \rrbracket) && (\mathcal{C}_3)
\end{aligned}
$$

185

$$\simeq \mathcal{M}_{\llbracket v \rrbracket \cdot \tilde{\mu} x. \langle x | s \rangle}(\llbracket e \rrbracket) \qquad \text{(Lemma A.3)}$$
$$\simeq \mathcal{M}_{\mathcal{M}(\llbracket v \rrbracket \cdot k; s)}(\llbracket e \rrbracket) \qquad (\mathcal{M}_9)$$
$$\simeq \mathcal{M}(\langle \llbracket e \rrbracket \mid \llbracket v \rrbracket \cdot k; s \rangle) \qquad (\mathcal{M}_{11})$$
$$\simeq \mathcal{M}_{k;s}(\mu \alpha. \langle \llbracket e \rrbracket \mid \llbracket v \rrbracket \cdot \alpha \rangle) \qquad (\mathcal{M}_6)$$
$$\simeq \mathcal{M}_{k;s}(\llbracket e\, v \rrbracket). \qquad (\mathcal{T}_4)$$

4. Case $e \simeq \pi_i\, e$:

$$\llbracket \mathcal{L}_k(\pi_i\, e) \rrbracket_s^* \simeq \llbracket \mathcal{L}_{\overline{\lambda v}.\text{let } x=\pi_i\, \overline{v} \text{ in } k\, @\, x}(e) \rrbracket_s^* \qquad (\mathcal{L}_6)$$
$$\simeq \mathcal{M}_{\overline{\lambda v}.\text{let } x=\pi_i\, \overline{v} \text{ in } k\, @\, x;s}(\llbracket e \rrbracket) \qquad \text{(IH)}$$
$$\simeq \mathcal{M}_{\pi_i\, \tilde{\mu} x. \llbracket k\, @\, x \rrbracket_s^*}(\llbracket e \rrbracket) \qquad (\mathcal{C}_2)$$
$$\simeq \mathcal{M}_{\pi_i\, \tilde{\mu} x. \langle x | k; s \rangle}(\llbracket e \rrbracket) \qquad \text{(Lemma A.3)}$$
$$\simeq \mathcal{M}_{\mathcal{M}(\pi_i\,(k;s))}(\llbracket e \rrbracket) \qquad (\mathcal{M}_8)$$
$$\simeq \mathcal{M}(\langle \llbracket e \rrbracket \mid \pi_i\,(k; s) \rangle) \qquad (\mathcal{M}_{11})$$
$$\simeq \mathcal{M}_{k;s}(\mu \alpha. \langle \llbracket e \rrbracket \mid \pi_i\, \alpha \rangle) \qquad (\mathcal{M}_6)$$
$$\simeq \mathcal{M}_{k;s}(\llbracket \pi_i\, e \rrbracket). \qquad (\mathcal{T}_5)$$

$\square$

The effect of the application of $\mathcal{M}$ can be achieved by applying μ-reductions in commands and η-expansions in coterms:

Lemma A.5. *For all terms, coterms and commands $t \in \Lambda_{\mu\tilde{\mu}}^{\mathrm{Q}}$ we have $t \equiv_\mu \mathcal{M}(t)$.*

Proof. By inspection of the relevant clauses in Definition A.1. The combination of the clauses $(\mathcal{M}_6)$ and $(\mathcal{M}_{11})$ corresponds to the reduction of μ-redexes. Coterms are η-expanded in the clauses $(\mathcal{M}_8)$ and $(\mathcal{M}_9)$. $\square$

Theorem 8.6. *For all terms $e \in \Lambda^{\mathrm{Q}}$, $\llbracket \mathcal{L}(e) \rrbracket^* \equiv_\mu \llbracket e \rrbracket$.*

Proof. By combining Lemmas A.4 and A.5. $\square$

Going Nowhere and Back: Is Trivialization the Same as Zero Execution?

Ivo Pezlar

Czech Academy of Sciences, Institute of Philosophy

pezlar@flu.cas.cz

Abstract

In this paper I will explore the question whether the Trivialization construction of transparent intensional logic (TIL) can be understood in terms of the Execution construction, specifically, in terms of its degenerate case known as the 0-Execution. My answer will be positive and the apparent contrast between the intuitive understanding of Trivialization and 0-Execution will be explained as a matter of distinct yet related informal perspectives, not as a matter of technical or conceptual differences.

1   Introduction and motivation

One of the most prominent features of transparent intensional logic (TIL, [9], [1], [8]), and simultaneously the source of much puzzlement for TIL newcomers, is the Trivialization construction, which more or less acts as a procedural analogue of a constant. For example, an abstraction term $\lambda x.x + 1$ of lambda calculus with constants corresponds to a Closure construction $\lambda x[x\,' + \,'1]$ of TIL, where the prime symbol "'" indicates Trivialization constructions, namely Trivializations of the

---

This work was supported by the Lumina quaeruntur fellowship awarded by the Czech Academy of Sciences (registration number: LQ300092101).

addition function + and of the number 1, respectively.[1] A lot has
been already written about Trivialization and its justification and I
do not wish to add to that literature. Instead I want the explore the
question whether Trivialization can be understood in terms of another
TIL construction known as Execution, specifically, in terms of its degenerate case known as the 0-Execution (read as "Zero Execution").
Although it is not an issue of great practical importance from the
viewpoint of natural language semantics, it touches on fundamental
theoretical aspects of TIL, which, I believe, make it a topic worthy of
a closer look.

The goal of this paper, and I wish to emphasize this point, is not
to argue for the removal of Trivialization from TIL, but to examine
whether it can be understood as Execution, specifically, 0-Execution.[2]
Ideally, it might also help TIL newcomers to more quickly grasp the
main idea of the construction denoted by the unnecessary-looking but
crucially important symbol ''.

Finally, it is worth mentioning that the idea of treating Trivialization as 0-Execution is not a new one and it has been floating around
the TIL community for some time (at least since [6]) but only recently
it gained more attention ([3], [4], [7]). In this paper, I want to reopen
this issue, revisit some of the previous arguments, and offer a new
perspective on the Trivialization vs. 0-Execution debate.

2 Trivialization vs. 0-Execution

Can we regard Trivialization construction in the terms of 0-Execution
construction? Or more bluntly put, is Trivialization the same construction as 0-Execution or are they distinct? To get a better grasp

---

[1]In the literature, Trivialization is most commonly denoted by the superscript
'0' not by ''. In this paper, however, I reserve the symbol '0' for denoting 0-
Execution. To ensure notational consistency I apply this convention to quotations
as well.

[2]I use this opportunity to revisit one of my earlier discussions with prof. Marie
Duží on the same topic. In the past, she was firmly against viewing Trivialization
as 0-Execution and with this paper I would like to, if not to change her mind, at
least to sow there a seed of doubt.

of the issue, let us assume that Trivialization and 0-Execution are indeed two distinct constructions. Let us denote the instances of the former as $'X$ and of the latter as ^0X where X is any entity whatsoever (within the scope of the ramified type hierarchy of TIL).

Our task will be to show that every role that is typically played by Trivialization can also be played by 0-Execution. I will investigate this issue from two aspects: technical and conceptual. From the technical aspect, we want to check if every instance of $'X$ can be replaced by ^0X (without affecting the procedural behaviour of the construction in the sense that it still produces the same output in the same way, i.e., that it is still the same construction). From the conceptual aspect, we want to check if all the standard intuitions typically associated with Trivialization can be preserved when dealing with 0-Execution. In other words, do we lose something if we view Trivialization as 0-Execution?

If no crucial discrepancies between Trivialization and 0-Execution of this sort are found, I believe we are warranted in claiming that Trivialization and 0-Execution can be understood as the same construction, just viewed from different perspectives, namely from the destination perspective and the path perspective, respectively (I will discuss these later).

For the purposes of this paper, from now on I will distinguish between Trivialization and Execution understood as different kinds of constructions, and Trivialization and Execution understood as specific instances of these kinds of constructions. I will write the former with capitalized first letters (Trivialization, Execution, ...) and the letter without capitalization (trivialization, execution, ...). We can think of this distinction as similar to the distinction between axiom schemata and their particular instances. For example, we can say that $'5$, $'Alice$, and $'['5 '+ '7]$ are all trivialization constructions, or trivializations for short. Or, alternatively, we can say that $'5$, $'Alice$, and $'['5 '+ '7]$ are specific instances of the Trivialization construction kind (distinct from other construction kinds such as Closure, Composition, etc.). Thus, when [2] says that "Trivializations are the one-step or primitive or atomic constructions of TIL" (p. 329) from our viewpoint he talks

189

about Trivialization as a general construction kind, but when [1] say phrases like "the meaning of 'cow' (here the Trivialization $'Cow)$" (p. 231), from our viewpoint they talk about a specific instance of the Trivialization construction kind, namely $'Cow$. In short, in the first case we are talking about Trivialization understood as a kind of construction, in the second case, we are talking about specific instances of thereof. This distinction might seem pedantic, but it will become useful later. Specifically, it will help us to better unpack some conceptual issues surrounding Trivialization and 0-Execution.

3   What is Trivialization?

Trivialization is an atomic construction, denoted as $'X$, and its main role is to pick definite objects that the compound constructions of TIL can operate on. Specific trivializations can be compared to constants, as they fulfil a similar role.

First, let us start by reviewing some of the standard specifications of Trivialization found in the literature. [9] originally specified it as follows:

> Where X is any entity whatsoever, we can consider the trivial construction whose starting point, as well as outcome, is X itself. Let us call this rudimentary construction the *trivialization* of X and symbolize it as $'X$. To carry $'X$ out, one starts with X and leaves it, so to speak, as it is. ([9], p. 63)

Compare with [1]:

> Trivializations match constants, by picking out definite entities in just one step. ([1], p. 9)

> When X is an object of any type (including a construction), the Trivialization of X, denoted $''X'$, constructs X without the mediation of any other constructions. $'X$ is the unique atomic construction of X that does not depend on valuation: it is a primitive, non-perspectival mode of presentation of X. ([1], p. 43)

> It constructs X without any change. ([1], p. 45)

And also with [7] (similarly in [8]):

> Every construction is a mode of presentation of a certain object [...] $'X$ presents X as it is (without any change of X) ([7], p. 52)

Thus, we could say that the key aspect of Trivialization $'X$ is that it does nothing with X, that it does not change X in any way.[3]

Note. It is worth pointing out that [9] considered even Trivializations and Executions of nonconstructions to have constituents ("Variables are the only *simple* constructions; all other constructions have constituent parts." [9], p. 63). This is not the case in [1] and later, where even Trivialization and Execution of nonconstructions are considered "partless", i.e., atomic ([1]):

> **Definition 2.17 (*atomic construction*)** A *construction* C is *atomic* if C does not contain any other constituent but itself.
>
> *Corollary.* A construction C is atomic if C is
>
> - a variable; or
> - a Trivialization $'X$, where X is an entity of any type, even a construction; or
> - an Execution ^1X or a Double Execution ^2X, where X is an entity of a type of order 1, i.e., a nonconstruction.
>
> An atomic construction of kind (i) or (ii) is v-proper for any valuation v. An atomic construction ^1X, ^2X, of kind (iii), is v-improper for any valuation v. In this case ^1X or ^2X does not v-construct anything, and $^1X \to \alpha$, $^2X \to \alpha$, for any type α, would constitute a type-theoretic mismatch. ([1], pp. 247–248)

At first, it might seem somewhat surprising that we can encounter executions that are atomic constructions as well as executions that are

---

[3]Later, I will argue that this naturally leads to viewing Trivialization as a degenerate case of Execution, specifically, 0-Execution.

nonatomic, i.e., compound constructions. I believe distinguishing between kinds of constructions and their specific instances (as discussed above) can help us here. Being atomic or compound construction is a property of executions understood as specific constructions, not of Executions understood as construction kinds. Thus, there is nothing extraordinary about the fact that we can have executions that are atomic, but also executions that are compound.

4 What is 0-Execution?

0-Execution is an atomic construction, denoted as ^0X, and it is understood as the limiting case of Execution construction kind which also includes Single Execution[4] and Double Execution, denoted ^1X, ^2X, respectively. The main idea is that while Single Execution tells us to execute X once, and Double Execution tells us to execute X twice, or more precisely, to execute X and then execute again the result we obtain (if any), 0-Execution tells us not to execute X, just to leave it as it is. It is simply the degenerate case of Executions when the number of consecutive executions equals 0. Its main role is to tell us what should not be executed.[5]

In the literature, we can find the following specifications of 0-Execution. For example, [3] introduces it as follows:

> Since ^1X and ^2X are Single and Double Execution, respectively, it would be natural if ^0X was known as *Zero Execution*. In fact, whereas 'Trivialization' gives the wrong idea about Trivialization, which is anything but trivial, 'Zero Execution' sums up what ^0X is all about: ^0X *displays* X. If X is a procedure, then ^0X does not proceed to executing X. This is in fact the gist of Tichý's original definition that ^0X produces X without any

---

[4] Single Execution would be also an interesting topic for a further investigation as it seems often overlooked for its apparent "simplicity". In this paper, I will, however, not discuss it further.

[5] Mind you, it does not mean that ^0X should not be executed, only that X itself should not be executed. However, using the standard conceptual framework of TIL, the X itself is "invisible", since 0-executions are considered to be atomic constructions (as are trivializations).

change of X. In this paper, however, I will stick to the original term 'Trivialization' for continuity. ([3], p. 1320)

Similarly [4]:

> If we should understand ^1X as 'execute X' (i.e., '1' = one execution) and ^2X as 'execute X and then execute its result X' (i.e., '2' = two executions), then, arguably the most natural reading of ^0X—if we have never heard of trivialization—is 'do not execute X' (i.e., '0' = zero executions). ([4], p. 203)

Thus, analogously to Trivialization, the key aspect of 0-Execution ^0X is that it does nothing with X.

5 Technical aspects

Assume that X is any entity whatsoever and that we have a trivialization of X, denoted $'X$, and a 0-execution of X, denoted ^0X. We want to show that the procedural behaviour of $'X$ and ^0X is the same. There are eight cases in total we have to consider: first, whether X is a nonconstruction or a construction, second, if it is a construction, what kind of construction. Although it might seem as unnecessary, I believe it will be instructive to present each case separately to better understand the relationship between 0-Execution and Trivialization.

1. X is a nonconstruction: $'X$ produces X, analogously ^0X produces X,

2. X is a construction:

    (a) X is a variable: $'x$ produces x, analogously ^0x produces x,

    (b) X is a closure: $'[\lambda x_1 \ldots x_m X]$ produces $[\lambda x_1 \ldots x_m X]$, analogously $^0[\lambda x_1 \ldots x_m X]$ produces $[\lambda x_1 \ldots x_m X]$,

    (c) X is a composition: $'[XX_1 \ldots X_m]$ produces $[XX_1 \ldots X_m]$, analogously $^0[XX_1 \ldots X_m]$ produces $[XX_1 \ldots X_m]$,

    (d) X is a trivialization: $''X$ produces $'X$, analogously $^{0'}X$ produces $'X$,

(e) X is a 0-execution: $^{\prime 0}X$ produces ^0X, analogously ^{00}X produces ^0X,

(f) X is a 1-execution (single execution): $^{\prime 1}X$ produces ^1X; analogously ^{01}X produces ^1X,

(g) X is a 2-execution (double execution): $^{\prime 2}X$ produces ^2X; analogously ^{02}X produces ^2X.

As we can see, both Trivialization and 0-Execution proceed in the same manner for all entities X: it just takes them and returns them without any change.

6 Conceptual aspects

So far it seems that the whole issue of Trivialization vs. 0-Execution is just a terminological dispute based on personal preference. Trivialization seems to be fully explainable, and thus replaceable, in terms of 0-Execution. This assessment would be, however, too rushed as it does not tell the whole story. Despite the considerations above, not all TIL researchers, prof. Marie Duží included, would agree that Trivialization can be viewed as 0-Execution.

What are the main objections against viewing Trivialization as 0-Execution? The five most important objections are, I believe, the following:

Objection 1: Trivialization binds free variables, Executions do not.

Reply: It is true that 1-Executions and higher do not bind free variables, but 0-Execution is not just another Execution. As we said, it is not only a limiting case but a degenerate case of Execution. And since degenerate cases exhibit qualitative differences from non-generate ones (e.g., a point can be considered as a degenerate case of a circle with radius 0), it should not be surprising that 0-Execution can have different properties than non-0-Executions. Especially, if these properties are necessary side effects of its specification. Recall that

194

0-Execution is essentially an instruction to "do nothing with X", to not change it in any way. If 0-Execution would not bind free variables, X might change with respect to a valuation or a substitution. And if so, it would no longer be 0-Execution, since something was done with X, it was not left as it was (see also [4], p. 205). Thus, the fact that 0-Execution binds free variables is not only unsurprising but to be expected.

Objection 2: Trivialization is always proper, Executions are not.

Reply: Analogous reasoning as above applies. In short, 0-Execution is a degenerate case of Executions and as such it exhibits different qualitative properties. The fact that 0-Execution is always proper construction can be explained with respect to its specification. The construction cannot fail, because, simply put, there is no possibility for it to do so: it just takes X and produces X back without any change.

Objection 3: Trivializations are always atomic constructions, but Executions can be atomic as well as compound.

Reply: True, but if anything, this fact rather supports the idea that Trivialization is a special case of Executions, since Executions appear to be a more general notion from this viewpoint: we can have atomic executions (including 0-executions) and compound executions, but we cannot have compound trivializations, only atomic ones.

Objection 4: Trivialization supplies entities for compound constructions, Executions do not.[6]

Reply: Analogous to replies to objections 1 and 2. In short, there is no reason why we cannot view the supplying of entities just as another

---

[6]See, e.g., "There are two atomic constructions that supply entities (of any type) on which complex constructions operate: *Variables* and *Trivializations*." ([1], p. 42) This, however, also brings up the question what is the purpose of the other atomic (although improper) constructions such as ^1X when X is a nonconstruction.

degenerate aspect of 0-Execution.[7]

Objection 5: Trivialization is a dual operation[8] to Double Execution: Trivialization raises context, while Double Execution decreases it and Trivialization cancels out Double Execution.[9]

Reply: Statements like "Double Execution suppresses the effect of Trivialization. More generally, Double Execution decreases the level of a context." ([1], p. 239) or "Unlike Trivialization, which is an operation of *mentioning*, Execution and Double Execution are operations of *using*." ([1], p. 239) are not incorrect, but they are somewhat imprecise.

Why is that? Because it is not the case that Double Execution always suppresses the effect of Trivialization or that Trivialization is always an operation of mentioning. As a counterexample to the first statement, consider a construction $^2{'}X$ where X is a nonconstruction: here ^{2} cannot suppress $'$, because if it did, we would end up with just X which is a nonconstruction. In other words, the "cancelling-out" process would transform a construction into a nonconstruction, which certainly cannot be correct.[10] As a counterexample to the second statement, consider a construction $'X$ where X is a nonconstruction. And since mentioning is defined only for constructions (see [1], p. 234), Trivialization cannot be here used for mentioning.

So, general statements like "Trivialization raises context" and sim-

---

[7]Furthermore, we should not forget that we can have compound constructions that contain no Trivialization, 0-Execution or Variable, e.g., a composition $[^15 {}^1{+}{}^17]$. True, it is an improper construction, but still a construction. So, even though 1-Execution of nonconstructions and higher cannot supply entities on which compound constructions can operate, they can in a way indirectly refer to them (e.g., ^{15} effectively tells us that "5 is not a construction") and be used to construct compound, albeit improper, constructions.

[8]I use the term "operation" in the sense of [1], i.e., an operation understood as a process, not in its more standard sense as a function/mapping.

[9]See, e.g., [3], p. 1320.

[10]The fact that $^2{'}X$ is an improper construction does not change anything: improper constructions are still constructions.

ilar are slightly misleading as they do not present the whole picture. The duality applies only to specific instances of Trivialization and Execution, namely to those of the form $'C$ and $^{2\prime}C$ where C is a construction. But these cases do not exhaust all possible instances of Trivialization and Execution. Hence, the raising/lowering of context is not an inherent property of Trivialisation/0-Execution construction kind, but only of their specific instances $'X/{}^0X$ where X is a construction. Thus, let us try to amend the original statements: trivializations of constructions raise context, while double executions decrease it and trivializations of constructions are cancelled by double executions.

Now, how can we explain this aspect of Trivialization, i.e., that it can *mention* constructions and thus give rise to hyperintensional contexts, in terms of 0-Execution? Again, the strategy is analogous as above: since 0-Execution is the degenerate case of Execution, special properties are to be expected. And, arguably, 0-Execution offers an even more intuitive explanation of its ability to mention constructions than Trivialization does. Consider, e.g., a construction $\lambda w \lambda t [{}^0 Calculates_{wt} \, {}^0 Alice \, {}^0[{}^0 5 \, {}^0 + \, {}^0 7]]$ presented as a result of a semantic analysis of the sentence "Alice calculates $5 + 7$". Since Alice is engaged in the calculation process itself and not in its result, the construction $[{}^0 5 \, {}^0 + \, {}^0 7]$ corresponding to this calculation should *not be executed* in the respective semantic analysis. And that is precisely what "${}^0[{}^0 5 \, {}^0 + \, {}^0 7]$" stands for, i.e., it tells us to "execute $[{}^0 5 \, {}^0 + \, {}^0 7]$ zero times".

In short, it seems there is no property typically associated with Trivialization that could not be associated with 0-Execution and explained by the fact that 0-Execution is the degenerate case of Execution, hence it is bound to have different properties from 1-Execution and higher. Moreover, we have shown that these degenerate properties naturally arise from the specification of 0-Execution (e.g., binding of free variables as a necessary side effect of the fact that 0-execution ^0X should not change X in any way).

So what is Trivialization? Is it just an alternative title for 0-Execution? 0-Execution definitely seems to have enough distinctive properties (binding, properness, supplying objects, ...) that would warrant giving it a special name such as "Trivialization", similarly as, e.g., a set with a single element is also given a special name of "singleton". But if we choose to view Trivialization this way, we have to keep in mind that there is still 0-Execution running (or rather, Execution not running) under the hood of Trivialization, so to speak.

So far I have only discussed reasons why we can regard Trivialization as 0-Execution, however, I have not said much about why should we, i.e., what are the advantages of this approach. This is intentional, as my goal is not to argue for getting rid of the notion of Trivialization and relying instead solely on 0-Execution. I only wanted to show that Trivialization can be replaced by 0-Execution, that there is nothing that would prevent us from doing so, both from the technical and the conceptual viewpoints. That being said, there are, I believe, some advantages of using 0-Execution, which I will briefly discuss in the concluding section.

7 Reconciliation

7.1 Two perspectives

In the previous section, we have seen that all the standard roles played by Trivialization can be also played by 0-Execution. Yet, declaring flatly that there is no difference between Trivialization and 0-Execution still does not seem entirely accurate, at least as far as the involved intuitions are concerned. I believe there is a difference between them, but it is a difference of informal perspectives that cannot be expressed within the conceptual framework of TIL.

To illustrate this difference, we will need to keep in mind two things: our treatment of 0-Execution as a degenerate case of Execution and Tichý's original informal explanation of Trivialization. Recall that according to Tichý, Trivialization is "the trivial construction whose starting point, as well as outcome, is X itself." ([9], p. 63). So its construction "path" goes from X to X. But going from X to X can be also understood as not going at all, i.e., as starting from X and simply staying there. The following diagram (see fig. 1) should

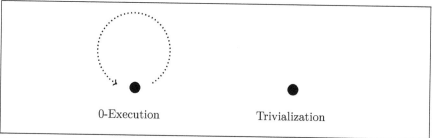

Figure 1: Difference between 0-Execution and Trivialization perspectives

make this distinction clearer.

Thus, we could perhaps best understand 0-Execution as giving us the following informal instruction:[11] "start at A and go to B and $A = B$". On the other hand, Trivialization tells us: "start at A and do not go anywhere", or alternatively "stay at A". In both cases we end up at A, but the ways how we got there are slightly different.

So, just as there is an intuitive difference between going out and returning to the same spot and not going out at all, so there is a difference between 0-Execution and Trivialization. It is not an important difference from a logical, or even a procedural point of view, but it can help us to explain the clash of intuitions associated with 0-Execution and Trivialization.

From this viewpoint, 0-Execution and Trivialization can be understood as capturing two different perspectives of the corresponding underlying construction typically denoted as ^0X. Furthermore, these two perspectives are not incompatible – we can easily imagine the "frivolous" path of 0-Execution as shrinking into the trivial path of Trivialization, which is just the starting point (similarly, as we can imagine a circle shrinking into a point). The following diagram (see fig. 2) depicts this process. Therefore, the difference between 0-Execution and Trivialization does not seem to be a simple matter of terminological preference but rather of different conceptual perspectives, each emphasizing different aspects of the construction known as

---

[11]In the style of Tichý's original informal explanation of Trivialization.

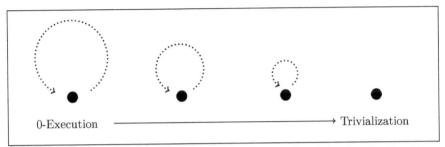

Figure 2: Transformation of the path perspective into the destination one

both 0-Execution and Trivialization. The former emphasizes its path, the latter its destination.

Are there some contexts where one perspective might be more suitable than the other? I believe so, e.g., when investigating foundational aspects of constructions it might be more beneficial to think of Trivialization in terms of 0-Execution. On the other hand, when dealing with more applied aspects of TIL, such as natural language analysis, opting for the Trivialization perspective might be more advisable, as in these cases we are not primarily interested in the inner workings of TIL constructions, they are mostly just a means to an end.

7.2 Pragmatic aspects

In this paper, I have focused on the technical and conceptual aspects but I have left out, arguably, the most decisive aspects determining the fate of 0-Execution – the pragmatic ones: habits, personal preferences, didactic considerations, and continuity. To these concerns I have little to say. If we prefer to use Trivialization instead of 0-Execution explicitly just for these pragmatic reasons, I have no issues with that. I just wanted to show that there is no technical or conceptual reason that would justify this choice. Aside from the practical considerations, we seem to be free in choosing and switching between 0-Execution and Trivialization.

However, I would like to briefly discuss the last mentioned concern, continuity. It is true that most of the TIL literature talks about "Trivialization", on the other hand, the notation itself does not change with the adoption of 0-Execution, it is just a matter of an informal retroactive interpretation whether we will read ^0X as 0-Execution or Trivialization. And most importantly, we do not need to get rid of Trivialization. We can keep it as it is – only with the added understanding that it is just a "disguised" Execution, specifically, its degenerate case of 0-Execution, and not a construction of its own kind. This newly acquired conceptual simplicity could lead to a system of TIL that is easier to learn (for example, TIL, as presented in [5], relies only on four basic constructions, namely, Variable, Closure, Composition, and n-Execution, instead of the standard six, i.e., Variable, Closure, Composition, Trivialization, Single Execution, and Double Execution). However, it might also have the opposite effect and produce a system that is more confusing, just as axiomatic systems with less axioms are not necessarily easier to understand and work with than systems with more axioms. Either way, this would be a matter for an empirical investigation.

8   Final remarks

Is Trivialization the same as 0-Execution? Based on the reasons given here, my answer is positive: yes, it is. There do not seem to be any convincing technical or conceptual reasons why Trivialization cannot be considered as a degenerate case of Execution. Of course, whether it is also a good idea to present it as such from a practical standpoint is another matter entirely. But if they are the same construction (i.e., given the same input, they produce the same output in the same way), how do we explain the fact that different intuitions seem to be associated with them? This, I believe, can be resolved by recognizing the two possible perspectives we can take in regards to 0-Execution/Trivialization, i.e., the path perspective and the destination perspective, with the former perspective being transformable into the latter: the "path" of 0-Execution can be contracted into a starting/ending point of Trivialization.

References

[1] Marie Duží, Bjørn Jespersen, and Pavel Materna. *Procedural Semantics for Hyperintensional Logic: Foundations and Applications of Transparent Intensional Logic*. Springer, Dordrecht, 2010. doi:https://doi.org/10.1007/978-90-481-8812-3.

[2] Bjørn Jespersen. Structured Lexical Concepts, Property Modifiers, and Transparent Intensional Logic. *Philosophical Studies*, 172(2):321–345, 2015. doi:10.1007/S11098-014-0305-0.

[3] Bjørn Jespersen. Anatomy of a Proposition. *Synthese*, 196(4):1285–1324, 2019. doi:https://doi.org/10.1007/s11229-017-1512-y.

[4] Ivo Pezlar. On Two Notions of Computation in Transparent Intensional Logic. *Axiomathes*, 29(2), 2019. doi:https://doi.org/10.1007/s10516-018-9401-7.

[5] Ivo Pezlar. Type Polymorphism, Natural Language Semantics, and TIL. *Journal of Logic, Language and Information*, forthcoming. doi:https://doi.org/10.1007/s10849-022-09383-w.

[6] Jiří Raclavský. Executions vs. Constructions. *Logica et Methodologica*, 7:63–72, 2003.

[7] Jiří Raclavský. *Belief Attitudes, Fine-Grained Hyperintensionality and Type-Theoretic Logic*. College Publications, London, 2020.

[8] Jiří Raclavský, Petr Kuchyňka, and Ivo Pezlar. *Transparentní intenzionální logika jako characteristica universalis a calculus ratiocinator*. Masaryk University Press (Munipress), Brno, 2015.

[9] Pavel Tichý. *The Foundations of Frege's Logic*. de Gruyter, Berlin, 1988.

What is the Message?

Martina Číhalová

Palacký University Olomouc, Department of Philosophy, Czech Republic

martina.cihalova@upol.cz

Marek Menšík, Adam Albert

VSB — Technical University of Ostrava, Department of Computer Science, Czech Republic

marek.mensik@vsb.cz, adam.albert@vsb.cz

Abstract

This chapter describes how Marek Menšík, Martina Číhalová, and Adam Albert met Professor Marie Duží a.k.a "Marie the Big" during their Ph.D. studies. It also introduces their mutual collaboration and how they were inspired by Marie in their research, for which they owe much to her. Last but not least, the appendix of this chapter includes a state diagram for effective communication not only with Marie but with anyone who is a fan of Transparent Intensional Logic.

Introduction

Professor Marie Duží is a personality who has had a very important influence on our lives not only professionally but also personally. Our contribution to this Festschrift for Marie contains a description of how we met Marie, how she has influenced our research, and last, but not least we provide instruction on how to communicate effectively with her.

1 How we met our 'mum'

In this chapter, it is stated from our point of view how we met our 'mum' a.k.a. 'Marie the Big' as Bjørn Jespersen calls her. It is because 'Duží' means *big* in Polish and Marie grew up in, and still lives in, Silesia close to the Polish border.

1.1 How Marek Menšík met our 'mum'

Once upon a time, in 1998, in the year of the greatest abundance of talented beings and in a distant galaxy called the Silesian University in Opava, I docked my spaceship-sized ego for a few years. I had to go through basic training to get my Master's degree. At that time I perceived it just as a necessary formality, which I had to accomplish with my left hand while drinking coffee with my right hand.

There were many training sessions, but one I found really unforgettable. It was the training in room B1. A large room was full of self-aware and know-it-all individuals proving how good they were. Not including myself, of course. That's where it all began. That's when I began my journey of initiation into the mystery and beauty of logic via a course named Introduction to Logic. Sitting in my seat, I was filled with anticipation and inner satisfaction that everything would be easy, and clear, just as it suited my mental state. I was waiting with the others for the arrival of the master of the art of logical argumentation. The door swung open and a female person entered. "Here we go", I said to myself. Her face was stern and uncompromising. Her arm was bent at the elbow and a purse full of secrets was hanging on it. My jaw dropped and suddenly it was clear to me that this wasn't going to be easy, this was going to be hard, and I did not think I was going to make it.

She introduced herself as Dr. Marie Duží and said she would be teaching the course *Introduction to the Logic* and also the exercises. She started off lightly. Everything seemed fine. Suddenly she began to explain some strange definition of the validity of an argument. I began to gasp for air. Everything collapsed, I could no longer understand what this strange definition was, followed by my 'Blue Death'. Then she explained the definition again, over and over, and then it gradually started to fall into place. I realized it was something I was interested in. That was the exact moment I realized who my master was and who would lead my journey to my dream degree.

My collaboration with Marie started slowly. I asked her if I could be her research assistant. I was approved and things started to go in the right direction. At first, I was only initiated into the HIT data model and then subsequently into Transparent Intentional Logic (TIL). However, my road was thorny and full of obstacles in the form of exams and credit tests. Not even my masterful escape from using the Logic textbooks and scripts helped me to avoid harming my health or damaging my so perfectly cultivated ego. Many other painful situations followed. They, unfortunately, required not only knowledge by heart but - more importantly -the understanding of the subject matter. And that hurt. But the vision of the Master's degree motivated me and made me study harder and harder. My master thesis, "Using Transparent Intensional Logic for Query Analysis and Processing", was the real beginning of our further scientific collaboration. This was followed by my application for Ph.D. studies, where I gradually met the rest of our team. At first, I met Martina Číhalová, who is a hypersensitive and clever person who does not like emails written in the form of bullet points. However, thanks to me, she had to learn to use them. And finally, I met Adam Albert who is too talkative but still a very talented student.

1.2   How Martina Číhalová met our 'mum'

I met Marie in 2006 only thanks to my superpower of confusion. I mistakenly thought I was supposed to apply for my Ph.D. after my state exams. My MA thesis supervisor, professor Štěpán from the Department of Philosophy at Palacký University, Olomouc, was surprised that I had missed the deadline to apply for Ph.D. studies. However, not long before that, he had gotten a call from Marie who was looking for a philosopher with experience in logical analysis of natural language for her project *Logic and Artificial Intelligence for Multi-agent Systems*. As I had passed Philosophy and Czech philology state exams, I could be a suitable candidate. He told me I could apply for Ph.D. studies at the Department of Computer Science at VŠB — Technical University of Ostrava. Being impractical and 'technically illiterate', I could not imagine myself as a doctor of computer science. However, I would have to wait a year to apply again at the Department of Philosophy anyway, so I called Marie to make an appointment, as I had nothing to lose. I will never forget how surprised I was by the depth of her voice. And then I was surprised for the second time when we met in Ostrava and she finished her cigarette in two puffs. Since then, I have constantly been amazed by her analytical thinking, her pertinent comments that never miss the point as well as her sense of humor. I finally got my Ph.D. degree in Ostrava after six years, participated in other projects with Marie, and even though I am back in Olomouc at the Palacký university, I am still cooperating with Marie and her 'dream team'.

1.3   How Adam Albert met our 'mum'

I met prof. Duží in 2019 during my Master's degree studies at VSB — Technical University of Ostrava. In my last two years of studies, I attended Mathematical Logic, Advanced Logic, and Transparent Intensional Logic taught by Marie. At that time, I had not yet developed a particular interest in logic, and I started to look for a job that I could do after finishing my studies. However, my Master's thesis supervisor, Marek Menšík, 'informed' me that I was going to study for my Ph.D. degree at the Department of Computer Science at my current university. Well, I did not know anything about Ph.D. studies, but Marek persuaded me, and I applied to study with Marie. In the last year of my Master's studies, I cooperated with Marek and Marie on my first papers, in which we applied TIL to supervised machine learning. It was not an easy task to explain our results to Marie with my vague and inaccurate explanations but thanks to those hard lessons I was able to improve my expression abilities and our articles. Little did I know that my explications of vague and inaccurate expressions will be the cornerstone of my dissertation thesis. Currently, I am in the third year of my Ph.D. studies and I study nearby her keen analytical mind. With her help and insights, Marek and I were able to complete several papers and I hope that we will work on many more.

2 How our research was inspired by Marie

Our first attempts at scientific papers were quite disastrous. When we consulted our outputs with Marie, her typical constructive criticism began with "What is the main message of your article?". This was closely followed by a criticism of our skills of expression based on not being afraid of vagueness. Marie has a special vagueness detector that never misses the point and its activation can be read from her expression. We soon learned that when Marie raised her eyebrows, we had to stop talking, start thinking properly and express ourselves explicitly, and possibly provide refinement of the crucial concepts. Coincidentally, our papers without the main message were about message communication between intelligent agents. The content of messages was specified in TIL which is the specification language with the greatest expressive power we have ever encountered. The main asset of using TIL is a fine-grained rigorous analysis and specification close to natural language. TIL enables us to distinguish three levels of abstraction, which means that it allows us to distinguish three types of context, namely extensional, intensional and hyperintensional.

2.1 At the beginning, there was Marie's idea of *refinement* applied to agents' communication

Marie Duží in [13] first proposed a general scheme of agents' messaging. Here is a bit adjusted Marie's proposal of the scheme of how to analyze different types of empirical messages and responses to them. This proposal and specification of types of message are modified according to Marie's later work.[1]

$$Message/(\iota\iota o_{\tau\omega})_{\tau\omega}$$
$$\lambda w\lambda t['Message_{wt}\ Who\ Whom\ \lambda w\lambda t['Type_{wt}\ What]]$$

where the types are $Who \rightarrow \iota$: construction of the sender of a message; $Whom \rightarrow \iota$: construction of the receiver of the message; $What \rightarrow \alpha_{\tau\omega}$: construction of the content of the message, which in the case of an empirical message is an intension of type $\alpha_{\tau\omega}$ (frequently a proposition of type $o_{\tau\omega}$); $Type/(o\alpha_{\tau\omega})_{\tau\omega}$.

There are three basic types of messages that concern propositions; these types are properties of propositions, namely $Order$, $Inform$, $Query/(oo_{\tau\omega})_{\tau\omega}$. In an ordinary communication act, we implicitly use the type $Inform$ affirming that the respective proposition is true. But using an interrogative sentence we ask whether the proposition is true ($Query$), or using an imperative sentence we wish that the proposition be true ($Order$). The content of a message is the construction of a proposition, the scheme of which is given by $\lambda w\lambda t['Type_{wt}\ What]$.

In what follows we specify in more detail typical types of messages. $Type = \{Seek,\ Query^{wh},\ Answer,\ Order,\ Inform,\ Unrecognised,\ Refine, \dots \}$; where $Type/(o\alpha_{\tau\omega})_{\tau\omega}$.

Examples of the content of a message:

- $['Inform_{wt}\ What]$; $What \rightarrow o_{\tau\omega}$
  Gloss. Informing that *What* is actually True.

- $['Order_{wt}\ What]$; $What \rightarrow o_{\tau\omega}$
  Gloss. Manage the proposition produced by *What* to be actually True (in a state of affairs w, t.).

- $['Seek_{wt}\ What]$; $What \rightarrow \alpha_{\tau\omega}$
  Gloss. Send me an answer, i.e. the actual α-value of *What* in a given state of affairs w, t of evaluation.

---

[1]In the original TIL, the simplest atomic construction that supplies entities to molecular constructions to operate on is the Trivialization. The original notation is ^0E, i.e., trivialization of the entity E. For the sake of typing simplicity, here we use the notation $'E$.

- $['Query_{wt} What]$; $What \rightarrow o_{\tau\omega}$
  Gloss. Send me an answer, i.e. the actual α-truth-value of the proposition produced by *What* in a given state of evaluation w, t.
  Remark. The analysis of wh-questions and their answering has been proposed by Marie (with her Ph.D. student Michal Fait) much later, in [12]. Briefly, each empirical wh-question is formalized by a construction of an intension of the form $\lambda w \lambda t [\lambda wh \ldots]$, where $wh \rightarrow \alpha$ is the variable the value of which the questioner would like to know.

- $[['Answer_{wt} What] = a/\alpha]$; where $a = [What_{wt}]$; the answer to a preceding query or seek.

Thanks to Marie's article [13], we soon understood that TIL enables agents to learn new concepts, due to the possibility of mentioning constructions on a hyperintensional level. Marie Duží proposed to use a new type of answer to messages of type *Unrecognized*. This answer consists in providing a compound construction that is a specification of the unknown concept. This specification is (or should be) an equivalent one with the unknown concept, i.e., it v-constructs the same entity for every valuation v.

Unrecognized is now a property of a concept, hence its type is $(o_n^*)_{\tau\omega}$. The subject of the message $What/_n^*$ is an unknown concept, i.e. a closed construction in its normal form. Thus, we have:

- $['Unrecognized_{wt} \, 'What]$; the unknown (usually atomic) concept *What* has not been recognized; a request for refinement.
  Gloss. Since *Unrecognised* is of type $(o^*n)_{\tau\omega}$, a property of a construction, the content of the message is not the entity constructed by *What*, but the construction *What* itself. The latter must be supplied by Trivialisation, hence $'What$.

The answer to such a message is of type $Refine/(*_n*_n)_{\tau\omega}$ an empirical attribute that associates the unknown concept with a new one.

- $[['Refine_{wt} \, 'What] = 'C]$: an answer to the message on the unrecognized concept, where $= /(o *_n *_n)$. The construction C is the respective molecular specification (definition) of *What*, i.e., C and *What* are equivalent, they v-construct the same entity for every valuation v.

For instance, the atomic empirical concept of a cat, i.e. '*Cat*, can be refined as a feline predator mammal. Hence, the answer to the message $['Unrecognized_{wt} \; ''Cat]$ would be

$$[['Refine_{wt} \; ''Cat] = '['Feline['Predator \, 'Mammal]],$$

where *Feline*, $Predator/((o\iota)_{\tau\omega}(o\iota)_{\tau\omega})$ are property modifiers; for more about property modifiers see, for instance [12]. For a mathematical example, consider the set of prime numbers. This set can be defined as the set of numbers with just two factors.

$$[['Refine_{wt} \; ''Prime] = '[\lambda x['Card \; \lambda y['Div \, x \, y] = '2]]],$$

where $x, y \rightarrow Nat$ (the type of natural numbers), $Div/(o \; Nat \; Nat)$ is the relation of being divisible by, $Card/ \; (Nat \; (oNat))$ is the number of elements of a finite set.

If an agent does not understand a concept, i.e., does not have this concept in their ontology, the agent sends a message of type *Uncerognized* and expects it to be refined via concepts the agent already has in their ontology. Marie's approach described above shows how the fine-grained semantics of TIL is suitable for analyzing agent communication in natural language, and how delivering a new type of message at the hyperintentional level can enable agents to learn new concepts.

Marie's analysis of the question types above was at the beginning of our story of cooperation on research. In the following sections, we lay out our individual perspectives on how our research was inspired and directed by Marie.

2.2 How Marek was influenced, instructed, and improved by Marie

My journey can be divided into several periods, all of which have something specific for me, and for 'Marie the Big', namely that it required a great effort from both sides to arrive at the desired goal.

Influenced. The effort to process natural language requires knowledge and, above all, determination and passion for the matter. Is it possible to somehow formalize natural language sentences appropriately into a formal system that can then be easily worked with? How to capture the semantics? Where to begin? It only took a few consultations with Marie to clarify that using TIL could be the right direction for my research activities. At the time, its logical notation was 'spilled tea' to me, as Martina put it. Lots of Greek letters, indexes, and a lot of bracketed terms seemed to me, not to form well-formed formulas. Gradually, however, I began to understand that parentheses are not just parentheses but labels of instructions specifying what to apply to what; and indexes are just shortcuts for further application of a function to an argument. Everything within TIL has its order, everything is in its proper place, and prefix notation is the best solution for future automated processing.

I was involved in a project dealing with multi-agent systems and agent communication. It was foundational research about designing systems and all the processes essential for agents to solve given tasks. My job was to work on controlling agents over a simulated infrastructure. Gradually, we ran into the problem of how to share information between the agents. We wanted to conceptualize it in a more general way, where not all agents may have the same knowledge. We need the given ontology to be distributed in a decentralized manner among the agents. Marie introduced us to the notion of *refinement* of concepts (constructions). This refinement was the starting point for our further research, which continues to the present day, where we use it in machine learning with a teacher. It inspired our specification of the concept of *explication* used to find appropriate textual resources or to search for suitable concepts matching a given explication. We use the term 'explication' in a broader sense than Marie for the result of the seeking process itself. The specification of *explication* is provided in more detail in section 2.3.

Instructed to learn a lesson. However, using terms like 'explication' and 'refinement', and all established terms in general, needs a clear definition that must be honored and followed. One cannot just

211

think intuitively that particular terms mean something and use them in that sense. And it took me a while to realize it. Why learn something if I have a 'clear idea' of what it means? One gets an idea and wants to share it with one's supervisor immediately. It has to be right now, and in the form it came to me. Why provide notes, think it through, and look at the subject from different angles? It was a waste of time for me. I have some ideas in my head, so sharing them with someone else can't be a problem. So I immediately used to go to Marie to tell her excitedly about my ideas. Today, in hindsight, I know what a crazy enterprise it was. The process was mostly the same and went as follows. I started talking. Thoughts unsorted and vague, I jumped from one thing to another and back again. The look on Marie's face, the horror in her eyes during my enthusiastic explanation! Marie's typical reaction was sentences like, "Once again, I don't understand you. What do you want to say? What do you mean? What is your main message?". And many of her questions followed, stemming from a misunderstanding of my vague interpretation, which was also often inconsistent. I left feeling misunderstood. But not only that. In my mind, I was thinking, "I don't know what I was going to say". I had only glimpses of ideas that had neither head nor tail. That's not the way it works. I have to think it through and then come forward with it.

Improved. Slowly, a space for collaborative exploration was beginning to emerge, where issues were addressed substantively and not about what I wanted to say. I was working on papers, going to conferences, and presenting results, and finally, I finished my Ph.D. We gradually formed a team and used the concept of *refinement* and further the concept of *explication* in machine learning. So this was my story about how one concept of refinement started a long trajectory of our mutual research and exploration.

2.3 History repeats itself, or else How Adam was influenced, instructed, and improved by Marie

In the not-so-distant past, as in the case with Marek, I was a student who didn't know exactly what he was going to do in his Master's studies after graduation. I was faced with the choice of a thesis project. Since I had no idea which topic I would be most interested in, I made a decision based on who would lead the project. My choice fell on a specific, very confident teacher Dr. Menšík who taught the course *Introduction to Theoretical Computer Science*. I liked the content of the course. The logic is based on following the rules. If one follows them, one runs no risk of making a logical error. Using strict algorithms of deduction, one can deduce what follows under given assumptions, whether a formula is logically true or whether it is a valid judgment. Why did I like it? Who knows? Perhaps because of the precise rules, proof procedure, and formulas transformations that brought order to things. Or was it satisfying some of my obsessive-compulsive disorder? The important thing was that I was accepted by Dr. Menšík for his thesis project.

My task was to study a symbolic machine-learning algorithm with a teacher. This is a type of machine learning in which the algorithm forms a symbolically written hypothesis describing some class of objects. This hypothesis is formed based on symbolically written descriptions of objects falling under the given class of objects or objects similar to them. Based on this work, together with Marek and Marie, I wrote my first paper (see [15]) in which we used TIL for symbolic representation. Already in the process of writing this paper, we intended to use the resulting hypothesis to describe atomic concepts. However, we ran into a problem. Our intention was to use the hypothesis as a definition of a given atomic concept. However, our hypothesis cannot be considered a definition. But what is it then? Marie came up with a solution. She likened the process of building a *refinement* to the Carnapian explication of a concept. In *Meaning and Necessity* [2], Carnap characterizes *explication* as follows:

> The task of making more exact a vague or not quite exact concept used in everyday life or in an earlier stage of scientific or logical development, or rather of replacing it by a newly constructed, more exact concept, belongs among the most important tasks of logical analysis and logical construction. We call this the task of explicating, or of

giving an explication for, the earlier concept [...]. [2, pp. 7–8]

Marie suggested the following specification of *refinement* based on the Carnapian approach to explication which is defined as follows:

Let C_1, C_2, C_3 be constructions. Let ^0X be a simple concept of an object X and let ^0X occur as a constituent of C_1. If C_2 differs from C_1 only by containing in lieu of ^0X an ontological definition of X, then C_2 is a refinement of C_1. If C_3 is a refinement of C_2 and C_2 is a refinement of C_1, then C_3 is a refinement of C_1.

Eureka! In [16] we used the hypothesis obtained here (molecular concept) as a refinement of the atomic concept. It is the process of refining a vague or inaccurate concept into an adequately precise concept. Using the constructions labeled as positive or negative examples, the algorithm iteratively builds the explication of the atomic concept. We refine the explication by adding more constituents into the molecular construction or we generalize the values of the constituents of the molecular concept with more general values. Positive examples are used for this purpose. We specialize the explication with negative examples by inserting additional constituents into the explication in a negated way. By those constituents, we differentiate the explication of the atomic concept from explications of similar concepts.

For example, let us explicate the atomic concept 'Cat' using positive examples such as 'Cats are mammals', 'Cats weight 5 kg', 'Cats weight 1.2 kg', and for a negative example, 'Cats bark'. The first positive example becomes a model (hypothesis in the refinement process), therefore the explication of cat (so far) is

$$\lambda w \lambda t [' Req \, 'Mammal \, 'Cat]$$

With the second example 'Cats weight 5 kg', the explication is refined by inserting a new constituent into the construction

$$\lambda w \lambda t [[' Req \, 'Mammal \, 'Cat \wedge [' Typ\text{-}p \; \lambda w \lambda t \lambda x [' = [' Weight_{wt} \; x]' 5]' Cat]].$$

Using the third example 'Cat weights 1.2 kg', the explication is generalized with a numerical interval spanning values in the model and the example.

$$\lambda w \lambda t [['Req\,'Mammal\,'Cat \wedge ['Typ\text{-}p\ \lambda w \lambda t \lambda x[['\le ['Weight_{wt}\ x]'11] \wedge ['\ge ['Weight_{wt}\ x]'1.2]'Cat]]$$

Finally, the explication is specialized with the negative example 'Cats bark' inserting constituent in 'negated way' into the construction.

$$\lambda w \lambda t [['Req\,'Mammal\,'Cat \wedge ['Typ\text{-}p\ \lambda w \lambda t \lambda x[['\le ['Weight_{wt}\ x]'11] \wedge ['\ge ['Weight_{wt}\ x]'1.2] \wedge ['\neg['Bark_{wt}\ x]]]'Cat]]$$

Types: Req/$(o(o\iota)_{\tau\omega}(o\iota)_{\tau\omega})$ the requisite relation;
Typ-p/$(o(o\iota)_{\tau\omega}(o\iota)_{\tau\omega})$ the typical property relation;
Cat, Mammal, Bark/$(o\iota)_{\tau\omega}$, Weight/$(\tau\iota)_{\tau\omega}$.

The resulting construction can be read in natural language as 'Cats are mammals that weigh between 1,2 to 11 kg and do not bark'.

Our paper [16] was the springboard for our further research, by which we worked with the concept of explication. In consulting this article with prof. Duží, I found myself in the situation that Mark once found himself in. Thanks to her and a few other experiences I have improved my skills of expression, and I was able to take my presentation and expression to a more precise level. We continue to work with the concept of explication and develop the system on the recommendation of information sources based on the obtained explication concept from these sources. Our ongoing research is documented in the following articles: [17, 1, 18, 19], and I hope there is still more to come.

2.4 How Martina's research was influenced and inspired by Marie

I have been inspired by Marie since the first moment I met her. In this section, however, I will focus on the two articles we wrote together, which followed many years of working together on various projects.

As a 'philosopher', asking questions is close to my heart. Hence, I suggested to Maria that we could write an article together that

addressed the issue of questions (see [11]). Marie had been dealing
with the issue of presuppositions for a long time, which she had been
addressing within the framework of TIL. I asked her if there were
different kinds of presuppositions, depending on the types of questions
(Yes-No questions, Wh-questions, and exclusive-or questions) and if
we could work it into a paper. Marie answered 'yes, of course' and
seemed intrigued.

Generally, in case the presupposition is not true, there is no unambiguous direct answer. In such a case an adequate complete answer
is a negated presupposition. Yet these simple ideas are connected
with a bunch of problems. First, it is important to distinguish between pragmatic and semantic presupposition and thus also between
presupposition and mere entailment. Second, Marie showed that the
common definition of a presupposition of a question as such a proposition that is entailed by every possible answer to the question is not
precise. She followed Frege and Strawson in treating survival under
negation as the most important test for presupposition. But a negative answer to a question is often ambiguous. The ambiguity consists
in not distinguishing between two kinds of negative answers, to wit the
answers applying either *narrow-scope* or wide-scope negation. Duží
[9] analyzed these two kinds of negation and showed that Russellian
wide-scope negation negates the entire proposition (it is not true that
the respective proposition has the truth-value **T**, hence it has the
value **F** or no truth-value) while Strawsonian narrow-scope negation
negates the respective property or relation; negation is propagated in.
While the former preserves presupposition, the latter seems to be presupposition denying. The paper shows that in order for the negative
answer to be unambiguous, instead of the *wide-scope negation* presumably denying presupposition, an adequate and unambiguous answer
is just the negated presupposition. Having defined the presupposition
of a question more precisely, we then examine Yes-No questions, Wh-
questions, and exclusive-or questions with respect to several kinds of
presupposition triggers. These include inter alia topic-focus articulation, verbs expressing termination of activity, factive verbs, the "whys
and how comes ", and past or future tense with reference time interval.

Our background theory was TIL, of course. We applied the general analytic schema for sentences with a presupposition that was introduced in [7, 8] together with the definition of the If-then-else function. The schema modified for the analysis of a question $Q \to \alpha_{\tau\omega}$ with a presupposition $P \to o_{\tau\omega}$ is this:

$$\lambda w \lambda t[\textit{If } P_{wt}\,\textit{then } Q_{wt}\,\textit{else } \neg P_{wt}]$$

Gloss: If the presupposition P is true in a given world w and time t (If P_{wt}) then evaluate Q_{wt} to provide the answer, else reply by negating the presupposition ($\neg P_{wt}$).

As an example, consider the question 'When did the Mayor of Ostrava visit Brussels?' If the topic of this question is the 'Mayor of Ostrava' then there is the existential presupposition that the office is occupied. The narrow-scope, unambiguous, negative, direct answer 'Never!' implies that the Mayor exists, and instead of the wide-scope interpretation of this negative answer an adequate answer is just the piece of information that the presupposition is not true: the Mayor does not exist (which in TIL means that the office of Mayor of Ostrava is vacant). Applying the general analytic schema for sentences with a presupposition to the question above, we obtain

$$\lambda w \lambda t[\textit{If } ['Exist_{wt}\ \lambda w \lambda t['Mayor\_of_{wt}\ 'Ostrava]]$$
$$\textit{then } \lambda t'[[t' < t] \wedge ['Visit_{wt'}['Mayor\_of_{wt}\ 'Ostrava\ 'Brussels]]$$
$$\textit{else } \neg ['Exist_{wt}\ \lambda w \lambda t['Mayor\_of_{wt}\ 'Ostrava]]].$$

Additional types: $Exist/(o\iota_{\tau\omega})_{\tau\omega}$: the property of an office of being occupied; $Visit/(o\iota\iota)_{\tau\omega}$; Brussels/$\iota$. Note that the answer is of type $(o\tau)$, that is, the set of past times t', namely those times when the Mayor visited Brussels.

While working on this article, most of the technical details and also the application of concepts of narrow-scope or wide-scope negation were handled by Marie. At the time I did not have much experience writing research articles, and the joint article with Marie taught me a lot. I am grateful that Marie did not give up on working with me and we instead started working on another joint article together in 2020. The main idea was inspired by the approach to event ontology developed in my dissertation and the article [3]. Here I proposed to analyze the activities of agents using the theory of valency frames and Sowa's thematic roles; (see [20, pp. 506–510]). The determination of the ontological category of concept was the differentiation between the *static* and *dynamic* parts of the respective domain of interest. The static part of the domain is made up of simple and non-decomposable unique object and their characteristics, and the dynamic part is made up of *activities* that form *events*.

The *static part* is made up of entities like individuals which can be characterized by their properties and attributes. I distinguished between *substantive* and *accidental* characteristics of individuals. On an intuitive level, I felt there was a difference, but I was unable to rigorously specify the difference between these two types of characteristics. My expressive difficulties were followed by Marie's lessons to eliminate my vague expressions. At the end of it, Marie proposed to specify substantive characteristics as those that individuals have *nomically* necessarily. Examples of substantive characteristics are, for instance, 'being a person', 'being a horse' or 'being a piece of fruit'. For instance, according to the laws of physics and biology, if an individual is born as a person, then during its lifespan it cannot become, say, a dog or a vase. These properties in ontology usually form so-called ISA relationships, such as 'Every apple is a piece of fruit'. On the other hand, accidental characteristics are possessed by individuals purely contingently, e.g., the property of being a student is accidental; one and the same person contingently becomes a student or stops being a student or contingently was never a student or will never become a student. Other accidental characteristics of the person-type individuals can be, for example, 'weight', 'height', 'age'.

The *dynamic part* is made up of entities like *activities* which are linguistically detected by some special types of verbs called episodic verbs according to Tichý's distinction in [21]. Tichý distinguished between episodic and attributive verbs. Episodic verbs (e.g., 'drive', 'tell') express the actions of objects or people, which we call *activities*. However, attributive verbs (e.g. 'is a dog', 'looks speedy') ascribe some empirical properties to individuals. They express *accidental* or *substantive* properties of individuals according to the proposed ontology above.

Each *activity* has an *actor* (who/what is doing the activity) and can involve other objects that are called participants. Thematic roles (participant), such as *Agent, Patient, Beneficiary, Destination, Instrument,* etc., expresses the role that a noun phrase plays with respect to the activity described by a governing verb. It can be specified as a relation-in-intension between an activity and the respective object.

So much for the fundamental concept of *activity* which is closely connected with concepts of *event* and *process*. In [4] I specify the concepts of *process* and *event* and the relation between them and in [5] I made it more precise. In a number of approaches, the concepts of process and event overlap, and these terms are treated as synonyms. However, I proceed from the approach of Sowa, who draws a difference between them. Each simple process consists of *state*1 (S1), *event* (*transition*), and *state*2 (S2). Processes could be compounded from two or more simple processes. A process can be defined as a sequence of states and events. A state as a particular state of affairs is a proposition. An event, as a change of states, can be also viewed as a proposition. For a simple example, consider this process.

State$_1$: Tom is standing;
Event: Tom starts running;
State$_2$: Tom is running.
This process can be dealt with as the sequence of three propositions:
$\lambda w \lambda t['Standing_{wt} \,'Tom]$, $\lambda w \lambda t['Start\text{-}Running_{wt} \,'Tom]$,
$\lambda w \lambda t['Running_{wt} \,'Tom]$.

Hence, the process as a whole can then be captured by a sequence of propositions. The measure of the granularity of the process depends on the aims of the application that the ontology serves. If we want to capture speed changes, say, we need to specify the process in more detail. Each speed change has to be captured by adding the actions of 'accelerate' and 'decelerate' to the ontology.

Linguistically, we express states and events by means of sentences, and these sentences are made up of at least a subject and a predicate plus they can contain other clause members based on the verb and noun valency. Hence, *state1* can be described in more detail by other characteristics of the individual and his or her or its activity (e.g., '30-year-old Peter is standing at the railway station'). There are two examples of processes as a state-event-state sequence below.

> Example: S1: Peter stands, *Event*: Peter starts running, S2: Peter runs.
> Example: S1: an/the apple is green, *Event*: an/the apple turns red, S2: an/the apple is red.

You can see there are two basic types of *activities* One is based on the actions of deliberative agents and the second is based on *passive events* which are not intentional. Hazal, Svátek, and Vacura [14] call this type of process *happening*. Note that the activity is the necessary condition for the occurrence of an event.

I suggested applying this framework for the process ontology to provide a classification of the logical types of Wh-questions for multi-agent systems and the logical analysis of such questions in TIL. In [6] me and Marie proposed a new classification of Wh-questions that matches the logical structure of agents' knowledge and the logical types of possible direct answers to Wh-questions. To this end, we distinguished questions about activities, their participants, substantive properties of objects, accidental characteristics, and, last but not least, the agents can ask for explication (refinement) of concepts themselves and thus learn new concepts and enrich their ontology. Marie managed to work out the specification of questions on participants of activities specified in different tenses with reference time and frequency when this or that activity happened or will happen. She elaborates on this tense specification in detail and further develops it (see [10]). By way of examples, we illustrated the application of the general analytic schema that takes into account presuppositions of such questions. Below is an example of an analysis of the question and answer concerning the participants of an activity.

Using a general place-holder π for the type of activity and $\alpha^{\text{Kind}-i}$ for an attribute/participant of a Kind-i, the type of *Do* is $(o\iota)_{\tau\omega}$, and the assignment of participants to the activity is then an entity $Asgn$ of type $(o\pi\alpha^{\text{Kind}-\text{i}})_{\tau\omega}$. To simplify the notation and make the formulas easier to read, we will use '$X^{\text{Kind}-\text{i}}$' instead of '$Kind - i\,X$' to signify that X belongs to the class of participants of the Kind-i. Thus, we obtain a general pattern for analysing an activity $P \to \pi$ with the actor A and participants $X_1^{\text{Kind}-1}, \ldots, X_n^{\text{Kind}-\text{n}}$.

$$\lambda w \lambda t [' Do_{wt} AP] \wedge [' Asgn_{wt} P' X_1^{\text{Kind}-1}] \wedge [' Asgn_{wt} P' X_2^{\text{Kind}-2}] \wedge \ldots$$
$$\wedge [' Asgn_{wt} P' X_n^{\text{Kind}-\text{n}}].$$

For instance, the analysis of the sentence *John goes to Brussel* by train amounts to this construction:

$$\lambda w \lambda t [' Do_{wt} \, ' John' Go] \wedge [' Asgn_{wt} \, ' Go' Train^{\text{Inst}}] \wedge [' Asgn_{wt} \, ' Go' Brussel^{Dir3}].$$

To summarize our joint paper, we provided a new detailed classification of Wh-questions in particular with respect to the analysis of dynamic activities of human and or software agents. This classification is apt for agents' reasoning and communication, and it is utilized in the rigorous logical analysis of agents' dynamic activities, including queries about the participants of the activities.

I am so grateful to have had the opportunity to work with Marie. I owe her not only for introducing me to TIL but especially for her excellent feedback and constructive criticism. I hope our collaboration will continue and we will do some more articles together.

How to communicate effectively not only with Marie

Vagueness is a source not only of misunderstanding but above all also of errors in reasoning. Marie continues to teach us how to avoid vagueness by explicitly specifying our ideas and refining them, and we will never cease to be grateful to her for this. Transparent Intensional Logic is a great tool for logical analysis of natural language which helps to reach this goal. The state diagram (see figure 1 in the appendix) shows how to communicate effectively, especially with Marie, but also with anyone who is familiar with TIL and who wants to avoid vagueness.

Acknowledgments

Finally, we would like to express our thanks to our 'mum', for her guidance, all the inspiration, and feedback she has given us throughout the years, and all the fun not only with Transparent Intensional Logic. And last but not least we have to express our special thanks to Bjørn Jespersen for his "pobjørnování" of this chapter.

References

[1] Albert, A., Duží. M, Menšík, M., Pajr, M., Patschka, V. (2020): Search for appropriate textual information sources. In: *Frontiers in Artificial Intelligence and Applications*. Amsterdam: IOS Press, 2020, s. 227-246. ISBN 978-164368140-5. ISSN 09226389. doi:10.3233/FAIA200832

[2] Carnap, R. (1947): *Meaning and Necessity*. Chicago University Press, Chicago.

[3] Číhalová, M. (2016): Event ontology specification based on the theory of valency frames. In *Frontiers in Artificial Intelligence and Applications, Information Modelling and Knowledge Bases XXVII*, T. Welzer, H. Jaakkola, B. Thalheim, Y. Kiyoki and N. Yoshida, eds, pp. 299–313. IOS Press, Amsterdam.

[4] Číhalová, M. (2021): Conceptual Framework for Process Ontology, In A. Horák, P. Rychlý, A. Rambousek (eds.) *Recent Advances in Slavonic Natural Language Processing (RASLAN 2021)*, 83–90, Brno, Tribun EU.

[5] Číhalová, M. (in review): Specification of the Fundamental Concepts in the Ontology of Processes; Event, Process, Activity. Special issue of *Organon F*.

[6] Číhalová, M., Duží, M. (2022): Modelling dynamic behaviour of agents in a multiagent world: Logical analysis of Wh-questions and answers. *Logic Journal of the IGPL*, vol. 30, No. 1, pp. 1–32. https://doi.org/10.1093/jigpal/jzab034

[7] Duží M. (2009): *Topic-focus articulation from the semantic point of view. Computational Linguistics and Intelligent Text Processing*, Springer, LNCS vol. 5449, pp. 220–232.

[8] Duží M. (2010): Tenses and truth-conditions: a plea for if-then-else. In the *Logica Yearbook 2009*, Peliš, M. (ed.), London: College Publications, pp. 63–80.

[9] Duží, M. (2014): How to Unify Russellian and Strawsonian Definite Descriptions. In *Recent Trends in Philosophical Logic, Trends in Logic*, Roberto Ciuni, Heinrich Wansing, Caroline Willcommen (eds.), vol. 41, pp. 85–101.

[10] Duží, M. (to appear): Specification of Agents' Activities in Past, Present and Future. Special issue of *Organon F*.

[11] Duží, M., Číhalová, M. (2015): Questions, answers and presuppositions. *Computación y Sistemas*, vol. 19, No. 4, pp. 647–659.

[12] Duží, M., Fait, M. (2021): A hyperintensional theory of intelligent question answering in TIL. In Loukanova, R. (ed.), *Natural Language Processing in Artificial Intelligence - NLPinAI 2020*, pp. 69-104, Springer book series Studies in Computational Intelligence (SCI, volume 939), Springer, ISBN 978-3-030-63787-3

[13] Duží, M., Vojtáš, P. (2008): Multi-Criterion Search from the semantic Point of View. *Frontiers in Artificial Intelligence and Applications*, Amsterdam: IOS Press, vol. 166, pp. 21–39.

[14] Hazal, T., Svátek, V., Vacura, M. (2016). Event categories on the semantic web and their relationship/object distinction. In R. Ferrario and W. Kuhn (eds.), *Formal Ontology in Information Systems* (pp. 183–196). Amsterdam: IOS Press.

[15] Menšík, M., Duží, M., Albert, A., Patschka, V., Pajr, M. (2019a.): Machine Learning Using TIL. In: *Frontiers in Artificial Intelligence and Applications*. Amsterdam: IOS Press, 2019, 344–362. ISBN 978-164368044-6. ISSN 0922-6389. doi:10.3233/FAIA200024

[16] Menšík, M., Duží, M., Albert, A., Patschka, V., Pajr, M. (2019b.): Refining Concepts by Machine Learning. *Computación y Sistemas*. 2019, 23(3), 943–958. ISSN 2007-9737. doi:10.13053/cys-23-3-3242

[17] Menšík, M., Duží. M, Albert, A., Patschka, V., Pajr, M. (2019c.): Seeking Relevant Information Sources. In: *INFORMATICS 2019 - IEEE 15th International Scientific Conference on Informatics*, Proceedings. 2019, s. 255-260. doi:10.1109/Informatics47936.2019.9119332

[18] Menšík, M., Albert, A., Patschka, V. (2020): Using FCA for seeking relevant information sources. In: *Recent Advances in Slavonic Natural Language Processing*. Brno: Virtual, 2020, s. 47-54. ISBN 978-802631600-8. ISSN 23364289. Also available from: `https://www.scopus.com/inward/record.uri?eid=2-s2.0-85103626231&partnerID=40&md5=420e25e2ed9b8bfc6bc6cf56a5e76c0d`

[19] Menšík, M., Albert, A., Patschka, V, Pajr, M. (2021): Improvement of

Searching for Appropriate Textual Information Sources Using Association Rules and FCA. In: *Frontiers in Artificial Intelligence and Applications*. Amsterdam: IOS Press, 2021, s. 204–214. ISBN 978-164368242-6. ISSN 09226389. doi:10.3233/FAIA210487

[20] Sowa, J. F. (2000): *Knowledge Representation (Logical, Philosophical, and Computational Foundations)*. Brooks Cole Publishing Co., Pacific Grove, CA.

[21] Tichý, P. (1980): The semantics of episodic verbs. *Theoretical Linguistics*, 7, pp. 263–296. Reprinted in (Tichý 2004: 411–446).

[22] Tichý, P. (2004): *Collected Papers in Logic and Philosophy*, V. Svoboda, B. Jespersen and C. Cheyne, eds. Prague: Filosofia, Czech Academy of Sciences, and Dunedin: University of Otago Press.

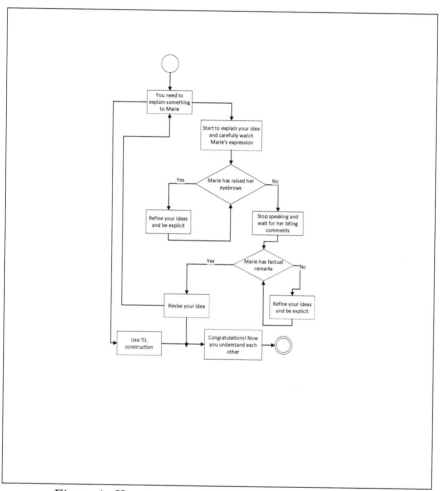

Figure 1: How to communicate effectively with Mary

How to Manage Cultural Differences?

HANNU JAAKKOLA
Tampere University, Information Technology and Communication Sciences
hannu.jaakkola@tuni.fi

1 Introduction

The trend is toward a global world, in which borders are becoming less important, both in an economic and social context. There are still nations that have isolated themselves, mostly for political reasons, but these are exceptions rather than the rule. Despite this progress there is still space for national values. How is globalization seen in practice? Traditionally, formal agreements are enablers for free trade, easy travelling, and cultural exchange. Nowadays, the importance of informal channels has become more important: the internet provides a platform for a wide variety of applications and social networking, which accelerate global access to the same services and functionalities. Simultaneously, the growth of economic welfare makes use of services, as well as the opportunity to travel for a growing share of the world population; long distances have become short both in the physical and cyber world, and digitalization has replaced physical products and services with their digital counterparts. Globalization provides opportunities, but simultaneously it creates dependencies between counterparts; national becomes international, cultural means multicultural and cross-cultural. *Business strategies* must consider

Title of Docent, University of Lapland; Visiting Professor, Keio University, Japan (1/2015-3/2015; 9/2016-3/2017; 11/2019-3/2020)

global and networking aspects: typical manifestations of the progress are outsourcing on national level (*subcontracting*), neighborhood level (*nearshoring* to the countries in the neighborhood), *offshoring* (worldwide distribution of the operations). As an example of nearshoring, a Finnish ICT company Tieto (currently Tietoevry) established a new branch in the Czech Republic, in the city of Ostrava (near to the VSB-TU Ostrava campus), in 2004. Currently it has 2.600 employees. The main reason to select (offshore to) this location was the available close collaboration with the University and a skilled, well-educated workforce. Typical phenomena also include recruiting experts from the international workforce instead of the national market and a global market for products. On a human level, typical manifestations are seen in increasing travel to foreign countries, globalization of entertainment services, organized student exchange programs among universities (foreign internship as part of study programs) and scientific collaboration, which extends the level of multicultural collaboration.

The progress of globalization is measured by the globalization index. Economists have developed different approaches to this topic. One of the most commonly referred to is the *KOF (Konjunkturforschungsstelle, ETH Zürich) Globalization Index*, which was introduced by Axel Dreher in 2006 [6]. A revised version has been published by Gygli [9]. In this index, globalization is divided into three categories: *economic, social, and political*. The globalization drivers are further divided into two main classes: *de facto* globalization measures actual international flows and activities (how it is seen in practice), and *de jure* globalization measures policies and conditions that enable, facilitate, and foster flows and activities. The *economic* category covers trade and financial aspects; social globalization covers interpersonal, informational, and cultural dimensions; *political* globalization deals with mainly official and organizational issues. The index data is maintained by ETH Zürich, and it covers data from 195 countries from 1970 until the present day [8]. The national level of globalization is measured by the index with a value of 0-100 and calculated from 42 separate factors. The tool (in [8]) allows analysis with different focuses and geographical areas, as well as comparisons between countries

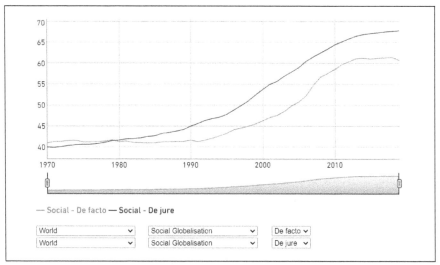

Figure 1: Social globalization in the world 1970-2019 [8]

and continents. Figure 1 illustrates the progress of *social globalization* (covering interpersonal, information, cultural); this focus was selected in view of the *goal of this paper*, which is to understand and manage cultural differences in multicultural, distributed global collaboration.

The fastest growth in globalization was triggered by the collapse of the Soviet Union and the liberation of the Eastern Bloc in Europe; the growth reversed at the end of the first decade of 2000 due to the economic crisis [21]. This was also the fastest period as the main *motivation* to reach a cheaper cost level. Nowadays the trend has become the opposite — *insourcing* activities back to the home country. For example, the latest available (2019) social globalization index of *Finland* is 85.84 and that of the *Czech Republic* 84.85. As a comparison, the index of *China*, 57.96, is reasonably low, although it has been one of the fast-growing targets of offshoring (in past decades). In the records of Statista [20], Finland was in position 9 (value 87.68) in the overall globalization index values among the fifty top countries in the world; the Czech Republic was in position 15 (value 84.85). China is ranked 85.

The discussion above is background and motivation for the actual topic of this paper. Despite acting as part of the global community, every nationality has its own *national culture*, which defines the behavioral patterns and interactions of individuals. In a *multicultural environment* this might be a source of conflicts and problems in daily situations — "Cultures Collide". Additional complexity comes from the distribution of work, and physical and temporal distance between the collaborating parties.

This paper is focused on *multicultural distributed collaboration* in a global environment — how to manage and recognize the problems related to it. The experience base comes from *software development*, which is the author's area of expertise. Some of the findings are derived from experiences based on over fifty years of co-operation with a wide base of *software companies* and on the author's research output (with international collaborators). *Finland* and the *Czech Republic* are used as case countries[1], in consideration of the context of the Publication Forum (and Professor Marie Duží); for the same reason some references to Japan as a country steeped in Asian culture are used. Despite the selections, the findings are scalable and transferrable to other sectors of business and societies, and partially to other cultures as well.

Distributed multicultural collaboration in a global context holds problems that are complex to manage, or even to recognize in time to avoid them. Because of the different behavioral patterns of the collaborating parties, misunderstandings exist, and people do not meet the expectations of others in typical collaboration situations. This fact comprises the *research problem* of this paper. Understanding cultural

---

[1]The topic of this contribution to the Festschrift for Marie Duží characterizes the long-standing rapport with her Japanese colleagues, in which Marie as a Czech and I as a Finn have been active partners. Additional European partners have come from Germany and two other Finnish universities. The collaboration started in 2000, when Marie was an invited speaker at the 10th European-Japanese Conference on Information Modelling and Knowledge Bases held in Finland. The Japanese partners represent several research institutes, among which the Keio University plays a key role. Through it several Asian partners joined the collaboration, which further increased the multicultural dimension.

differences helps to avoid problems in cultural complexity and aid adaptation in daily situations. From this starting point the following research questions are derived.

RQ1: What are the key elements of (national) culture and are there other dimensions (apart from nationality) to consider?

RQ2: Are there any frameworks applicable for understanding cultural differences?

RQ3: If there are such frameworks, how to apply these in practice?

The rest of the paper is structured in three sections, according to the research questions. Section 2 discusses the characteristics of the concept "culture". Section 3 responds to RQ2 and introduces two selected frameworks of cultural differences. Section 4 deals with some additional aspects on culture. Section 5 summarizes the paper.

2 What is Culture?

2.1 Culture Defined

The concept of culture has a wide range (hundreds, if not thousands) of definitions. In this paper we have adopted the approach of Geert Hofstede, who is one of the pioneers in the "practical" research of cultures. Hofstede *et al.* [12] define (national) culture as "*The collective programming of the mind which distinguishes the members of one human group from another — Software of the Mind*". He started the work on cultures in the 1960s in the HRM organization of IBM, which employed people all around the world. To illustrate the role of culture as part of the human mind, Hofstede introduced two (meta)models: the pyramid and onion models of culture (Figure 2).

The pyramid model (Figure 2, left side) consists of three layers. The lowest layer (*Human nature*) is common to all human beings. It is inherited and universal— culture independent. Hofstede uses the term "Operating System of the Mind". The Culture layer is collective and learned by a specific group or category (nation, geographical

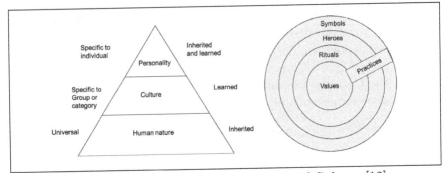

Figure 2: Pyramid and Onion Models of Culture [12].

area) of people. It is something that human beings learn from their parents and from the surrounding society. Hofstede calls it "*Collective program of the Mind*". The top layer is *personality*, which is specific to individuals, partially inherited and partially learned. In this part, *personal mental programs*, the social environment plays an important role.

The onion model (Figure 2, right side) illustrates some of the key elements of culture — how *practices* (behavior, way of thinking, attitude) as manifestations of a culture are built. People in the same culture typically share the same (learned) *values*, which are the core of the culture and the preferred states over others. *Rituals* are collective activities, which are based on the values and essential in a culture (e.g., attitude to religion). *Heroes* are highly prized persons of the culture, which emphasize togetherness and are typical to one culture (these would be persons such as national poets, admired successful sportsmen, etc.). *Symbols*, the outermost layer of the onion, are words, gestures, and objects that are common to those who share the culture (e.g., religious symbols).

2.2 Culture is Stored and Dynamic

Culture is *learned* under the influence of the surrounding society. However, human beings are learning all their life and perceiving new things. Culture is

- re-learned and incremental: on the individual level, new experiences "update" and enrich the culture

- clustered and multidimensional: the same person has different culture roles

- dynamic: on macro-level temporal changes appear in the environment and modify the culture

- adapted: external pressure, e.g., a long stay in a foreign culture, modifies the culture on individual level.

In summary, the culture of an individual is the 'stock' of *lifetime experiences*.

Multidimensional and clustered culture

The *clustered character* of the culture may have different dimensions. Based on the area where people live, this may be regional, subregional, tribe, and family culture, which represent minor variations of the national culture. People are also representatives of their active life environment, which is seen as the cultural aspects from an organization, sub-organization, site, project, and team. A sub-culture may also be functional, with dependence on education or school, and profession. This multidimensional character allows an individual to appear in different *roles* in different situations. The same person may be a strong manager (company culture), a soft family man (family culture), and a schoolmate (school culture) with different values and practices depending on the respective role and the context of the activity.

Temporal dynamics of the national culture

The World Values Survey (WVS) is an international research organization conducting studies of the social, political, economic, religious, and cultural values of people in a worldwide scope [23]. WVS has collected worldwide data on people's values to see the ongoing cultural changes and the persistence of distinctive cultural traditions on the national level. The data indicates the long-term temporal *dynamics of national cultures*; the dynamics have been visualized by Ronald Inglehart and Christian Welzel in a two-dimensional map called the *Inglehart-Welzel Cultural Map*. Countries are located in a two-dimensional space with the dimensions Traditional values versus Secular-rational values, and Survival values versus Self-expression values.

A low score in the *y-dimension* indicates the importance of *traditional* values, which cover tight parent-child ties, deference to authority, family and traditions, high levels of national pride, and a nationalistic outlook. A high score in contrast indicates *secular-rational values (materialism)*, which are opposite preferences to the traditional values. These are typical in countries with a long history of social democratic or socialist policy, and where the population has a high level of education.

A low score in the *x-dimension* indicates the importance of *survival values*, which cover an emphasis on economic and physical security. It is linked with a relatively ethnocentric outlook and low levels of trust and tolerance. A high score in turn indicates the importance of *self-expression values*. These cover high priority given to environmental protection, high tolerance of foreigners and unusual phenomena, gender equality, demands for democratic participation in decision-making in economic and political life. These are characteristics of Western world democratic countries and liberal post-industrial economies in the Western world.

The current map (Figure 3) reports the situation in 2020 ([23], in the subsection "Findings and Insights"). In addition, a visualization of temporal changes in the period of 1981-2015 is given there (also in https://youtu.be/ABWYOcru7js). It indicates the changes in 35 years and provides an opportunity to analyze the reasons behind the changes.

Typical factors keeping a country in the *lower left* corner are low welfare and tight control by the authorities. Traditions (e.g., strong rule-based religion) are a means to survive in these conditions. These countries are emergent with future opportunities, in case obstructive

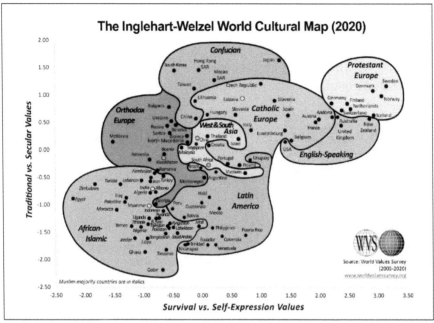

Figure 3: Cultural map - WVS wave 7 (2017-2021). [23], subsection "Findings and Insights").

factors will disappear. The alternative path is toward chaos, as seen in (too) many cases. Typical examples in this category are African-Islamic countries.

Countries in the *upper left* corner have high education and acceptable welfare, but the society still has some less-developed features, like an authoritarian administration, limited human rights, lack of equality, or corruption. Welfare is acceptable but life is still somewhat controlled, compared to well-developed democratic societies. Examples in this segment are some former Eastern Bloc countries. Societies in the lower right corner have a reasonably high degree of freedom, are flexible but still controlled by traditions instead of rational policies. People may suffer from an unpredictable unstable political environment and disorderly conditions of life. Despite that, the degree of freedom is high. Examples are Latin-American countries.

The societies in the *upper right* corner are mature and have high welfare, the degree of freedom is high, and administration is well-organized and democratic. Traditions dominating peoples' lives are minimal, if any. Typical countries in this category are European Protestant countries, Australia, and USA.

3 How to Increase Understanding of Cultural Differences?

3.1 Culture stereotypes

Culture has been defined as " *the collective programming of the mind which distinguishes the members of one human group from another".* The common features typical to a culture build a stereotypical approach to culture, which builds the impression that all the members of the same culture behave in the same way. However, culture is not a stable concept; continuous learning changes it, it is clustered and multidimensional. Personal characteristics reinforce differences between individuals. Are stereotype-based generalizations useful at all for recognizing cultural differences?

Stereotypes do not fit everybody. The behavior of two members of one national culture may differ from each other more than the behavior of representatives of two different cultures. However, stereotype models are the results of systematic analysis of the selected properties typical in a culture, measured from big masses of people, and organized in the form of an applicable framework. Despite their shortcomings, these generalizations provide a means to understand the features that are somewhere in the *culture kernel* of individuals. An essential part of human behavior is bound to the social control of the society and its institutions and becomes true latest for a big mass of people. Good examples can be found in everyday occasions: shaking hands in Europe vs. making a deep bow (Japan), how to greet a woman, the role of members in a delegation (age, position), interpretation of silence (no response). Also, spoken and written language (characters, reading direction) has implications for the formation of concepts and structures (Asian visual vs. Western linear), as well as the use of body language in communication differs between cultures.

Culture has implications for numerous everyday phenomena. Typical sources of issues in the work context appear in:

- *Communication*: misunderstanding, language barrier, creation of concepts and structures varies, the way of communication is different (hidden meanings).

- *Leadership* practices: people management, providing feedback, mentoring, need for control, solving conflicts.

- *Management*: organizing work, responsibilities, decision making practices.

- *Trust creation*: building up internal and external relationships, the role and importance of trust and trusted relationships.

- *Management of change*: attitude to changes, resistance and ability to accept changes.

- *Motivation and loyalty*: motivation factors, loyalty to stakeholders.

- *Competence differences*: existence and ability to see competence differences in an organization, self-evaluation skills, truthfulness of the individual's skill profile (trying to look better than one really is).

The reasons in the background of the issues are manifold: attitude to more powerful colleagues, acceptance of equality, fear of losing face, acceptance of a flexible truth. Differences between emergent and mature societies exist, and the role of church and organized religion in society also has an important role. Where problems are ultimately simple solutions are also relatively easy to realize, provided that the root causes (characteristics typical to cultures) are understood.

To provide a means for understanding and analyzing the differences between cultures, the rest of this section introduces two stereotype-based models of culture:

- *Hofstede Six Dimension Model*, which is based on the calculated scores of six cultural dimensions for the (national) culture (sub-section 3.2).

- *Lewis model*, which defines three basic culture stereotypes and combined cultures based on these (sub-section 3.3).

These models have been selected because they are the ones most commonly used and referred to. Additional models, like *Trompenaars* Seven Dimension Culture Model or the GLOBE model of *Robert House* do not bring any remarkably added value to the discussion. Both Hofstede and Lewis have made studies in organizational culture; these are not handled here.

3.2 The Hofstede Six Dimension Model

Geert Hofstede is a Dutch social psychologist who started to collect systematic data from the employees of IBM in forty countries; currently the geographical coverage is high. Hofstede noticed systematic differences between nations. He classified the differences in measurable and organized classes, called *cultural dimensions*, in a framework (model) providing a multidimensional culture profile of national cultures. The first version of the model had four dimensions, two were added gradually later. The close connection to IBM and white-collar workers has been a source of criticism against the model. However, the model is supplemented by data from other sources to make it organization-independent and to increase its overall validity.

The model is based on culture dependent scores given to each dimension. The scale of score values is from 0 to 100 with 50 in the middle; values below 50 mean low and above it a high score in each dimension. The absolute score value is not meaningful in the model, which is developed to compare countries to each other; the score scale is *relative*, not arithmetically linear (the distance of two values is not the difference between them). The model details are available in the book [12] and in the resource pages [13, 14].

The six dimensions in the Hofstede model are (cited and modified from the original sources):

- *Power Distance (PDI)*: the extent to which power differences are accepted in society.

- *Individualism/Collectivism (IDV)*: the extent to which a society emphasizes the individual or the group.

- *Masculinity/Femininity (MAS)*: the dominance of general values in society — hard vs. soft values.

- *Uncertainty avoidance (UAI)*: the degree to which the members of a society feel uncomfortable with uncertainty and ambiguity.

- *Long-term/Short-term orientation (LTO)*: the extent to which the delayed gratification of material, social, and emotional needs is accepted.

- *Indulgence/Restraint (IND, earlier IVR)*: acceptance of enjoying life and having fun vs. controlling life by strict social norms.

A more detailed description of the dimensions is available in Hofstede Insight [13].[2]

What can we read from the score values? Examples of practical implications are:

- PDI: Democratic (low) vs. hierarchical, centralized (high) decision making; flat (low) vs. hierarchical organization (high); extent of bureaucracy - complex (high), flexible (low); need for supervision and control (high) vs. independence (low); power to make decisions - need confirmation of higher authorities (high) vs. delegated (low).

- IDV: collective (low) or individual (high) decisions; fear of losing face as a member of the collective group (low); importance of the collective (subjective) sources of data (low) in decision making instead of facts (high); collectivism (low score) combined with a high UAI indicates sensitivity to losing face and fear of taking responsibility.

---

[2]Direct access: `https://hi.hofstede-insights.com/national-culture`.

- MAS: dominance of hard values (high)— money, visible welfare, visible power - vs. soft values (low)— family life over work, importance of leisure time, money not important; in decision-making economic aspects (high) vs. societal consequences (low); deviance from normal accepted (low) or not accepted (high); high MAS combined with low LTO supports short-term hard values in decision making (e.g., discharging employees instead of other adjustments and carrying social responsibility); equality (low) vs. inequality (high) in the society; femininity (low score) correlates to the welfare of the society— see Inglehart-Wenzel map.

- UAI: need for rules and norms (high); merged with high PDI indicates bureaucracy and management by power; resistance to change (high); inability to accept differences and exceptional behavior (high); risk avoidance (high).

- LTO: traditions dominate in decision making (high); ability to wait for profits in business and long-term development of companies (high); short-term positive consequences dominate in decision making (low); high LTO is typically combined with low IDV, high MAS, and high PDI.

- IND: acceptance of visible happiness and joy (high); dominance of norms and rules in everyday life (low); optimistic (high) vs. pessimistic (low) attitude; work (low) vs. leisure (high); importance of religious values (low).

In addition to national cultures, Hofstede has studied the role of national cultures in organizations on both organization and team level.

The Hofstede model provides score values in six dimensions for national cultures. The combination of values provides a means for looking into a *national culture profile*, especially for comparing the differences of selected cultures. The differences recognized and understood support success in multicultural collaboration and communication. Hofstede Insight [13] provides a *Country Comparison Tool*

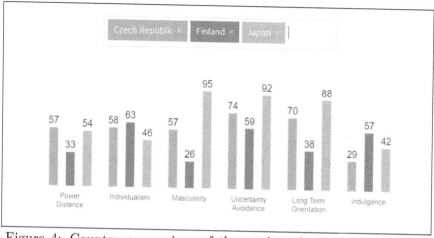

Figure 4: Country comparison of three selected countries, Hofstede model [13].

to conduct such comparisons. A similar tool is also available for mobile devices — Hofstede Insights (former Culture Compass) for both Android and IOS. The resource pages [13, 14] provide a lot of model related readings and reports, which are worth studying.

The respective national cultures of Czech Republic, Finland, and Japan are compared in Figure 4. *Finland* is an individualistic country (IDV) having a democratic work culture (PDI). At work, a low level of supervision is needed (PDI), employees can modify work plans and guidelines (PDI, IDV), and are willing to participate in important decisions (PDI). A high level of equality exists in the society and workplace (MAS). In leadership, open and direct feedback is accepted (PDI, IDV). People will act in a well-organized environment (UAI) and avoid lack of order.

The Czech Republic is also an individualistic country (IDV), in which some level of authoritarian management culture exists. People are used to working in a well-organized environment (UAI) and accept some level of inequality both at work and in society (MAS). In leadership, open feedback is accepted (IDV) but is not desirable (UAI). The level of organizational control over peoples' work is higher than in

Finland (PDI). The long-term orientation (LTO) value is high, which indicates readiness for long-term plans to reach the goal. According to the Inglehart-Wenzel temporal animation, the Czech Republic has been floating around its current location having all the time a high secular value score, first growing self-expression score (closer to the Protestant Europe island), but recently moving backwards slightly on the self-expression score. The nearness to the average European culture base is clear, however.

Both Finland and the Czech Republic have a lot of similarity in their cultures. The author of this paper as well as the inspiration of the publication (Professor Marie Duží) have a long-standing experience collaborating with the Japanese research community. Japan is a strongly collectivist culture (importance of the social group). An authoritarian and multi-layer decision culture dominates; respect for more experienced, older, and wealthier people is visible and present in decision making. Inequality (MAS) is acceptable and exists. Japan is very bound to rules and organizations (UAI, PDI). Conflicting situations with European partners may appear in leadership (fear of losing face), decision making (individualistic vs. hierarchical collective), organizing the work (linear, chunk based), responsibilities, and flexibility at work in relation to rules.

3.3   The Lewis Model

Richard Lewis is a British cross-cultural communication researcher and cosmopolite, who has lived in several countries, including Finland and Japan. He speaks half a dozen languages, and his experience base comes from actions in practice. The *Lewis culture model* is based on an analysis of working patterns, first in the IT branch but gradually extended to cover experiences in organizational cultures in general. The focus of the model is on *communication, organizing the work, handling facts, and data.* Lewis has published details and applications of his model in a wide variety of books; the kernel is the book by Lewis [17] which covers the general aspects of the model. It is supplemented by a country analysis, books on culture and organizations, etc. The resource page [4] provides general information about the model and

Linear-active	Multi-Active	Reactive
introvert	extrovert	introvert
patient	impatient	patient
plans ahead	plans grand outline	looks at general principles
does one thing at a time	does several things at once	reacts
punctual	not punctual	punctual
compartmentalizes activities	one activity influences another	sees whole picture
sticks to plans	changes plans	makes slight changes
sticks to facts	juggles facts	statements are promises
information from official sources	prefers oral information	information from official and oral sources
follows correct procedures	pulls strings	networks
completes action changes	completes human transactions	reacts to partners
likes fixed agendas	interrelates everything	thoughtful
uses memoranda	rarely writes memos	plans slowly
dislikes losing face	has ready excuses	must not lose face

Table 1: Characteristics of cultural stereotypes according to the Lewis model [17, modified].

the CultureActive page [5] has access to detailed material on national cultures.

The Lewis model is based on three basic stereotypes (descriptions cited and modified from the original):

- *Linear-active*: task-oriented, technical competence is important, highly organized societies; level-headed, factual, and decisive planners; fact-based data sources.

- *Multi-active*: extrovert, human force is an inspirational factor, doing many things at one time in an unplanned order, people-oriented; warm, emotional, loquacious, and impulsive.

- *Reactive*: people-oriented, dominated by (wide scale, even subjective) knowledge, patience and quiet control, listening before reacting; courteous, amiable, accommodating, compromiser, and being good listeners.

Examples of typical features of each basic stereotype are listed in Table 1. The model itself is simple: it is a triangle with the basic stereotypes as the corners (Figure 5).

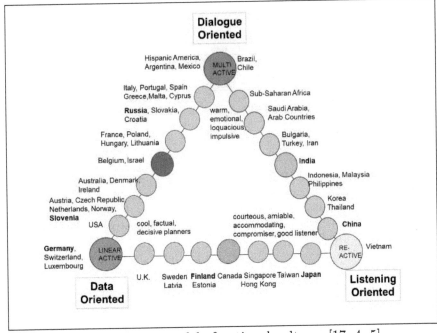

Figure 5: Lewis model of national cultures [17, 4, 5].

The national cultures in the Lewis model are located either at the corners (representatives of the pure basic stereotypes) or on the sides as representatives of mixed cultures. The national culture of *Finland* is strongly linear-active, but it has some features (e.g., listening, silence) of the reactive behavior. The Czech Republic is also a strongly linear-active culture but is located on the side towards the dialogue-oriented culture. Japan (as with many Asian countries) is strongly reactive but has some features of linear-active culture.

A detailed description of each culture is available in the books by Lewis (e.g., [17]) and on the resource page [5], which has unfortunately restricted access (paid registration needed). Every culture description covers general facts about the country, data on the values and core beliefs (iceberg), culture in general (classification and overview), concept of time (organizing the tasks), concept of space, communication pattern, listening habits, and interaction (e.g., behavior in meetings).

The descriptions include text and visualizations.

Based on the Lewis culture analysis, *Finland* is linear-active and data oriented (fact-based sources). **Concept of time**: Tasks are executed in linear order. Finns are punctual; they value good time-keeping; time is divided and used for maximum efficiency. **Listening**: Finns are good listeners (do not speak much) and react when sufficient grounds for a decision exist (ref. Hofstede UAI). **Communication**: Meeting behavior is based on silence, listening, minimal speech, and saying only that which is necessary (UAI). They repeat, summarize, and decide. **Concept of space**: They feel uncomfortable by any attempt to limit their personal space; face-to-face communication distance is at least 1.2 meters; this rule is also followed in bus queues. Equality and personal independence are important.

The Czech Republic is also a linear-active culture, but towards a dialogue-oriented instead of listening culture (Finland). This indicates a lot of similarities but, being "around the corner," also differences between the countries. **Concept of time**: Basic structure is linear, similarly to Finland. Czechs come to appointments on time and tasks are executed in a linear order. **Listening**: Dutiful listeners, polite, and courteous. They do not interrupt, and their feedback is minimal. They do not like digressive discussions. **Communication**: Soft speakers, they communicate in a thoughtful manner. Meetings are prepared internally to anticipate difficulties. They do not show their feelings; their humor is dry and black. Seek compromise, reasoned decisions. **Concept of space**: Hand shaking is mandatory. Distance of comfort is over 1 meter. Bus queues are disciplined and orderly.

Both case countries are ultimately very similar. When approaching the *dialogue-oriented* corner in the triangle, behavioral patterns change. Feelings affect behavior, having visible external implications, dialogues with people and data from unofficial sources replace facts, tasks are executed in no particular order— in reorganized dynamic chunks instead of a predefined order. Even meetings may become chaotic, and the concept of time is fluid. When approaching the *listening-oriented* corner (Japan, for instance), politeness, diplomacy, and harmony beat rationality and sometimes also the truth. *Trust creation* is important— serious collaboration is possible with *trusted partners*. Data sources are rich and also cover subjective sources. Time has value. Typical of collective cultures, in which people are used to living together, short distances are accepted. Organizing tasks is similar to dialogue-oriented cultures but executed in a more organized way.

There are some similarities between the Hofstede and Lewis models, but mainly, because of the different viewpoints, they supplement each other. Whereas the Hofstede model is based on general aspects of a culture, Lewis provides a means for everyday practices in communication and collaboration. Lewis also provides a tool to compare the difference between individual (personal) characteristics and the national generalizations, including sixteen culture related factors.

4 Additional Approaches and Aspects of Culture

4.1 The Hall Model — High and Low Context Cultures

In his book Hall [10]), the American anthropologist Edward Hall introduced an idea about differences in communication between cultures. He defined two cultural classes and the corresponding styles of communication — *high and low context cultures*. This complements the models introduced above, especially from the point of view of verbal and written communication, although it cannot be classified in the category of stereotype models for cultures.

In a high-context culture many things are left unsaid. Messaging is implicit, and less formal information is given in written and spoken form; part of the information is hidden, wordless, "between the lines". Understanding the message is based on internalized understanding of the message context by the target group. This is based on the long-term relationship between the communicating parties. Knowledge is situational and relational, depending on the context that is recognized only by the insiders. In a low-context culture, everything is included in the message and said directly (explicit messaging). People play by

	Low context	High context
Example countries	USA, UK, Canada, Germany, Denmark, Norway	Japan, China, Egypt, Saudi Arabia, France, Italy, Spain
Business outlook	Competitive	Cooperative
Work ethic	Task-oriented	Relationship-oriented
Work style	Individualistic	Team-oriented
Employees desires	Individual achievement	Team achievement
Relationship	Many. looser, short-term	Fewer, tighter, long-term
Decision process	Logical, linear, rule-oriented	Intuitive, relational
Communication	Verbal over non-verbal	None-verbal over verbal
Planning horizons	More explicit, written, formal	More implicit, oral, informal
Sense of time	Prenset/future-oriented	Deep respect for the past
View of change	Change over tradition	Tradition over change
Knowledge	Explicit, conscious	Implicit, not full conscious
Learning	Knowledge is transferable	Knowledge is situational

Table 2: High and low context cultures compared [16].

external rules, known by everybody. Knowledge is codified, public, accessible, and transferable. No additional information is needed to understand the meaning of the message.

Table 2 lists the differences between high and low context cultures. It is based on Gisela Schmalz' (non-published) presentation and modified by the authors of the reference [16]. Looking at the row of example countries, it is easy to find a correlation between this classification to Hofstede's IDV dimension and Lewis' linear-active culture (low context). Both Finland and Czech Republic are low context cultures.

4.2 Adapted and Extended National Culture

Katharina Reinecke is one of the pioneers in studying the consequences caused by staying under the strong influence of a foreign culture. She defines the term "*user's extended national culture*" for the storage of the influence of foreign culture on the (original) national one when a person is living in a foreign culture. Experiences from a foreign culture are gradually adopted as changes and enrichments to the existing cultural patterns; the original national culture remains the kernel, but the outer layer gradually adapts to the features of the foreign

247

Interface aspect	Linked to	Low	Medium	High
Information density	LTO	To-do items provide little information at first sight, requiring a user to click before seeing more information	To-do list shows all information at first sight	Complex version that additionally presents encoded information with big icons
Navigation	PDI	Tree menu and to-dos in a list view, allows nested sorting	Flat navigation and list view, or tree menu and icon-represented to-do list	Flat navigation and icon-represented to-do list
Accessibility of functions	PDI	Functionalities are always accessible but grayed out if not needed	Functionalities appear on mouse-over	Functionalities are always accessible
Guidance	UAI	While users enter a dialog, all other information on the UI retains visible and accessible	Information other than the current dialog is still visible, but inaccessible	Unnecessary information is hidden in order to force users to concentrate on a currently active dialog
Structure	PDI	Minimum structure: Different elements of the UI are only structured through alignment	Elements are separated and each color-coded for better distinction	Maximum structure: Elements are bordered and affiliations between information is accentuated across elements
Colorfulness	IDV	Many different colors	A medium number of colors	The UI is homogeneously colored
Saturation	MAS	Pastel colors with little saturation	Medium saturation and contrast	Highly contrasting, bright colors
Support	UAI	On-site support with the help of short tool-tips	The UI offers question mark buttons that expand into help bubbles	An adaptive wizard that is always visible

Table 3: Hofstede's dimensions and User Interface Preferences [9].

culture as a function of time. The changed and enriched culture is also called the *adapted culture.*

The adaptation has been measured by testing user preferences in the use of information systems. The paper [19] introduces the test environment MOCCA, which provides 39, 366 variations of the same information system. Changes in user preferences are measured over a long time frame in a group of multicultural users. Changes in the user preferences are registered by the system's internal user model, in which the adaptation to a foreign culture is measured in Hofstede dimension scores. In long-term use, the changes in scores indicate the adaptation of the user.

The correlation between UI properties and national cultures are introduced in Table 3. Although the origin of this table comes from the test environment described above, it points out the importance of culture as part of the information system itself, in addition to the information system development, which is based on collaboration between the developers and other interest groups.

Although the work of Reinecke concerns the use of information systems and cultural adaptation of the users, its results can be generalized to include the effects of foreign culture as a change factor even at a general level.

4.3 Non-verbal Signs of Culture and Culture Independent Interaction

Colors (even the extent of the color map) and images may have hidden meanings: their level of relevance and semantics varies between cultures. A discussion of the correlation between colors and emotions is available in several texts[3]. Understanding these differences is important when designing user interfaces, web content, and advertisements. Mental connections to the human mind may also appear in *symbols*. These dependences are stronger in Asian (nature religions and philosophies) and Southern American and European (strongly bound to religion) than in Western liberal cultures. The role of colors as symbols of cultures has been studied by Professor Kiyoki (one of our collaborating parties) and his research group at Keio University, Japan. The semantics of color (emotional experience: 'Kansei' in Japanese) is analyzed by means of semantic computing (based on comparisons of the semantic space of colors and conceptual structure) and has been reported in several papers, e.g., [15]. In Kiyoki's work, colors are also used as a common (intermediate) language to transfer knowledge from one context to another.

In communication, the encoded concept structures are transferred between the communication parties. However, *the formation of concepts and concept structures* varies between cultures, which might cause misunderstanding of the message. The reasons vary — one might be the written language — visual vs. linear writing and read-

---

[3]Color map: [1]. Meanings of colors: [2]; Color psychology: [18]; Color preferences: [22]; Colors in design: [3].

ing order (from left -from right; up-down). Duží addresses this topic in her papers [7] and [11]. She argues the thesis that human communication is based on *procedurally structured abstract concepts* that people learn, execute, and discover. This process plays a central role in human communication in all cultures, although its external manifestation may look different; people tend to build their inner world of thought driven by external expression. To substantiate her thesis, Duží compares three different (visual) communication structures — Egyptian hieroglyphs, pictures, and Inca knot writing (khipu), which has different ways to encode meanings. The "character set" of hieroglyphs consists of 5000 characters, which covers the expressive power of written language of concrete concepts; abstract concepts are left out (at least partially) of the written language, but of course, not from human communication. Khipu (Inca knot-writing) indicates the message bound to it in binary coded form, as units of structured information in a similar way to the computer of today. The coding system cannot be analyzed within the traditional approach to semantics. A third example comes from pictures: a picture is seen as the whole of its parts, in which parts of it are not separable and cannot be interpreted as separate concepts. Based on the analysis of these three very different approaches in communication, Duží draws the conclusion that these very different writing systems encode *procedurally structured concepts*, which consist of a finite number of constituents executable in any possible world at any time. Duží finds similarity between these structures and the key elements of *Transparent Intensional Logic (TIL)*. TIL forms a culture- and language-independent approach for specifying and understanding conceptual structures. In several papers, she has introduced the use of TIL in *Natural Language Processing (NLP)* as a culture-independent intermediate form for communication and messages (concept structures).

TIL is an example of the interpreted formal languages that can be seen as a means for *culture-independent communication*. People are used to communicating in natural languages — using their native language or a commonly accepted foreign language. If an exact transfer of information is needed, natural language is not the best option — it is imprecise and requires long-winded expression to specify the contents exactly. The use of formal or semiformal languages (like UML) is commonly used in a software engineering context, especially in requirements engineering, but also in the transfer of common understanding between different interest groups. The use of standardized messages — a format that consists of predefined components — represents the same approach and is commonly used, for example, in outsourced helpdesk services in the ICT sector.

The paper [11] also covers a discussion of the use of *icons* to support multicultural communication without depending on one specific culture. Generally known icons are traffic signs that are more or less the same around the world. However, culture dependency exists even here. Warnings of kangaroos or koalas are not valid in Finland and the Czech Republic, whereas warnings about moose and deer might be. Heimbürger introduced the concept of the *"intelligent icon (I-icon)"* to support multicultural communication with icons. I-Icons have application domain sensitive areas and can include computational functions such as linking to collections of Web resources, searching and discovering deeper cross-cultural knowledge. The paper also introduces an extension of UML using icons to be used in requirements engineering.

5 Conclusions

Globalization indicates an increasing need for communication and collaboration independently across geographical borders and nationalities (cultures) between the collaborating parties. Even in the academic world, research projects are funded to an increasing degree by international sources and based on collaboration between partners from different countries. The experience base of this paper has reported on three-way collaboration between the Czech Republic, Finland and Japan; in fact, partners from Germany and Indonesia were also part of the collaborating team. The interest in understanding cultural differences inside this multicultural collaborating team triggered the author's culture studies.

The purpose of this paper was to report some findings of the author's work, to provide the means and preparation to survive and
thrive in a multicultural global context. Work having similar goals
has been done by the collaborating partners of the author in the Czech
Republic, Germany and Japan. The paper looked at the progress of
globalization and pointed out, as a *research problem*, the growing requirements needed to act in this continuously changing environment.
The resulting research questions were:

- RQ1: *What are the key elements of (national) culture, and
  are there other dimensions (apart from nationality) to consider?*
  The characteristics of the concept were handled in Section 2.

- RQ2: *Are there any frameworks applicable for understanding
  cultural differences?* Two frameworks and additional aspects
  were introduced in Section 3.

- RQ3: *If there are such frameworks, how to apply these in practice?* The examples of application practices were embedded in
  the discussions throughout the paper.

Naturally, the models introduced only provide support for surviving in
the global multicultural environment. It remains for everybody who
acts in such an environment to decide how to use this information in
a beneficial way.

The author (with collaborating partners from different cultures)
has dealt with this subject in a dozen peer-reviewed publications,
which focus on different aspects of information system development
and usage. While the role of information systems in modern society
has grown, simultaneously the complexity of the systems has grown
(exponentially, in fact), especially in point of development. Instead of
one system the goal is to achieve a network of collaborating systems of
systems, developed for the global market and for an unknown client
base. The implementation work of these systems is made by distributed multicultural teams, they include open-source components
(on a black-box basis) and are distributed over the cloud, without
physical contact to the end user. The key element in these systems

of systems is fluent collaboration between its individual components by communicating via open interfaces in an expected way toward the common goal. The analogy between human collaboration in system development and the integrity of such a complex artifact is clear.

Cultural differences have been discussed in this paper. The difference between two cultures could be defined by the term *cultural difference*. It is an abstract concept and hard to turn into an absolute value. On abstract level it is easy to accept that a bigger cultural difference in the models handled means higher complexity in collaboration and communication. The *Hofstede model* provides scores as a measure in six dimensions. The values are relative, and the scale is not valid for arithmetic comparisons. However, the visualized profile of six dimensions in a country comparison provides some means to see a profile difference between countries. The *Lewis model* indicates that nearby cultures have a shorter cultural distance, and that distance increases remarkably when the partner is on another side of the triangle. In *combined analysis* (integration of analysis results from different sources) visualizations like a *Kiviat graph* would be helpful and provide a two-dimensional culture profile for comparison purposes.

Culture is dynamic — changes happen inside the culture but especially because of external influences. Nowadays data is global, cultural offerings are global, and foreign language studies open a window to foreign cultures. Intercommunication with foreigners is both easy and inevitable. What is the future — will cultures unify or will they remain separate? The latter alternative is more plausible. Simultaneously with globalization progress, national values, the cornerstones of culture-related values, are becoming more important. People want to know their roots, not as a citizen of the world, but as a member of the culture in which their roots lie.

References

[1] Allison R. (2017), Colors represent different things in different cultures. Retrieved from https://blogs.sas.com/content/sastraining/2017/06/29/colors-in-different-cultures/ on April 27th, 2022.

[2] Alida D. (2022), Color Meanings in Different Cultures. Retrieved from https://study.com/academy/lesson/color-meanings-in-different-cultures.html on April 27th, 2022.

[3] Cousins C. (2012). Color and Cultural Design Considerations. Retrieved from https://www.webdesignerdepot.com/2012/06/color-and-cultural-design-considerations/ on April 27th, 2022.

[4] CrossCulture (2022). Know Culture for Better Business. Richard Lewis Communications. Retrieved from https://www.crossculture.com/on April 21st, 2022.

[5] CultureActive (2022). CultureActive— Cultural Assessment and Team Development Tools. Retrieved from https://secure.cultureactive.com/index.html 1stPanel on April 20th, 2022. Access to the key material needs registration.

[6] Dreher, Axel (2006). Does Globalization Affect Growth? Evidence from a new Index of Globalization. Applied Economics 38 (10), pp. 1091-1110.

[7] Duží, Marie (2014). Communication in a Multi-Cultural World. International Journal of Analytical Philosophy, Organon-F 21 (2), pp. 198-218.

[8] ETH Zurich (2022), KOF Globalization Index. Retrieved from https://kof.ethz.ch/en/forecasts-and-indicators/indicators/kof-globalisation-index.html on April 19th , 2022.

[9] Gygli, Savina, Florian Haelg, Niklas Potrafke and Jan-Egbert Sturm (2019). The KOF Globalisation Index— Revisited. Review of International Organizations, 14 (3), pp. 543-574. Retrieved from https://doi.org/10.1007/s11558-019-09344-2 on April 19th, 2022.

[10] Hall, Edward T. (1976). Beyond Culture. Anchor Books, New York.

[11] Heimbürger, A., Duží, M., Kiyoki, Y., Sasaki, S., Khanom, S. (2014). Cross-cultural communication with icons and images. In Tokuda T., Kiyoki Y., Jaakkola H., Yoshida N. (eds), Information Modelling and Knowledge Bases XXV. IOS Press. pp. 306-321.

[12] Hofstede, Geert, Hofstede, Geert Jan, Minkow, Michael (2010). Cultures and Organizations: Software of the Mind: Intercultural Cooperation and Its Importance for Survival. New York: McGraw-Hill.

[13] Hofstede Insights (2022). Compare Countries. Retrieved from https://www.hofstede-insights.com/fi/ product/compare-countries/ on April 20th, 2022.

[14] Hofstede, Geert, Hofstede, Geert Jan (2022). Hofstede Globe. Retrieved

from http://geerthofstede.com/landing-page/ on April 20th, 2022.

[15] Itabashi, Yoshiko, Sasaki, Shiori, Kiyoki, Yasushi (2014). An Explorative Cultural-Image Analyzer for Detection, Visualization, and Comparison of Historical-Color Trends in Thalheim Bernhard, Jaakkola Hannu, Kiyoki Yasushi, Yoshida Naofumi (eds.), Information Modelling and Knowledge Bases XXVI, pp. 152 - 171. IOS Press, Amsterdam. DOI: 10.3233/978-1-61499-472-5-152.

[16] Jaakkola, H. and B. Thalheim (2018). Web Information Systems for High and Low Context Cultures. Information modelling and Knowledge Bases XXIX. Sornlertlamvanich V., P. Chawakitchareon, A. Hansuebsai et al. (eds.). Amsterdam, IOS Press, pp. 299-320.

[17] Lewis, R. D. (2011). When Cultures Collide. Leading Across Cultures. Third Edition. London, Nicholas Brealey International.

[18] Racoma B. (2019), Color Symbolism— Psychology Across Cultures. Retrieved from https://www.webdesignerdepot.com/2012/06/color-and-cultural-design-considerations/on April 27th, 2022.

[19] Reinecke, Katharina, Bernstein, Abraham (2013). Knowing what a user likes: A Design Science approach to interfaces that automatically adapt to culture. MIS Quarterly, 37(2), 427-453.

[20] Statista (2022). Top 50 countries in the Globalization Index 2021. Retrieved from https://www.statista.com/statistics/268168/globalization-index-by-country/ on April 19th, 2022.

[21] Taskinen, Kristian (2020). Kääntyykö globalisaatiokehityksen suunta? [Whether the direction of globalisation trend reverts?]. In Finnish. Statistics Finland. Retrieved from https://www.stat.fi/tietotrendit/artikkelit/2020/kaantyyko-globalisaatiokehityksen-suunta/ on April 19th, 2022.

[22] Viková, Martina, Vik, Michal , Kania, Eva (2015). Cross-cultural variation of color preferences. 21st International Conference Light Světlo 2015. DOI 10.13140/RG.2.1.1882.1601.

[23] WVS (2022). World Values Survey. Retrieved from https://www.worldvaluessurvey.org/ WVSContents.jsp on April 19th, 2022.

ARE DEONTIC MODALS HYPERINTENSIONAL?

VLADIMÍR SVOBODA

Department of Logic, Institute of Philosophy Czech Academy of Sciences, Czech Republic

svobodav@flu.cas.cz

Let me begin with a personal memory. Many years ago I presented a conference paper devoted to the problems of deontic reasoning. As an example of inferential steps that agents involved in deontic language games can make I put forward two model inferences:

IA1   *Mary, feed the sheep and the goats!*
       Mary, feed the goats!

IA2   *Mary, feed the sheep!*
       Mary, feed the sheep or the goats!

I then asked the people in the audience to reflect on their intuition as concerns the validity of the inferences, i.e. I asked them whether the conclusion is, in their view, entailed by the premise.[1]

My work on this paper was supported by grant No.20-18675S of the Czech Science Foundation. I am grateful to Jaroslav Peregrin and Bjørn Jespersen for their valuable critical comments.

[1]It is, of course, somewhat controversial to speak about *inferences* in connection with compounds consisting of imperative sentences. The debates among philosophers and logicians on whether such compounds deserve to be called logical inferences begun with Jørgensen [14]. They are rather complex and persistent. We can, with some simplification, say that those who deny that such compounds deserve to be seen as inferences think that there is nothing like the logic of imperatives (this position is sometimes called the *imperativological skepticism*) while those who are open to the idea that it is worthwhile to study the logic of imperatives can be called the *imperativological optimists*. Non-philosophers, however, usually don't have any problem understanding the point of the question concerning the validity of simple inferences of this kind.

As I expected, nearly all the people present (they were predominantly philosophers) were ready to acknowledge (though some with a bit of hesitation) that the first inference is valid. As concerns the second — IA2 — the audience was more divided. Most members of the audience adhered to the view that the inference is incorrect while some were hesitant to take a clear stand. There was one person, however, who vigorously defended the view that the inference was valid. That person was Marie Duží.

As my intuition was that the inferential step presented in the second inference is rather clearly invalid,[2] I tried to undermine her position. I suggested, that those who imagine themselves in the position of an addressee of the command/instruction "Feed the sheep or the goats!" are very likely to understand it so, that they are allowed to choose which animals they are to feed and as the command "Feed the sheep!" apparently excludes any choice of this sort it cannot entail the choice opening command. I argued that an average lazy person, upon receiving the command "Feed the sheep or the goats!", is likely to consider whether it is easier to feed the sheep or the goats and will then choose to undertake the less demanding task (an assiduous person can, of course, choose to follow the command by feeding both the sheep and the goats). If, for example, there are only two goats on the farm and fifty sheep, a typical addressee is going to react with something like "Well, then I am going to feed the goats" (to make cooperatively clear which of the tasks is going to be completed by them). Marie conceded, sort of, that such an understanding makes sense but she still insisted that the inferential step in IA2 must be correct as the content of the command in the conclusion is logically weaker than the content of the command in the premise. The inferential step, according to her conviction, must be analogous to the step taken from "Mary will feed the sheep" to "Mary will feed the sheep or the goats". I was surprised by how strong her intuition seemed to be and how vigorously she defended it.[3]

---

[2]I share this intuition with large number of scholars — in fact, most of logicians find this kind of inference, which is commonly called Ross paradox, highly problematic. (We will return to Ross paradox later.)

[3]In fact, I shouldn't have been so surprised — vigorousness, as all who know

I remembered this occasion when I, relatively recently, read a paper written by a scholar who can be regarded as a follower of Marie — at least in the sense that she is significantly younger than Marie, they have been in touch for many years and they both tend to look at any issue related to logic through the prism of transparent intensional logic (TIL). The person is Daniela Glavaničová and the paper was her "Hyperintensionality of Deontic Modals: an Argument from Analogy" [11]. It struck me that Glavaničová's perspective on deontic reasoning must be, in its spirit, very similar to Marie's and that the paper perhaps indicates why their approach to (deontic) logic is so different from mine.[4]

In this paper I will try to get a firmer grasp on the kind of outlook that is, I think, behind the specific approach to deontic reasoning which appears to me to be common for the two scholars and perhaps also for other people who insist on hyperintensionality of deontic modals.[5] First, I will devote attention to Glavaničová's argument from analogy which is meant to support the view that deontic modals create hyperintensional contexts. Subsequently, I will focus on Faroldi's argumentation, which shares the same goal. I will try to show that his arguments in favour of hyperintensionality of deontic modals are unconvincing and the logical system which he presents is unsatisfactory. In the last part of the paper, I will turn attention to a more general issue — the different views on the nature and mission of logic. I will formulate hypotheses that are meant to explain why the stances of adherents of TIL towards logic in general and on deontic logic in particular are so disparate from those which I hold. I will distinguish two positions concerning the nature of logic — the Platonist one and

---

Marie can confirm, belongs among the quintessential characteristics of her personality. (In fact, I suspect that Marie is one of the few people who would choose to feed a herd of sheep instead of feeding two goats if she were offered a choice between the tasks — just for the fun of accomplishing a more challenging mission).

[4]In fact, this idea should (or at least could) have occurred to me earlier when I read earlier papers by Glavaničová [8, 9], to which I reacted (see [26]), but I didn't at that point think of the possible relationships between their outlooks.

[5]I should perhaps note that I don't mean to suggest that all 'hyperintensionalists' are bound to share the mentioned Marie's view on the validity of IA2 or that she still sticks to the view which she spontaneously adopted many years ago.

the Protagorean one — and I will explain how their adoption in my view affects our strategies concerning the formation of logical systems and assessing their acceptability and other related qualities.

As I have suggested, the core idea defended in Glavaničová's paper that inspired my deliberations is that deontic modals create hyperintensional contexts. The idea is not brand new but it is of fairly recent vintage. Let's cite what she says:

> "Recently, hyperintensionality of deontic modals (expressions such as it is *obligatory that, it is permitted that, it is forbidden that, (an agent) ought to...*, *(an agent) is allowed to...*, and so forth) has been brought into the focus (cf. [1, 5, 6, 8, 10]. Paradoxes of deontic logic and the failure of substitution of classical equivalents have been enlisted as the main motivation for going hyperintensional in deontic logic." [11, p. 653]

Glavaničová, in her paper, formulates a new argument for the hyperintensionality of deontic modals which "is based on an over-looked analogy between epistemic logic and deontic logic" (ibid).

The initial idea, namely that epistemic modals like *(an agent) knows that...*, *(an agent) believes that...*, *(an agent) excludes that...*, create hyperintensional contexts, is not controversial. Proponents of TIL have devoted a lot of attention to the logical analysis of sentences of this kind (see [4, 18, 29]) and I gladly admit that TIL is probably the most sophisticated available tool for the logical analysis of sentences that express different propositional and notional attitudes, including the epistemic ones.

Compared to the attention that was devoted to the analysis of these kinds of sentences, the attention which adherents of TIL devoted to the analysis of sentences and inferences containing imperatives or deontic modals is negligible. To the best of my knowledge, Pavel Tichý — the creator of TIL — never considered analyzing special logical features of sentences containing the deontic modals (let's call them *deontic sentences*).

He also neglected other sentences which are characteristic of deontic discourse — sentences that articulate commands and prohibitions as well as those sentences that are suited for issuing permissions or granting privileges, i.e. sentences in which phrases like *you may...*, *you are hereby granted the privilege to...* appear. The disregard for imperative sentences among adherents of TIL has relatively clear reasoning behind it. As Duží, Jespersen, and Materna point out in their classical book from 2010, Tichý adopted the view that we don't need a special logic of imperative sentences[6] as their logic is essentially the same as the standard logic of indicative sentences and the difference is not on the semantic but only on the pragmatic level (c.f. [4, p. 351]). The logic of imperatives is thus seen as (at best) parasitic on the logic of indicatives, i.e. on the (traditional) logic of propositions.[7]

The disregard of deontic sentences that lasted until Glavaničová's contributions doesn't have so straightforward an explanation. Deontic sentences clearly have their specific features and questions like: "Does (the proposition expressed by) the sentence[8] *It is forbidden that Tom sells his house* (logically) entail the sentence *It is permitted that Tom doesn't sell his house?*" or "Does the sentence *Tom ought to sell his car or his house* entail the sentence *If Tom won't sell his house he should sell his car or his yacht?*" seem legitimate and worth answering.

There is, I guess, one plausible explanation of the tricky status of sentences of this sort. While we can assume that the sentence *It is forbidden that Tom sells his house* is, after all, i.e. when it is properly completed,[9] true or false when we suppose that the sentence

---

[6]The same holds for interrogative sentences. Still, the pivotal figures of TIL devoted some attention to the analysis of interrogative discourse (see [17, 28]). A comprehensive but in my view somewhat too skeptical discussion of the logic of imperatives can be found in Hansen [12].

[7]The idea that the logic of imperatives is parasitic on the logic of indicatives was articulated by Dubislav in the period which from the present perspective can be called the *prehistory of deontic logic* (see [3]).

[8]I will hereinafter neglect the — undoubtedly important — difference between indicative sentences and propositions which those of them that are meaningful express, and I will simply talk about sentences assuming that such a simplification won't cause any misunderstandings.

[9]We take it, for example, as a shorthand for the sentence *It is forbidden by Canadian law that Thomas Bernard, born 11.9.1948, sells his house in Toronto, at 75 Bolzano Avenue, after April 21, 2022.*

is used within a particular legal discourse, in case of the sentence *Tom ought to sell his house* where the "ought" is interpreted as prudential or moral we can hardly take for granted that the sentence bears a definite truth value. In some cases/contexts we may conclude that the (suitably complemented) sentence is clearly true or false[10] but in many, perhaps even most, cases sentences in which "ought" or a similar deontic expression plays a central role are correctly understood as expressing recommendations or instructions. Recommendations, as well as instructions, can be wise, silly or controversial but they are not true or false — their role is to *guide actions* not to *state facts*, they are *prescriptive* not *descriptive*.

From a logical viewpoint we can (and in a sense have to) choose between three basic strategies concerning deontic sentences: a) we will disregard the prescriptive use/interpretation of deontic sentences and take them as a specific category of statements, b) we will disregard the descriptive use/interpretation of deontic sentences and take them as a specific category of prescriptions (action-guiding sentences) or c) we will decide to take both the descriptively and prescriptively interpreted deontic sentences as legitimate objects of logical study. Let us call strategy a) *the descriptivist strategy*, strategy b) the *prescriptivist strategy* and strategy c) the *two-fold strategy*.

If we adopt strategy c), we open space both for the *logic of deontic statements* and for the *logic of prescriptive judgements*. (We can then take this latter logic as coinciding or overlapping with the *logic of imperatives* or as a different field of study.)

As we can see, logicians interested in deontic discourse must make "strategic" choices that are not easy or, more precisely, which may be quite difficult for those who want to have their conception of deontic logic well justified. Among logicians who were aware of the intricacy of the choices was the main founder of deontic logic, G. H. von Wright, who struggled with them for most of his professional life (see [31, 32, 33, 34, 35]). Many logicians, however, don't perceive the

---

[10]For example when a kibitzer comments on a chess game, saying "If white wants to win she should castle" when castling is the only way for white to prevent check mate.

choices as being too difficult. A number of them tacitly (or perhaps even unconsciously) accept the descriptivist strategy. In this sense, the mainstream of deontic logic, which takes the so-called standard deontic logic (SDL) as a prototypical system (or at least as a benchmark system), can be — if we simplify a bit — classified as descriptivist. [11]

I am not quite sure whether Marie Duží would resolutely adhere to the descriptivist strategy but there are good reasons to ascribe it to Daniela Glavaničová, who clearly treats deontic sentences as truth-bearers in her texts.[12] Adoption of the descriptivist approach is in a sense convenient as we don't need to bother about the so-called Jørgensen's dilemma, which raises the question of whether expressions like imperatives and prescriptively interpreted deontic sentences, which are (according to the prevailing view) not truth-bearers, can be related by the relation of logical consequence. The descriptivist position is, of course, legitimate providing that it is adopted consistently and those who hold it carefully avoid casual digressing to the prescriptive interpretation of deontic sentences (such digressing may, unsurprisingly, be tempting in many contexts).

Let us in the following paragraphs presume that we have adopted the descriptivist approach or the two-fold strategy with a focus on the logic of deontic statements. We can then concentrate on the question of whether deontic modals seen in this way are hyperintensional — whether expressions like *it is obligatory that...*, *it is permitted that...*, *it is forbidden that...*, *(an agent) ought to...*, *(an agent) may ...* create contexts which are like the contexts created by epistemic phrases or whether they contribute to the meanings of the relevant sentences in significantly different ways.

As I have indicated, authors like Glavaničová and Faroldi are convinced that they do create hyperintensional contexts. What are their

---

[11]One should be aware of the fact that the decision to view deontic sentences as descriptive, i.e. true or false (if we accept the law of excluded middle) has important implications which by far not everybody is ready to fully accept.

[12]It is, however, possible that she is open to the adoption of a version of the two-fold strategy. In Glavaničová [10], we come across formulations that indicate that she might be open to accept the two-fold strategy, to this point she has, however, focused only on the logic of deontic statements.

reasons? Besides Glavaničová's argumentation that points to the alleged analogy between epistemic and deontic modal terms, the crucial motives are not difficult to understand. Both authors are convinced that if we consequently treat the deontic modals, resp. the relevant logical constants O, F, and P as we know them from SDL (c.f. [13])[13] as creating hyperintensional contexts, we easily avoid most or perhaps even all the paradoxes which trouble SDL and similar systems of deontic logic.

Such a strategy makes sense: it is quite obvious that if we raise the standards for what can be substituted for what in the contexts of deontic operators (in particular if we disallow substituting of co-extensional terms within the contexts), we significantly weaken the original system and hence it is likely that we won't be forced to accept inferences which seem incorrect or implausible as valid. This kind of strategy, which we might call *better safe than sorry* amounts to putting stress on the natural and highly relevant requirement that logic shouldn't authorize inferences that (according to our intuitions) lead from true premises to a false conclusion.

The strategy, unsurprisingly, also has its limitations: if we put all the weight on this requirement we might conclude that the best (safest) system of logic is one that does not approve any inferences as valid. It is, however, not difficult to see that a logical system of this kind wouldn't be satisfactory as the possible problems just "get swept under the carpet". In fact, a system of this kind would hardly deserve the title "logical".

We thus should aspire to build logical theories that are not only safe but also ambitious. A logical theory should allow us to classify as logically correct as many inferences (arguments) which a) are intuitively correct and b) are such for logical reasons — i.e. their correctness is based on the meaning of the logical constants (resp. on the formal features of the relevant language) as being possible. Only

---

[13]It should be clear that SDL is not standard in the sense that most logicians interested in deontic logic accept it. It is a proper name of a system with which hardly anyone is satisfied but which serves as a kind of common reference point in discussions about deontic logic.

then the theory will be useful.

One can try to put pragmatic requirements like ambitiousness, simplicity, usefulness and user-friendliness aside and simply require that we have to identify the right/correct logical theory — the one that captures logical relations as they really are. This, seemingly natural, requirement however breaks down when we seriously consider the question of how we might get a clear and reliable insight into the realm in which logical relations are distinct and definite. Platonic recollection appears somewhat unreliable and we can also hardly build on something like the mystic visions of those who dare claim that they have a good insight into the realm of logical relations. Thus we are, I am afraid, left with intuitions of fallible humans and with balancing reliability and ambitiousness.[14]

Let us now take a look at how the idea that deontic modals are hyperintensional fares from this perspective. We can start with simple deontic arguments like the following one:

DA1   *No one is allowed to drive any motor vehicle on Bolzano Road.*
    Any electric pick-up is a motor vehicle
    Jim is not allowed to drive an electric pick-up on Bolzano road.

This argument seems straightforwardly correct and we can be quite sure that any policeman who is convinced that its first premise is true would be unwilling to enter into an argument as to whether Jim is, on this basis, forbidden to drive an electric pick-up on Bolzano road or not. But if we insist that the deontic term "allowed" creates a hyperintensional context, there is, as far as I can see, little chance to classify it as correct. The fact exposed in the second premise, namely that any electric pick-up is (actually and perhaps even by necessity) a motor vehicle, is unessential when we cannot "make use of it" in the — allegedly hyperintensional — context created by the deontic terms like "allowed" or "forbidden". Thus we can see that insisting on the claim that deontic contexts are hyperintensional can get us into

---

[14]More about the related issues can be found in Peregrin and Svoboda [21].

trouble if, for example, the police employ a stronger (more ambitious) logic than we do.

Let us consider an analogous argument in which an "epistemic operator" plays a central role, namely:

EA1   *Jim believes that he didn't drive any motor vehicle on Bolzano road.*
      Any electric pick-up is a motor vehicle
      Jim believes that he didn't drive an electric pick-up on Bolzano road.

In this case, I presume, we would have a reasonable chance to convince a policeman that Jim can be, at least in principle, so ignorant that he is not aware of the obvious fact that driving an electric pick-up amounts to driving a motor vehicle and thus the conclusion is not substantiated by the two premises. (Still, this is unlikely to save Jim from a potential fine if he did drive the electric pick-up on Bolzano road as we all know that ignorance, and less so blatant ignorance, is no excuse.)

Let us consider another argument:

DA2   *Jim ought to visit all his cousins*
      Tim is one of Jim's cousins
      Jim ought to visit Tim

I think that hardly anyone would wish to deny that it is an intuitively valid inference. If we try to identify its logical form, the following two options seem plausible:

DAF2
$$O_j \forall x (C(x,j) \rightarrow V(j,x))$$
$$\underline{C(t,j)}$$
$$O_j V(j,t)$$

DAF2*
$$O \forall x (C(x,j) \rightarrow V(j,x))$$
$$\underline{C(t,j)}$$
$$OV(j,t)$$

266

The first one corresponds more closely to DA2 as it straightforwardly speaks about Jim's obligation (which is indicated by the subscript to the operator O). The second corresponds rather to the argument whose first premise reads: *It ought to be the case that if any individual is a cousin of Jim then Jim visits the individual (in the near future).* In any case, it seems reasonable to expect that respectable systems of deontic logic should classify both DAF2 and DAF2* as (logically) valid (if they occur in their language). Nevertheless, it is difficult to imagine how we could recognize their validity if we assume that phrases like "is obliged" or "it is obligatory that", resp. the operators O_j and O, create hyperintensional contexts — the relevant information about the kinship relation can't be easily retrieved for the sake of argumentation.

Let us again compare the argument with its "epistemic analogue":

EA2   *Jim believes that he has visited all his cousins*
        Tim is one of Jim's cousins
        Jim believes that he has visited Tim

I am convinced that most logicians, as well as laymen, would agree that this argument is incorrect and its form, let us say

$$EAF2 \quad \frac{B_j \forall x (C(x,j) \to V(j,x))}{B_j V(j,t)} \quad C(t,j)$$

should be classified as invalid by any reasonable epistemic/doxastic logic.

Let us, finally, consider the following two statements:

DS1   *No one is allowed to draw a square on this blackboard.*

DS2   *Tom may draw a regular quadrilateral on the blackboard.*

It seems obvious that the two statements are incompatible for logical reasons — they should be, if properly analyzed, logically inconsistent. But once again, if we take the phrases ... *is (not) allowed to* ... *and* ... *may*... as forming hyperintensional contexts, it is difficult to see how the formulas that are going to represent the logical structure of the sentences might be shown to be incompatible, notwithstanding that we explicitly adopt a statement like

DS3   *Necessarily, any square is a regular quadrilateral (and vice versa)*

as our premise. This is, once again, an unfortunate effect of the assumption that deontic terms standardly create hyperintensional contexts.

Now it is, I believe, quite clear what I want to suggest: I wish to claim not only that the analogy between epistemic and deontic modal terms on which Glavaničová wants to build her argument is not, by far, as convincing and robust as she supposes. In fact, I want to cast doubt on the very idea that deontic modals create hyperintensional contexts.

The controversial idea that they do is also promoted by Federico Faroldi [6], who not only presents arguments designed to show the failure of conceptions within which deontic modals are conceived as non-hyperintensional (conceptions allowing for the substitution of classical logical equivalents in the contexts created by deontic modals), but he also outlines a theory which is meant to surpass the controversial deontic theories.

Before I turn my attention to Faroldi's ideas, I should perhaps mention a problem that is characteristic not only of Faroldi's account but also of other approaches to deontic logic. I have already suggested that we must (or at least we should) choose how we want to conceive our project when we want to deal with issues covered by the general term "deontic logic". I mentioned that essentially we have three options: a) the *descriptivist strategy*, b) the *prescriptivist strategy*, and c) the *two-fold strategy*. Unfortunately, many authors writing about deontic logic don't clearly say which approach they have chosen. This doesn't seem like a too big problem as we can expect that their choice can be recognized from the way in which they talk, but this is not always the case — often the indications in the texts are conflicting and hence confusing.

This is, unfortunately, also the case with Faroldi. He seems to oscillate between the descriptivist approach and the two-fold approach. He, on the one hand, clearly takes deontic sentences as truth-bearers, but, on the other hand, he apparently takes them as items that have (when suitably used) an action-guiding force — a force which is typical of, for example, proclamations contained in legislative codes. We are thus left in doubt about whether he is just a somewhat careless descriptivist or a special sort of logician adopting the two-fold strategy — one that is convinced that distinguishing between descriptive and prescriptive understanding of deontic sentences is futile, or at least unimportant, because the logical principles guiding their logical behavior are exactly the same (or strictly parallel) in both cases. This kind of view might be called the *both-at-one-blow policy in deontic logic*. I have argued elsewhere that this policy is — no matter whether it is adopted deliberately or unconsciously — a crucial source of confusion in deontic logic (see [27]).

Let me show how the ambiguity which concerns the "strategic approach" to the study affects one of Faroldi's arguments in favour of the view that deontic modals create hyperintensional contexts.

Let us carefully read Faroldi's argumentation. He writes:

"Consider the following:
(1) You ought to drive.
(2) You ought to drive or to drive and drink.

The prejacents of (1) and (2) are logically equivalent, because A is logically equivalent to $A \vee (A \wedge B)$. But (1) and (2) cannot be considered equivalent obligations, as many legal systems scrupulously remind us. The fact that driving and [driving or (driving and drinking)] have the same truth value in all situations is not enough to make them interchangeable in ought contexts, preserving all normatively relevant features. Besides these intuitive considerations, where these two obligations are equivalent, we should take normative systems to be either largely irrational or very bad at guiding action effectively, since driving and drinking would be an acceptable (even required) course of action to satisfy one's obligation to drive." [6, p. 389]

From what Faroldi says here, one would guess that he either accepts the prescriptivist strategy or he is an adherent of the two-fold strategy focused on the logic of prescriptive judgments — the two sentences he mentions in his example are clearly understood as "guiding actions", i.e. prescriptive. But when he later begins to use a formal apparatus we can see that he works with sentences of this kind as if they express declarative statements which are either true or false. The best explanation for this seems to be that he accepts the *both-at-one-blow policy* and this allows him to freely oscillate between viewing one and the same sentences once as expressing an action-guiding prescription and another time as a deontic statement. But if he does this, he should declare this openly and defend the chosen approach as it is far from uncontroversial.

But let us pass over this issue and focus on assessing how convincing Faroldi's argument is. He points out that "driving and [driving or (driving and drinking)] have the same truth value in all situations". This claim, I am afraid, does not make much sense — driving (similarly as swimming, sitting or sleeping) clearly does not have any truth value. Types of actions, resp. generic acts, simply don't have any truth values. This point was made already in von Wright's seminal paper *Deontic Logic* [31].

If we disregard this problem and suppose that Faroldi's "prejacents" are more appropriately expressed by the sentences *You drive* and *You drive or you drive and you drink* we will easily see that the problem which he points out is only a variant of the well-known *Ross paradox*,[15] which indicates that disjunction in imperative sentences (or

---

[15]In his paper from 1941, Alf Ross turned his attention to the implausibility of the inference from *Mail the letter!* to *Mail the letter or burn it!*. (To be accurate — Ross says that "from: slip the letter into the letter-box! we may infer, slip the letter into the letter-box or burn it!" (p. 62) which is, in his view, obviously a wrong step.) The inference turns out to be correct if we adopt Dubislav's convention mentioned above. (It is worth noting that Ross takes sentences occurring in the inference in question as action-guiding, no matter what grammatical form they have (see [24]).

more generally in action-guiding sentences) behaves differently than in ordinary declarative statements. There are, however, many logical theories that more or less successfully deal with the problem without involving means so "heavy-weight" as hyperintensionality.

If we consider the problem of the substitution of (classical) logical equivalents within prescriptive sentences without a prejudice we will see that the substitution is not controversial if we avoid using "or". The substitution may yield somewhat strange-sounding sentences but the arguments don't seem wrong at all. Let us consider for example the following equivalents of Faroldi's (1):

3. *You ought to drive and (to) drive and (to) drive.*

4. *You ought to drive and if you don't drive then [you ought to] drive.*

Both the sentences which are formed analogously to Faroldi's (2) sound somewhat strange but they appear to say/require the same thing as (1) and thus the substitution step does not seem too controversial and still much less evidently wrong.

Another strange thing about Faroldi's argument is his surprising conclusion that "driving and drinking would be an acceptable (even required) course of action to satisfy one's obligation to drive". I don't understand what Faroldi is after here. If I have an obligation to drive, I can satisfy it by a complex action that consists in driving and singing or in driving and smoking or in driving and drinking (either non-alcoholic or alcoholic beverages), etc. Nothing, of course, guarantees that all of the combinations of actions are acceptable — if I am obliged to avoid smoking then driving and smoking is clearly an unacceptable course of action[16] but this seems quite uncontroversial. Similarly unproblematic is the case with driving and drinking: driving and drinking water is fine unless this combination of actions is forbidden by the relevant code and the situation is quite analogous when by 'drinking' we mean 'drinking alcoholic beverages'.

From what I have said it is, I believe, clear that I find the argument of Faroldi's that I have just outlined above utterly unconvincing. His

---

[16]I can't see what might make one think that it is "even required".

next argument is quite bizarre as it concerns the biblical Eve eating infinitely many apples, and I will skip it. His third argument invites us to consider two deontic sentences

5. *It ought to be the case that the pope shakes hands with Shakira.*

6. *It ought to be the case that Shakira shakes hands with the pope.*

complemented by an assumption that "it is the case that the pope shakes hands with Shakira if it is the case that Shakira shakes hands with the pope" [6, p. 390]. Faroldi then emphatically states that if "ought" were to be analyzed non-hyperintensionally, we would have to conclude that the two deontic sentences come out as equivalent. But this, in his view, is patently false.

Once again I don't understand the gist of his argument. The *requirement* suggesting that the relevant event of shaking hands is something that ought to take place can be, in my view, expressed by both the sentences and in this sense the sentences are equivalent. It is, in my view perfectly natural to understand (5) and (6) as sentences *describing* (correctly or incorrectly) a certain deontic situation and they are, I believe, true in exactly the same situations.

Faroldi disagrees. He points out that the pope might have a (shaking) obligation towards Shakira, without it being the case that Shakira has a (shaking) obligation towards the pope. I gladly agree with his point, but I think that it doesn't concern his example and so it is not relevant. Neither (5) nor (6) speak about an obligation of a certain person. Such obligations would be properly expressed by sentences like *Shakira ought to (see to it that she) shake(s) hands with the pope.* But then it is clear that no substitution within the reach of the deontic modal (which might be, as I have indicated, symbolically expressed by indexing the deontic operator — let say Os) can turn Shakira's obligation into an obligation of the pope, and hence Faroldi's argumentation fails.

Even if Faroldi's arguments in favour of hyperintensionality of deontic modals are unconvincing, one might be ready to admit that if the system of hyperintensional deontic logic that he proposes (HDL) satisfies the crucial requirements: i) it is not affected by paradoxes, and at the same time ii) it is ambitious/fruitful enough (i.e. allows us to classify inferences that are intuitively logically valid as valid according to the system), then his point is generally convincing. And Faroldi apparently believes that the propositional deontic logic that he proposes is satisfactory or at least a promising candidate for a respectable system of deontic logic. He briefly shows that HDL does not validate some paradoxical inferences that affect SDL. He, however, doesn't try to convince the reader that this success is not achieved at the expense of the ambitiousness/fruitfulness of his logic. This is, as I have suggested, a natural worry that arises when we "go hyperintensional". In the case of a logical system like HDL, which contains as its *only* rule the rule of the substitution of hyperintensional equivalents, such a worry is surely more than well substantiated.

I don't want to subject HDL to a complex critique here. I will only outline two critical points. The first one I have already mentioned. The language of HDL is a propositional language which allows for "mixed" formulas like $A \vee OB$ or $\neg OA \to B$, i.e. it allows for expressing sentences like *Ostrava is the capital of Moravia or Bjørn ought to drive* or *If it is not obligatory that Bjørn fasts then (it is true that) Pavel is a philosopher.*[17] It is quite obvious that language of this kind is not suitable for the formalization of sentences that are action-guiding. The deontic sentences expressible in this language are either true or false, otherwise they couldn't be connected by classical connectives with sentences expressing common statements.

If we appreciate this, it is obvious that Faroldi's system is quite detached from his argumentation in which he clearly talks about prescriptively interpreted (action-guiding) sentences. Faroldi might perhaps connect the system and the argumentation if he successfully defends the acceptability (or better appropriateness) of a version of the "both-at-one-blow policy", but he doesn't try to do that.

The second critical point concerns his axiom

---

[17]HDL doesn't contain the connective of implication, but I assume that it can be defined in the usual way. Of course, using material implication in the formalization of sentences about conditional obligations is associated with a number of problems, and it is hard to believe that at least some of them don't affect the plausibility of HDL.

13. $O(A) \vee O(B) \approx_H O(A \vee B)$.[18]

The fact that this formula is valid straightforwardly suggests that, for example, the sentence *Jim ought to mail the letter or Jim ought to burn the letter* is, as far as its meaning goes, equivalent to the sentence *Jim ought to mail the letter or burn it*. The step from the left side to the right one is, as we have seen above, quite controversial — it is a version of Ross' paradox. I don't want to exclude that the reasonability of such a step might be plausibly defended, but the defense is far from straightforward.

Unfortunately, the right-to-left direction is controversial too. It apparently validates, for example, the inference from *Jim ought to buy oranges or tangerines* to *Jim ought to buy oranges or Jim ought to buy tangerines*. Intuitively, the first sentence suggests that Jim is obliged to buy some fruits, specifically oranges or tangerines, while it is not specified which of the two — he is apparently in a situation where he can choose. The second, which is supposedly equivalent, nevertheless quite clearly suggests that there is no room for choice — Jim is obliged to buy a specific kind of fruit, we just don't learn from the sentence whether he ought/is obliged to buy oranges or he ought/is obliged to buy tangerines. In any case, the sentence instantiating the left side of axiom 13 is true only if Jim is not free to choose which fruits to buy.

Once again, it is perhaps possible to defend the step in some way, but then we can't help but recognize that our language is unsuited to appropriately capture the common situation which arises when the addressee/subject of an obligation can choose in which way he will comply with a command. (We have come across this situation in the very beginning in connection with the command *Mary, feed the sheep or the goats!*) A language that is incapable of describing such a common type of deontic situation is, I am afraid, substantially flawed. So I dare to conclude that the consequences of axiom 13 are unacceptable.

I don't want to categorically exclude the possibility of building a satisfactory system of deontic logic in which deontic operators are

---

[18]The symbol '$\approx_H$' represents hyperintensional equivalence (see [6, p. 396].

treated as creating hyperintensional contexts but I don't think that Faroldi has succeeded in completing this project. The argumentation he presents is, in my view, unconvincing, and he has not succeeded in either of his two tasks, namely a) to subvert the plausibility of non-hyperintensional deontic theories and b) to build a satisfactory deontic logic. As I have indicated, I am not very optimistic about the prospects of building an acceptable logical system of this sort — if we "go hyperintensional" in deontic logic, the resulting theory is likely to be safe (providing that we build it carefully), so it will not approve as valid any argument forms which are instantiated by incorrect arguments but it will, I am afraid, hardly be useful.

Let me close this paper with a general consideration that turns attention to the ways in which our philosophical positions affect our perception and evaluation of logical systems in general and systems of deontic logic in particular. I assume that Marie and Daniela share a general view of the nature of logic which, quite naturally, determines also their outlook on deontic logic. This is the picture according to which logic "investigates logical objects and ways they can be con- structed" and according to which findings of logic "apply regardless of what people do with those objects: whether they exploit them in asserting, desiring, commanding, or questioning" Tichý, [1978, p. 278; 2004, p. 298].[19]

My hypothesis is that it is exactly this picture of logic that, if taken as seriously as Marie tends to take it, almost inevitably leads its advocates to take a stand that is quite unfavourable to the speci- ficities of reasoning that involves sentences characteristic of deontic and in particularly prescriptive discourse. Her principled approach thus doesn't allow her to accept certain inferential steps that com- mon language users consider clearly correct and the correctness of which can be recognized on the level of the form, as truly logical. If

---

[19]In the passage, Tichý refers to Fitch, who claims: "[W]e do not need a special 'logic of imperative statements', 'logic of performative statements', and so on, as logic over and beyond, or basically different from the standard logic of propositions" (see [7, p. 40]). I should perhaps mention that I find the concept of imperative statement difficult to comprehend, but I won't go into the details here.

we insist that on the semantic level there is no difference between the sentences *Jim drives Tim's car, Does Jim drive Tim's car?* and *Jim, drive Tim's car!*, and all the difference in their meaning is relegated from the sphere of semantics to the sphere of pragmatics, then there is no space for admitting that the individual sentences might occupy different places in the relevant "logical webs".[20] There is also no space left for admitting that *ought* sentences interpreted prescriptively or imperative sentences might require a specific logical analysis that would reflect their specific features.

The position of Marie Duží (as I conceive it) is rigorous: If we have made some discovery concerning logical principles, then we have to bite the bullet and stick with it even if it appears counterintuitive in a specific case. This "obligation" stems from the outlook within which a) logic is meant to reveal the true logical reality as it is and b) all logically relevant features of sentences are reduced to their (narrowly conceived) semantic characteristics.[21] From such a perspective it is natural to insist that if *Mary feeds the goats or the sheep* is entailed by *Mary feeds the sheep* we must bite the bullet and accept that the prescription *Mary feed the goats or the sheep!* is entailed by *Mary feed the sheep!* (providing that the sentences are at all logically interconnected).

The position of Daniela Glavaničová is in my view somewhat less

---

[20]That there are such specific webs is, in my view, convincingly demonstrated in a number of publications that focus on the logical features of questions (see e.g. [2]) or on the logical features of imperatives (see e.g. Vranas [36, 37]).

[21]When Pavel Materna cooperated with A. Svoboda and K. Pala in the 1970s he flirted with the idea of distinguishing between the *external pragmatics* (including the speaker, the context of her utterance etc) and the *internal pragmatics* which "operates within language 'dead', or in other words, constitutes part of a language user's apparatus preconditioning the actual use of a given language" [25, p. 208]. (The phrase "language 'dead' " apparently refers to a language seen as a system lacking any dynamics.) From such a perspective the difference between the three sentences about Jim and Tim belongs to internal pragmatics. Admittedly, the borderline between semantics and internal pragmatics is, to a large extent, optional. From such a perspective, the differences connected with the grammatical mood of sentences might turn out to be logically relevant. This suggests that Tichý's (Fitch's) conception of the logically relevant features of our language is by far not the only possible one even from the perspective of an adherent of TIL.

rigorous — she obviously takes into account intuitions that are associated specifically with employing sentences for prescribing and in this way she implicitly takes them as relevant from a logical point of view. In this way, she apparently diverges from the standard TIL position according to which mood is a matter of pragmatics (and hence irrelevant for logical considerations). She, however, seems to be ready to sacrifice ambitiousness/fruitfulness of a logical system when its safety/reliability might be endangered. It is, I guess, the main reason why she is ready to adopt the unfortunate consequences associated with conceiving deontic modals as creating hyperintensional contexts.

My position as concerns the nature of logic is quite different from the one favoured by Marie and Daniela. In my view, there is no fixed logical reality to be described. Still, logic is concerned with a kind of reality — with the, by its very nature fuzzy/indistinct, reality of our common normatively established practices that are called argumentation, (overt) reasoning or proving. Their indistinctness is, of course, connected with the fact that their natural media are natural languages.[22] Any nontrivial language, be it natural or artificial, is by its nature a "growth medium" for such practices, and being a competent speaker of the language involves (some would perhaps say amounts to) the ability to recognize (though not infallibly) which simple steps in reasoning are correct, which are incorrect and which are somewhere in between (almost always correct, correct or incorrect depending on a given context, strictly speaking incorrect, etc). The ambition of logic is to identify those kinds of inferential steps which are (or can be) employed no matter what we talk about (i.e. which are independent of a particular discourse topic), "extract" them from the language (the extraction involves "purifying" and schematizing) and systematize them.[23]

Unsurprisingly, the process of extracting, purifying and systematizing may result in approving steps that are controversial. If they

---

[22]We may speculate about whether these practices are characteristic only of the reasoning of us humans or of any possible reasoning (e.g. reasoning practiced by extraterrestrial civilizations), but such speculations are likely to be sterile as we, I am afraid, cannot look at our concept of reasoning from the outside.

[23]More about the principles of such processes can be found in [21].

are too controversial then they are likely to subvert the proposed logical system which approves them. If they are controversial only to some extent we may decide to tolerate such a minor discrepancy between our intuition and our systematization and adopt the system in spite of its imperfection as deserving the label "logic" without any provisos.[24]

The history of deontic logic can serve as a paradigmatic example of different attempts at suitable extracting, purifying and systematizing. Most of the theories proposed during the last century failed because the controversial inferences approved by the theories turned out to be unacceptable for (most of) the community of involved logicians.[25] Others have failed and may fail because the resulting theory is too weak — doesn't allow us to classify inferences that seem clearly correct for logical reasons as correct within the theory.[26] One can, of course, presume that there is a "genuine" deontic logic to be discovered. Such a conviction is safe in the sense that there is no way in which it might be convincingly refuted. One can only insist that those who adopt it try to make clear how can we "measure" the *appropriate-likeness* (an analogue of truth-likeness or verisimilitude of scientific theories) of the individual theories and how we might recognize that we have identified the one and perfect (deontic) logic.

Marie Duží, as far as I know, strongly holds to the "realistic" outlook on logic just outlined, and I believe that she would keep holding to it if she turned her attention to deontic logic — she would, I believe, tend to think that there is a genuine deontic logic to be captured by our theory. I also presume that she would be unwilling to admit that it might be reasonable to distinguish between the logic of *deontic*

---

[24]For example, many logicians are ready to tolerate 'the paradoxes of material implication', as the systematization provided by classical propositional logic is so valuable, elegant and useful for various purposes. Others, of course, may find the paradoxes intolerable and search for another systematization — see, e.g., already [15].

[25]A good example is the very first modern system of deontic logic, proposed by Ernst Mally (see [16]). Mally didn't foresee all the unacceptable consequences which stemmed from his axioms, which seemed to him to be unquestionable.

[26]This is, in my view, likely to be the fate of Glavaničová's Δ-TIL or Faroldi's HDL.

statements (descriptively interpreted deontic sentences) and *deontic judgments* (prescriptively interpreted deontic sentences). I am, however, not sure how she would react if asked to provide criteria for measuring the appropriate-likeness of the individual theories. (Perhaps she will take a stand on these issues in the future.)

I suppose that Daniela Glavaničová, as a devoted follower of Pavel Tichý's legacy, also tends to lean towards the "realistic" outlook with respect to logic (including deontic logic). As in Marie's case, I am uncertain what would be her attitude toward the "appropriate-likeness measuring task". I only know that she believes that building deontic logic is a project worth devoting energy to — she has proved it by her activities in this field.

I also believe that the three of us don't differ (at least not too much) as for the criteria that we tend to almost automatically employ for assessing which systems of deontic logic are better and which are worse. Still, there is a principled difference between the conception of logic shared by Marie and Daniela which can be called Platonic, and the one that I have defended for many years (often together with Jaroslav Peregrin, who has adhered to it for most of his professional career)[27] which might be perhaps called Protagorean as it suggests a specific kind of man-as-measure doctrine for logic. Within this doctrine, logic is primarily a tool that we — humans — create and use to achieve different goals related to our communication.[28]

Frankly, I tend to think that even those who adhere to conceptions that are overtly radically Platonic tend, after all, to measure the "logicality" of individual logical theories as well as their comparable acceptability by pragmatic human measures which are based on common considerations. Good logical theory should be consistent, safe (free of counterintuitive consequences), fruitful, useful, sophisticated and yet elegant. Those who adopt the Protagorean conception of logic have the advantage that they can openly adjust their criteria of logicality so that they reflect the purpose of building individual logical theories. A Protagorean logician can openly do trade-offs be-

---

[27]See Peregrin [19, 20].

[28]We have defended this picture in Peregrin and Svoboda [22, 23].

tween various "qualities" of logical theories: sometimes (relative to some purposes), she can take as a crucial quality of a logical theory the extent to which it allows for fine-grained analysis, in other cases she can put stress on simplicity and user-friendliness. In one case she can, for pragmatic reasons, give preference to the safety/reliability of the theory in question and reprobate a system that has only slightly counterintuitive consequences, while in the other case she may highly value inferential fruitfulness of the logical theory which she considers and assesses.

The defenders of the Platonic conception are, I am afraid, determined by their creed to search for *the* genuine logic or, if they are pluralists, for *the* genuine logics. They cannot openly say: "The system that I propose is not a true deontic logic — I am sure that it is not appropriate in the sense of capturing the (relevant) real deontic logic, but I propose that you take it as such because it may be useful". (Well, they can do this, but they would have to feel awkward doing so. Or at least they should feel awkward.)

As I have suggested, my impression is that scholars like Duží, Glavaničová (and probably also Faroldi) prefer safety and fine-grainedness to ambitiousness and inferential fruitfulness. This is, in a way, natural — if you aspire to capture the logical relations as they truly are in a realm that we can call the *third realm* or the *Platonic realm* then the worst mistake one can make while building a logical system is that the theory will approve some inferential relations that do not exist in this special realm. In such a case, the theory is simply wrong and hence it is only a failed attempt at building a logic. If one, on the other hand, presents a logical theory that does not capture all the relations but identifies at least some of them correctly then the theory is not perfect but it is not wrong in any strong sense, it is only not as good/perfect as it might/should be.

Also, if one assumes that we should aim at *revealing/discovering* the true logical structure which certain sentences of natural language exhibit, then their natural ambition is to capture the structure as it is — in its full richness/fine-grainedness. This then naturally leads one to prefer as an analytic tool a logic that allows for a really fine-grained, and ideally exhaustive, analysis, even though it may not be user-friendly. Practicality and user-friendliness are likely to be inessential to a dedicated logical Platonist. If the reality of the realm of logical relations is complex and difficult to untangle, our logic should be correspondingly complex.

If, on the other hand, one views logical theories as a human tool — a tool that is primarily designed to allow for a clearer formulation of our assertions (or of our instructions) and for minimizing the space for mutual misunderstandings everywhere where clarity and understanding are of utmost importance — one is likely to be more open to introducing convenient simplifications when this enhances the usefulness of the theory. Such an approach also leads one to highly value the fruitfulness of the theory as one always has in mind that it should be useful as *our* tool (and possible confusion can be prevented by introducing explicit provisos of which everyone is then aware).

I, however, don't want to say that adhering to a version of logical Platonism has only disadvantages. Adoption of certain philosophically grounded principles and limitations can be challenging and inspiring and also the quest for unity and comprehensiveness which is (or should be) characteristic of Platonist conceptions can catalyze the formation of worthwhile and insightful logical theories. However, I think that the cons of the position prevail over its pros. Moreover, I believe that deontic logic is the area in which we are in danger of paying particularly dearly for our preconceptions, no matter how well they fare in different areas of logical research.

References

[1] Anglberger, A. J. J., Faroldi, F. L. G. & Korbmacher, J. (2016): "An Exact Truthmaker Semantics for Permission and Obligation". In: O. Roy — A. Tamminga — M. Willer (eds), *Deontic Logic and Normative Systems: 13th International Conference, DEON 2016, Bayreuth, Germany, July 18 – 21, 2016.* London: College Publications, 16–31.
[2] Ciardelli, I., Groenendijk, J., and Roelofsen, F. (2019): *Inquisitive Semantics.* Oxford University Press.

[3] Dubislav, W. (1938): "Zur Unbegrundbarkeit der Forderungssätze". *Theoria* 3, 330–342.

[4] Duží, M., Jespersen, B., and Materna, P. (2010): *Procedural Semantics for Hyperintensional Logic. Foundations and Applications of Transparent Intensional Logic.* Berlin: Springer Series Logic, Epistemology, and the Unity of Science 17.

[5] Faroldi, F. L. G. (2019): *Hyperintensionality and Normativity.* Cham: Springer.

[6] Faroldi, F. L. G. (2019): "Deontic Modals and Hyperintensionality". *Logic Journal of the IGPL* 27 (4), 387–410.

[7] Fitch, F. (1971): "On kinds of utterances". In: P. Kurtz (ed.), *Language and Human Nature*, St. Louis: Waren H. Green.

[8] Glavaničová, D. (2015): "K analýze deontických modalít v Transparentnej intenzionálnej logike". *Organon* F 22 (2), 211–228.

[9] Glavaničová, D. (2016): "Δ-TIL and Normative Systems". *Organon* F 23 (2), 2016, 204–221.

[10] Glavaničová, D. (2017): "In Defence of Δ-TIL". *Organon* F 24 (1), 105–113.

[11] Glavaničová, D. (2019): "Hyperintensionality of Deontic Modals: An Argument from Analogy". *Filozofia* 74 (8), 652-662.

[12] Hansen, J. (2013): "Imperative Logic and Its Problems". In: Gabbay et.al. (eds.),*Handbook of Deontic Logic and Normative Systems*, College Publications, 137-192.

[13] Hilpinen, R. (ed.) (1971): *Deontic Logic: Introductory and Systematic Readings.* D. Reidel, Dordrecht.

[14] Jørgensen, J. (1937–1938): "Imperatives and Logic", *Erkenntnis* 7, 288–296.

[15] Lewis, C. I. (1918): *A Survey of Symbolic Logic*, University of California Press, Berkeley.

[16] Mally, E. (1926): *Grundgesetze des Sollens. Elemente der Logik des Willens*, Graz: Leuschner & Lubensky. Reprinted in Ernst Mally: *Logische Schriften. Großes Logikfragment — Grundgesetze des Sollens*, K. Wolf, P. Weingartner (eds.), Dordrecht: Reidel, 1971, 227–324.

[17] Materna, P. (1981): "Question-like and non-question-like imperative sentences". *Linguistics and Philosophy* 4, 393–404.

[18] Materna, P. (1998): Concepts and Objects. *Acta Philosophica Fennica*, vol. 63, Helsinki.

[19] Peregrin, J. (1995): *Doing Worlds with Words.* Dordrecht: Kluwer.

[20] Peregrin, J. (2019): *Philosophy of Logical Systems,* New York: Routledge.

[21] Peregrin, J., and Svoboda, V. (2017): *Reflective Equilibrium and the Principles of Logical Analysis: Understanding the Laws of Logic,* New York, Routledge.

[22] Peregrin, J., and Svoboda, V. (2021): "Moderate anti-exceptionalism and earthborn logic". *Synthese* 199, 8781–8806.

[23] Peregrin, J., and Svoboda, V. (2022): "Logica Dominans vs. Logica Serviens". *Logic and Logical Philosophy,* Special Issue: Logics and Their Interpretations. H. Antunes — D. Szmuc (eds.) Vol. 31 No. 2, 183-207.

[24] Ross, A. (1941): "Imperatives and Logic". *Theoria* 7, 53–71 (reprinted in: A. Ross, "Imperatives and Logic". Philosophy of Science 11, 1944).

[25] Svoboda, A. (1976): "Apropos of internal pragmatics". *Brno Studies in English,* 12, 187–224.

[26] Svoboda, V. (2016): "Δ-TIL and Problems of Deontic Logic". *Organon F* 23 (4), 539–550.

[27] Svoboda, V. (2018): "A Lewisian taxonomy for deontic logic". *Synthese* 195, 3241–3266.

[28] Tichý, P. (1978): "Questions, answers, and logic". *American Philosophical Quarterly* 15, 275–284. Reprinted in (Tichý 2004: 293—304).

[29] Tichý, P. (1988): *The Foundations of Frege's Logic.* De Gruyter, Berlin — New York.

[30] Tichý, P. (2004): *Collected Papers in Logic and Philosophy.* V. Svoboda, B. Jespersen, and C. Cheyne (eds.), Filosofia, Prague and University of Otago Press, Dunedin.

[31] von Wright, G. H. (1951): "Deontic Logic". *Mind* 60, 1–15.

[32] von Wright, G. H. (1969): "On the Logic and Ontology of Norms". In: J. W. Davis — D. J. Hockney — W. K. Wilson — D. Reidel (eds), *Philosophical Logic,* Dordrecht.

[33] von Wright, G. H. (1971): "A New System of Deontic Logic". In: R. Hilpinen (ed.) (1971), 105–120.

[34] von Wright, G. H. (1985): *A Pilgrim's Progress, Philosophers on Their Own Work* vol. 12. Peter Lang, Bern.

[35] von Wright, G. H. (1991): "Is there a Logic of Norms?". *Ratio Juris* 4, 265–283.

[36] Vranas, P. B. M. (2011): "New Foundations for Imperative Logic: Pure

Imperative Inference". *Mind* 120, 369–446.

[37] Vranas, P. B. M. (2016): "New foundations for imperative logic III: A general definition of argument validity". *Synthese* 193 (6): 1703-1753.

Currying Order and Restricted Algorithmic Beta-Conversion in Type Theory of Acyclic Recursion

Roussanka Loukanova
Institute of Mathematics and Informatics,
Bulgarian Academy of Sciences, Sofia, Bulgaria
rloukanova@gmail.com

Abstract

In this paper, I introduce the type-theory of acyclic recursion. I present some of its algorithmic expressiveness to represent semantic phenomena that has not been handled adequately by Montague's IL. I investigate some of the differences between denotational and algorithmic semantics in the formal language of Moschovakis type theory of acyclic recursion (TTAR). I show that, while denotational beta-conversion holds in TTAR, algorithmic beta-conversion is not valid in TTAR, in general. A limited, restricted version of algorithmic beta-conversion is valid in TTAR.

1 Introduction

The formal theory L_{ar}^{λ} used in the technique of syntax-semantics interface in this paper is based on development of the theories of recursion introduced by Moschovakis [24, 25]. The formal languages of full recursion include terms constructed by adding a recursion operator along with the typical λ-abstraction and application. The resulting theories serve as a powerful, computational formalisation of the abstract, mathematical notion of algorithm with full recursion, which, while operating over untyped functions and other entities, can lead to calculations without termination. The languages of full recursion were then extended to a higher order theory of acyclic recursion L_{ar}^λ, see Moschovakis [26]. The formal language L_{ar}^λ is more expressive, in sense of adding higher-order types and typed terms denoting typed algorithms restricted to computations that close-off. I.e., the class of languages L_{ar}^λ, and their corresponding calculi, represent abstract, functional operations (algorithms) that terminate after finite number of computational steps. Moschovakis [26] introduced higher-order, typed, acyclic recursion by targeting algorithmic semantics of human language. In this paper, I introduce the theory of L_{ar}^λ, with an extended calculus, by presenting possibilities for computational syntax-semantics interfaces, in computational grammars of formal and human languages.

Detailed introduction to the formal language L_{ar}^λ of Moschovakis acyclic recursion, its syntax, denotational and algorithmic semantics, and its theory, is given in Moschovakis [26] and Loukanova [20]. The formal theory L_{ar}^λ is a higher-order type theory, which is a proper extension of Gallin's TY_2, see Gallin [6], and thus, of Montague's Intensional Logic (IL), see Montague [29].

Denotational and Algorithmic Semantics The denotational semantics of L_{ar}^λ follows the structure of the L_{ar}^λ-terms, in a compositional way. The reduction calculi of L_{ar}^λ, including the extended γ^*- and γ-reduction calculi, of terms to canonical forms, play the key role in the algorithmic semantics of L_{ar}^λ, see Loukanova [21, 20].

The algorithm for computing the denotation $\mathsf{den}(A)$ of a meaningful L_{ar}^λ-term A, is determined by its canonical form[1]:

$$\mathsf{cf}(A) \equiv A_0 \ \text{where} \ \{p_1 := A_1, \ldots, p_n := A_n\} \tag{1}$$

Intuitively, the denotation $\mathsf{den}(A)$ is computed recursively, by computing $\mathsf{den}(A_i)$ and saving it in the recursion variable p_i, step-by-step, according to recursive ranking. In this paper, I use the ex-

---

[1]The symbol "$\equiv$" is a meta-symbol, which is external, i.e., not in the vocabulary of L_{ar}^λ. We use it to specify literal identity between expressions of L_{ar}^λ and for definitional notations.

tended γ-reduction, which uses the (γ)-rule, formulated in the next sections. For more details and on mathematical properties of L_{ar}^λ, see Loukanova [20].

1.1 Related Work

Musken [27] introduced Relational Montague Grammar, which is a significant generalisation of Montague's PTQ, based on Musken type theory TT_2, an improvement of Gallin type theory TY_2, see Gallin [6]. A conjecture is that the type system TT_2 is properly included in L_{ar}^λ, which is not in the subject of this paper, and I have not formally carried out this. Musken [28] is another contemporary and significant work on higher-order Intensional Logic that formalizes similar ideas of algorithmic senses as computations of references, also tentatively properly included in, or related to L_{ar}^λ.

There have been applications of Moschovakis' theories of algorithms. E.g., Hurkens et al. [7] uses the untyped language of recursion, Moschovakis [25, 24], for reasoning about mutually recursive identities; Van Lambalgen and Hamm [32] for application to logic programming in linguistics and cognitive science, for representing temporal information in discourse. Loukanova [9, 10, 11, 13, 12, 16, 15, 14], uses L_{ar}^λ for representation of fundamental semantic notions in human language. Loukanova [17] is on applications of L_{ar}^λ on the important topic of quantifier scope ambiguities, while [19] is on formalisation of binding argument slots across mutual recursion assignments in recursion terms and modeling neural receptors.

1.2 Contribution

I shall use the γ-reduction calculus of L_{ar}^λ, which is introduced in Loukanova [20] and extends the calculus and theory of recursion L_{ar}^λ introduced by Moschovakis [26].

The γ-reduction calculus of L_{ar}^λ employs an additional reduction rule, i.e., the (γ) rule. For sake of self-containment of this paper, the γ-rule and the reduction calculus based on it, are presented in Sect. 3.

The contributions of this paper are in demonstrating the usefulness of the formal language L_{ar}^{λ} of Moschovakis acyclic recursion L_{ar}^{λ} and its extended variants by [21, 20], in particular:

(1) I develop the original type theory L_{ar}^{λ} by Moschovakis [26], by using the additional (γ) rule and its corresponding reduction calculus, in algorithmic, i.e., computational, semantics of formal and human languages

(2) By the (γ) rule, it is possible to significantly simplify semantic representations of human language expressions

(3) The (γ) rule can reduce extraneous λ-constructs denoting constant functions, which do not depend essentially on the values filling up the argument slots introduced by the relevant λ-abstract. Such extra λ-constructs can be introduced via reduction sequences that apply the λ-rule of L_{ar}^{λ}

(4) I explicate a preferable order of currying coding of arguments of relations that have at least two arguments. While, in general, the currying order in functional terms of λ-calculi can be a matter of subjective preferences of authors and developers, it is also dependent on applications of selected λ-calculus

The currying order influences the algorithmic computations of semantic representations in syntax-semantics interfaces. For example, Constraint-Based Lexicalized Grammar (CBLG) of human language with syntax-semantics interfaces can employ L_{ar}^{λ}-terms in canonical forms. The formal models of the syntax of CBLG is presented in Loukanova [18]. For the first time ever, syntax-semantics in CBLG, by using L_{ar}^{λ}, is introduced in Loukanova [10].

In this paper, I demonstrate the use of the preferable currying, with respect to the syntactic rules in certain varieties of computational grammar, including CBLG. More importantly, by using L_{ar}^{λ}-terms as logical forms for semantic representations, the syntactic parse of human language expressions can provide a vehicle for compositional syntax-semantics interface, by the highly expressive type-theory of algorithms L_{ar}^{λ}.

(5) I show that algorithmic β-conversion is not valid in TTAR, in general. A limited, restricted version of algorithmic beta-conversion is valid in TTAR.

As in L_{ar}^{λ}, β-reduction does not hold in procedural semantics of hyperitensional Transparent Intensional Logic (TIL), as shown by work of Marie Duzi. For example, Duží [1, 2] promotes procedural semantics of TIL, in close lines of my work on computational semantics.

It is my great honour to underline the significant work carried on by Marie Duží, who continues to contribute to a line of work on procedural semantics. It is my strong conviction that I share passion for research and work, in similar directions on the importance of computational intension for adequate coverage of semantic phenomena that are still widely open. Among the most significant works by Duží, some of which in collaboration, are [1, 3, 4, 5, 22, 23, 2, 8]. It all began by Pavel Tichý [30, 31].

2 Syntax and Semantics of Typed-Theory of Acyclic Recursion

2.1 Syntax

Types: The set Types of the set of the type terms is the smallest set defined recursively, by using the Backus-Naur Form (BNF) notation style as follows:

$$\theta :\equiv e \mid t \mid s \qquad (BasicTypes)$$
$$\mid \sigma \mid (\tau \to \sigma), \quad \text{for } \tau, \sigma \in \text{Types} \qquad (2)$$

The basic type e is the type of the entities in the semantic domains and the expressions denoting entities; t of the truth values and corresponding expressions, s of the states. The types $(\tau \to \sigma)$ are the types of functions from objects of type τ to objects of type σ, and of expressions denoting such functions. For instance:

I commonly use the following types and useful abbreviations in L_{ar}^{λ}-terms, especially in semantic representations of expressions in human language by rendering them in L_{ar}^{λ}-terms.

Notation 1. *The type in* (3b) *is for state-dependent individuals; in* (3c) *for state-dependent truth values (i.e., type for representing state-dependent propositions); and in* (3d) *for state-dependent objects, of type* $\tilde{\tau}$, $\tau \in$ Types, *with the notation introduced by [26], the types of:*

$$e \to t, \quad \text{characteristic functions of sets of entities} \tag{3a}$$

$$\tilde{e} \equiv (s \to e), \quad \text{state dependent entities, i.e., "concepts"} \tag{3b}$$

$$\tilde{t} \equiv (s \to t), \text{state dependent truths,} \tag{3c}$$
$$\text{i.e., state dependent propositions}$$

$$\tilde{\tau} \equiv (s \to \tau), \quad \text{types of state dependent objects of type } (s \to \tau) \tag{3d}$$

The vocabulary of L_{ar}^{λ} consists of (pairwise disjoint sets):

Constants: For every $\tau \in$ Types, $K_{\tau} = \{c_0^{\tau}, \ldots, c_k^{\tau}, \ldots\}$ is the set of the constants of type τ

$$K = \bigcup_{\tau \in \text{Types}} K_{\tau}$$

Pure variables: For every $\tau \in$ Types, $\text{PureV}_{\tau} = \{v_0^{\tau}, v_1^{\tau}, \ldots\}$

$$\text{PureV} = \bigcup_{\tau \in \text{Types}} \text{PureV}_{\tau}$$

Recursion variables: For every $\tau \in$ Types, $\text{RecV}_{\tau} = \{r_0^{\tau}, r_1^{\tau}, \ldots\}$

$$\text{RecV} = \bigcup_{\tau \in \text{Types}} \text{RecV}_{\tau}$$

Definition 1 (Terms). *The set* Terms *of* L_{ar}^{λ} *is defined by recursion. In this work, I express it by using a typed style of a variant of BNF, with the type assignments given either as superscripts or with column sign:*

$$A :\equiv c^{\tau} \mid x^{\tau} \mid [B^{(\sigma \to \tau)}(C^{\sigma})]^{\tau} \mid [\lambda v^{\sigma} (B^{\tau})]^{(\sigma \to \tau)} \mid \qquad (4a)$$

$$[A_0^{\sigma_0} \text{ where } \{ p_1^{\sigma_1} := A_1^{\sigma_1}, \ldots, p_n^{\sigma_n} := A_n^{\sigma_n} \}]^{\sigma_0} \qquad (4b)$$

where the meta-variables in the rules (4a)–(4b) *are as follows:*

- *for* $n, m \geq 0$, $c^{\tau} \in K_{\tau}$ *is a constant*

- $x^{\tau} \in \text{PureV}_{\tau} \cup \text{RecV}_{\tau}$ *is a pure or recursion variable,* $v^{\sigma} \in \text{PureV}_{\sigma}$ *is a pure variable*

- $A, B, A_i^{\sigma_i} \in$ Terms *(i = 0, ..., n) are terms of the respective types*

- $p_i \in \text{RecV}_{\sigma_i}$ *(i = 1, ..., n) are pairwise different recursion variables*

- *the subexpression* $\{ p_1^{\sigma_1} := A_1^{\sigma_1}, \ldots, p_n^{\sigma_n} := A_n^{\sigma_n} \}$, *in the expressions of the form* (4b), *is a sequence of assignments that satisfies the* Acyclicity Constraint:

Acyclicity Constraint For any terms $A_1 : \sigma_1$, ..., $A_n : \sigma_n$, and pairwise different recursion variables $p_1 : \sigma_1$, ..., $p_n : \sigma_n$ $(n \geq 0)$, the sequence $\{p_1 := A_1, \ldots, p_n := A_n\}$ is an *acyclic system of assignments* iff there is a ranking function:

$$\text{rank} : \{p_1, \ldots, p_n\} \longrightarrow \mathbb{N} \qquad (5a)$$

such that, for all $p_i, p_j \in \{p_1, \ldots, p_n\}$:
$$\text{if } p_j \text{ occurs freely in } A_i, \text{ then } \text{rank}(p_j) < \text{rank}(p_i) \qquad (5b)$$

The terms of the form (4b) are called *recursion terms*.

Notation 2. *Often, we shall use abbreviations for sequences, e.g.:*

$$\vec{p} := \vec{A} \equiv p_1 := A_1, \ \ldots, \ p_n := A_n \quad (n \geq 0) \tag{6a}$$

$$H(\vec{x}) \equiv (\ldots H(x_1)\ldots)(x_n) \quad (n \geq 0) \tag{6b}$$

$$\lambda(\vec{v})(A) \equiv \lambda(v_1,\ldots,v_n)(A)$$
$$\equiv \lambda(v_1)(\ldots\lambda(v_n)(A)\ldots) \quad (n \geq 0) \tag{6c}$$

Definition 2 (Immediate and Pure Terms). *The set of the* immediate *terms consists of all terms of the form* (7), *for* $p \in \mathsf{RecV}$, $u_i, v_j, \in$ PureV $(i = 1,\ldots,n,\ j = 1,\ldots,m)$ $V \in \mathsf{Vars}$:

$$T :\equiv V \mid p(v_1)\ldots(v_m) \mid \lambda(u_1)\ldots\lambda(u_n)p(v_1)\ldots(v_m),$$
$$\text{for } m, n \geq 0 \tag{7}$$

Every term A that is not immediate is pure.

3 Extended Gamma Reduction Calculus of $\mathrm{L}_{\mathrm{ar}}^{\lambda}$

In this section, I give the formal definitions of the reduction rules of $\mathrm{L}_{\mathrm{ar}}^{\lambda}$, from Moschovakis [26], extended by the additional reduction rule, (γ) rule. The (γ) rule and the induced γ-reduction calculus of $\mathrm{L}_{\mathrm{ar}}^{\lambda}$ are introduced formally, by details, in Loukanova [20].

Definition 3 (Congruence Relation). *Two terms* $A, B \in \mathsf{Terms}$ *are* congruent, $A \equiv_c B$ *iff one of them can be obtained from the other by renaming bound variables and reordering assignments in recursion terms.*

3.1 Reduction Rules of Extended $\mathrm{L}_{\mathrm{ar}}^{\lambda}$

Congruence The congruence relation is supported by the reduction relation:

$$\text{If } A \equiv_c B, \text{ then } A \Rightarrow B \tag{cong}$$

Transitivity

$$\text{If } A \Rightarrow B \text{ and } B \Rightarrow C, \text{ then } A \Rightarrow C \tag{trans}$$

Compositionality Replacement of sub-terms with correspondingly reduced ones respects the term structure by the definition of the term syntax.

$$\text{If } A \Rightarrow A' \text{ and } B \Rightarrow B', \text{ then } A(B) \Rightarrow A'(B') \qquad \text{(rep1)}$$

$$\text{If } A \Rightarrow B, \text{ then } \lambda(u)(A) \Rightarrow \lambda(u)(B) \qquad \text{(rep2)}$$

$$\text{If } A_i \Rightarrow B_i, \text{ for } i = 0, \ldots, n, \text{ then} \qquad \text{(rep3)}$$

$$A_0 \text{ where } \{\, p_1 := A_1, \ldots, p_n := A_n \,\}$$
$$\Rightarrow B_0 \text{ where } \{\, p_1 := B_1, \ldots, p_n := B_n \,\}$$

Head Rule If $p_i \neq q_j$ and p_i does not occur freely in B_j, for all $i = 1, \ldots, n$, $j = 1, \ldots, m$, then

$$(A_0 \text{ where } \{\, \overrightarrow{p} := \overrightarrow{A} \,\}) \text{ where } \{\, \overrightarrow{q} := \overrightarrow{B} \,\}$$
$$\Rightarrow A_0 \text{ where } \{\, \overrightarrow{p} := \overrightarrow{A}, \ \overrightarrow{q} := \overrightarrow{B} \,\} \qquad \text{(head)}$$

Bekič-Scott Rule If $p_i \neq q_j$ and q_j does not occur freely in A_i, for all $i = 1, \ldots, n$, $j = 1, \ldots, m$, then

$$A_0 \text{ where } \{\, p := (B_0 \text{ where } \{\, \overrightarrow{q} := \overrightarrow{B} \,\}), \overrightarrow{p} := \overrightarrow{A} \,\}$$
$$\Rightarrow A_0 \text{ where } \{\, p := B_0, \overrightarrow{q} := \overrightarrow{B}, \ \overrightarrow{p} := \overrightarrow{A} \,\} \qquad \text{(B-S)}$$

Recursion-Application Rule If p_i does not occur freely in B, for all $i = 1, \ldots, n$, then

$$(A_0 \text{ where } \{\, \overrightarrow{p} := \overrightarrow{A} \,\})(B)$$
$$\Rightarrow A_0(B) \text{ where } \{\, \overrightarrow{p} := \overrightarrow{A} \,\} \qquad \text{(recap)}$$

Application Rule If B is a proper term and b is a fresh recursion (memory) variable, then

$$A(B) \Rightarrow A(b) \text{ where } \{\, b := B \,\} \qquad \text{(ap)}$$

λ-Rule If, for every $i = 1, \ldots, n$, p'_i is a fresh recursion (memory) variable, and A'_i is the result of the replacement of all the free occurrences of $p_1, \ldots, p_n$ in A_i with $p'_1(u), \ldots, p'_n(u)$, respectively, i.e.:

$$A'_i \equiv A_i\{p_1 :\equiv p'_1(u), \ldots, p_n :\equiv p'_n(u)\} \tag{9a}$$

$$\equiv A_i\{\overrightarrow{p} :\equiv \overrightarrow{p'(u)}\} \tag{9b}$$

then

$$\lambda(u)\,(A_0 \text{ where } \{\, p_1 := A_1, \ldots, p_n := A_n \,\}) \\ \Rightarrow \lambda(u)\, A'_0 \text{ where } \{\, p'_1 := \lambda(u)\, A'_1, \ldots, p'_n := \lambda(u)\, A'_n \,\} \tag{λ}$$

$$\equiv \lambda(u) A_0\{\overrightarrow{p} :\equiv \overrightarrow{p'(u)}\} \text{ where } \{ \\ \overrightarrow{p'} := \overline{\lambda(u) A\{\overrightarrow{p} :\equiv \overrightarrow{p'(u)}\}} \,\} \tag{10}$$

γ-Rule The (γ) rule is defined by:

$$A \equiv \quad A_0 \text{ where } \{\, \overrightarrow{a} := \overrightarrow{A}, \;\; p := \lambda(v)P, \;\; \overrightarrow{b} := \overrightarrow{B} \,\} \quad (\gamma)$$
$$\Rightarrow_\gamma A'_0 \text{ where } \{\, \overrightarrow{a} := \overrightarrow{A'}, \; p' := P, \qquad \overrightarrow{b} := \overrightarrow{B'} \,\}$$

given that:

- the term $A \in$ Terms satisfies the γ-condition (in Def. 4) for the assignment $p := \lambda(v)P$
- $p' \in \mathsf{RecV}_\tau$ is fresh for A
- $\overrightarrow{X'} \equiv_c \overrightarrow{X}\{p(v) :\equiv p'\}$ is the result of the free replacement $X_i\{p(v) :\equiv p'\}$, i.e., of *all of the occurrences of $p(v)$ by p', for the corresponding free occurrences of p in X_i,* in all parts X_i, for:

$$X_i \in \{\, A_i \mid i = 0, \ldots, n \,\} \cup \{\, B_i \mid i = 0, \ldots, k \,\}$$

modulo congruence with respect to renaming the scope variable v

294

Definition 4 (γ-condition). *Any given $A \in$ Terms satisfies the γ-condition, with respect to an assignment $p := \lambda(v)P$ in its where-scope, if and only if A is of the form (12), and the clauses (g:1)–(g:2) hold:*

$$A \equiv A_0 \text{ where } \{ \overrightarrow{a} := \overrightarrow{A}, \; p := \lambda(v)P, \; \overrightarrow{b} := \overrightarrow{B} \} \qquad (12)$$

for

- $v \in \mathsf{PureV}_\sigma, \quad p \in \mathsf{RecV}_{(\sigma \to \tau)}$

- $P \in \mathsf{Terms}_\tau$, *and thus,* $\lambda(v)P \in \mathsf{Terms}_{(\sigma \to \tau)}$

- $\overrightarrow{A} \equiv A_1, \ldots, A_n \in \mathsf{Terms}$,
  $\overrightarrow{a} \equiv a_1, \ldots, a_n \in \mathsf{RecV}$ *($n \geq 0$), of corresponding types*

- $\overrightarrow{B} \equiv B_1, \ldots, B_k \in \mathsf{Terms}$,
  $\overrightarrow{b} \equiv b_1, \ldots, b_k \in \mathsf{RecV}$ *($k \geq 0$), of corresponding types*

(g:1) The term $P \in \mathsf{Terms}_\tau$ does not have any free occurrences of v in it, i.e., $v \notin \mathsf{FreeV}(P)$

(g:2) Each of the free occurrences of p in any of the term parts A_0, A_i, B_j, i.e., in $\overrightarrow{A}$ and $\overrightarrow{B}$, is an occurrence in a subterm $p(v)$ that is in the scope of $\lambda(v)$ (modulo congruence with respect to renaming the scope variable v)

In such a case, we also say that the assignment $p := \lambda(v)P$ (and the recursion variable p) satisfies the γ-condition for A in (12), and that A satisfies the γ-condition for p in its where-scope.

Definition 5 (Term Irreducibility). *A term $A \in$ Terms is irreducible if and only if*

$$\text{for all } B \in \mathsf{Terms}, \text{ if } A \Rightarrow B, \text{ then } A \equiv_c B \qquad (13)$$

The following theorems are major results, which are essential for algorithmic semantics. For details, see Loukanova [20].

Theorem 1 (Criteria for γ-irreducibility). *With respect to the extended set of reduction rules, which includes the (γ) rule:*

(1) *All constants and variables are γ-irreducible, i.e.: for all $A \in$* Consts $\cup$ Vars, *A is γ-irreducible*

(2) *An application term $A(B)$ is γ-irreducible if and only if B is immediate and A is explicit and γ-irreducible (and thus, irreducible)*

(3) *A λ-term $\lambda(x)A$ is γ-irreducible if and only if A is explicit and γ-irreducible (and thus, irreducible)*

(4) *A recursion term $A \equiv A_0$* where *$\{ p_1 := A_1, \ldots, p_n := A_n \}$ ($n \geq 0$) is γ-irreducible if and only if*

    (a) *All of the parts A_0, $\ldots$, A_n are explicit and γ-irreducible (and thus, irreducible), and*

    (b) *A does not satisfy the γ-condition with respect to any of the assignments in its* where-*scope*

Proof. By structural induction on terms and checking whether the γ-reduction rules are applicable. $\quad\square$

Theorem 2 (Canonical Form Theorem). *See Moschovakis [26], § 3.1 and Loukanova [20].*
    Existence and uniqueness of the canonical forms: *For each term A, there is a unique, up to congruence, irreducible term C, denoted by* cf(A) *and called the canonical form of A, such that:*

1. cf$(A) \equiv A_0$ where $\{ p_1 := A_1, \ldots, p_n := A_n \}$,

    for some explicit, irreducible terms A_1, $\ldots$, A_n ($n \geq 0$)

2. $A \Rightarrow$ cf(A)

3. *if $A \Rightarrow B$ and B is irreducible, then $B \equiv_c$* cf(A), *i.e.,* cf(A) *is the unique, up to congruence, irreducible term to which A can be reduced*

4. *A constant* c $\in K$ *occurs in* cf$_\gamma(A)$ *if and only if it occurs in the term A*

5. *A variable $x \in$ Vars occurs freely in* $\mathsf{cf}_\gamma(A)$ *if and only if it occurs freely in the term* A.

Note 1. *In this paper, when we refer to reductions, canonical forms, algorithmic equivalence, and related notions, we assume that they are with respect t the extended γ-reduction calculus, see Sect. 3. Thus, we shall write* $\mathsf{cf}(A)$ *for* $\mathsf{cf}_\gamma(A)$.

The mathematical notions of algorithmic and referential equivalence, i.e., synonymy, in any given semantic structure $\mathfrak{A}$, are based on the mathematics of algorithmic, i.e., referential intention and natural isomorphism between functional recursors in $\mathfrak{A}$. For more details, see Moschovakis [26], especially, § 3.4–3.20.

Definition 6 (The Algorithmic, Referential Synonymy in $\mathfrak{A}$). *Assume that $A, B \in$ Terms. $A \approx B$ if and only if one of the following cases holds:*

(1) A and B are immediate, A and B have the same denotations:

$$\mathsf{den}(A) = \mathsf{den}(B) \qquad (14)$$

(2) or, A and B are proper terms, i.e., algorithmically meaningful, A and B are denotationally equal, and their denotation is computed by the same algorithm, i.e., A and B have naturally isomorphic referential intentions:

$$\mathsf{alg}(A) \cong \mathsf{alg}(B) \qquad (15)$$

Theorem 3 (Algorithmic / Referential Synonymy Theorem). *Assume that $\mathfrak{A}$ is a given semantic structure, with G the set of all variable valuations g in $\mathfrak{A}$.*

Two terms A, B are algorithmically, i.e., referentially, synonymous, $A \approx B$ in $\mathfrak{A}$, if and only if

(A) A and B are immediate, and:

$$\mathsf{den}(A)(g) = \mathsf{den}(B)(g),$$
$$\textit{for all variable valuations } g \in G \textit{ for } \mathfrak{A} \qquad (16a)$$
$$\mathsf{FreeV}(A) = \mathsf{FreeV}(B) \qquad (16b)$$

(B) *A and B are proper and there are explicit, irreducible terms of corresponding types, $A_i : \sigma_i$, $B_i : \sigma_i$ $(i = 0, \ldots, n)$, such that:*

$$A \equiv A^{\sigma_0} \Rightarrow_{\mathsf{cf}} A_0^{\sigma_0} \text{ where } \{\, p_1 := A_1^{\sigma_1}, \ldots, \qquad (17a)$$
$$p_n := A_n^{\sigma_n} \,\}$$

$$B \equiv B^{\sigma_0} \Rightarrow_{\mathsf{cf}} B_0^{\sigma_0} \text{ where } \{\, p_1 := B_1^{\sigma_1}, \ldots, \qquad (17b)$$
$$p_n := B_n^{\sigma_n} \,\}$$

$$\text{for all } i = 0, \ldots, n$$

$$\mathsf{den}(A_i)(g) = \mathsf{den}(B_i)(g), \text{ for all } g \in G \qquad (17c)$$

$$\therefore \ \mathsf{den}(A_i) = \mathsf{den}(B_i) \qquad (17d)$$

$$\mathsf{FreeV}(A_i) = \mathsf{FreeV}(B_i) \qquad (17e)$$

$$\mathsf{Consts}(A) = \mathsf{Consts}(B) \qquad (17f)$$

Proof. The detailed proof is provided in Moschovakis [26] § 3.4–3.20. It is based on the mathematical notions of algorithmic and referential equivalence, i.e., synonymy, in any given semantic structure $\mathfrak{A}$.

Alternatively, the proof follows directly from Def. 6.   □

4 Currying Order of Renderings of Constants

Formally, the currying order in λ-calculi is a matter of convention and can be considered as non-essential. Specific choices of order of the currying of the terms denoting multi-argument functions can be translated from one to another.

When λ-terms with currying are used for semantic representations of human language, the order can depend on the choice of the computational grammar providing syntactic parsing and syntax-semantic interfaces. The same phrase, e.g., "Kim hugs Maja" can be rendered to two different $\mathrm{L}_{\mathrm{ar}}^{\lambda}$-terms (18a) or (19a), depending on the order of currying coding. For instance, the rendering (19a) is popular in logic.

In (18a) and (19a)–(19c), the application order is inverted.

Rendering Φ to the term in (18a) or (18c), is more natural with respect to syntax-semantics interface in certain computational grammars of human language. CBLG, see Loukanova [18], provides syntactic parsing, which is linguistically motivated by shared phrasal structure across syntactic categories, most prominent in verbal phrases (VPs). Typically, the syntactic arguments, that are the complements of the head verb, are saturated at first. E.g., a verb like "hug" in a sentence like (18a), gets its syntactic argument of the complement, e.g., a NP, such as a proper name "Maja", before its syntactic argument of the subject, a NP, e.g., "Kim". Thus, a rendering that follows compositional syntax-semantics can be provided by the λ-term in (18a).

Computational syntax-semantics interace between syntax and semantics of human language is not in the subject of this paper. Such syntax-semantics interace provides computational connection between compositional, syntactic analisys of human language and semantic representations of the corresponding syntactic structures.

Denotational and Algorithmic Meanings in Specific Semantic Structures

Let's assume that $\mathfrak{A}$ is a given semantic structure, with G the set of all variable valuation functions g in $\mathfrak{A}$.

Assume that the English word "hugs" is rendered to a constant $hug \in \mathsf{Consts}$, in (18a)–(18c), e.g., by syntax-semantics interface that fills up at first the complement NP and then the subject NP:

$$\Phi \equiv \text{Kim hugs Maja.} \xrightarrow{\text{render}} A \qquad (18a)$$

$$A \equiv hug(maja)(kim) \qquad (18b)$$

$$\Rightarrow_{\mathsf{cf}} \ \mathsf{cf}(A) \equiv hug(m)(k) \ \text{where} \ \{ \qquad (18c)$$
$$m := maja, \ k := kim \ \}$$

Thus, (18a)–(18c) is a rendering of the sample sentence "Kim hugs Maja" denoted by Φ, to a $\mathsf{L}_{\mathrm{ar}}^{\lambda}$-term A, or its canonical form $\mathsf{cf}(A)$. In it, the constant $hug \in \mathsf{Consts}$ denotes a two-argument function, via currying, and applies:

(1) at first to a term M that designates the individual who is the hugged one

(2) and after that, to a term K that designates the individual who does the action of hugging

Alternatively, the same word "hugs" can be rendered to a constant hug_1, which applies:

(1) at first to a term K that designates the individual who does the action of hugging

(2) and after that, to a term M that designates the individual who is the hugged one

Thus, in the rendering (19a)–(19c) of the same sentence Φ to a $\mathrm{L}_{ar}^{\lambda}$-term B, or its canonical form $\mathsf{cf}(B)$, the order of the functional application is reverse to that in (18a)–(18c):

$$\Phi \equiv \text{Kim hugs Maja.} \xrightarrow{\text{render}} B \qquad (19\text{a})$$
$$B \equiv hug_1(kim)(maja) \qquad (19\text{b})$$
$$\Rightarrow_{\mathsf{cf}} \ \mathsf{cf}(B) \equiv hug_1(k)(m) \text{ where } \{\, m := maja,\ k := kim \,\} \qquad (19\text{c})$$

The currying order has significance, in the interpretations (21a)–(21c), and the corresponding denotations in (22a)–(22d), the constant $hug_1 \in \mathsf{Consts}$ has the reverse currying order to that of the constant $hug \in \mathsf{Consts}$.

Recall, that here, we assume a given semantic structure, $\mathfrak{A} = \mathfrak{A}(\mathsf{Consts})$, with a fixed interpretation function $\mathcal{I} = \mathcal{I}_{\mathfrak{A}}$. The corresponding denotation function $\mathsf{den} = \mathsf{den}_{\mathfrak{A}}$ is defined on the basis that $\mathcal{I} = \mathcal{I}_{\mathfrak{A}}$. In $\mathfrak{A}$, the set $\mathbb{T}_{\tilde{e}}$ is the semantic domain of objects of type $\tilde{e} \equiv (\mathsf{s} \to \mathsf{e})$.

We assume the understanding that these interpretation functions are for selected semantic structure $\mathfrak{A}$, such that (20a)–(20c), and respectively (21a)–(21c):

For all $a, b \in \mathbb{T}_{\tilde{e}}$:

$$[\mathcal{I}_{\mathfrak{A}}(hug)](b)(a) = [\mathcal{I}_{\mathfrak{A}}(hug_1)](a)(b) \qquad (20\text{a})$$
$$[\mathsf{den}_{\mathfrak{A}}(hug)](b)(a) = [\mathsf{den}_{\mathfrak{A}}(hug_1)](a)(b) \qquad (20\text{b})$$

$\therefore$ for all $K, M \in \mathsf{Terms}_{\tilde{e}}$, $K, M : \tilde{e} \equiv (\mathsf{s} \to \mathsf{e})$

$$\mathsf{den}_{\mathfrak{A}}(hug(M)(K)) = \mathsf{den}_{\mathfrak{A}}(hug_1(K)(M)) \qquad (20\text{c})$$

More generally, assume a class of semantic structures $\mathfrak{A}$ that have the same interpretation functions for the two constants, $hug, hug_1 \in$ Consts, as in (21a)–(21c). We can skip the subscript for the class of these $\mathfrak{A}$.

For all $a, b \in \mathbb{T}_{\tilde{e}}$:

$$[\mathcal{I}_\mathfrak{A}(hug)](b)(a) = [\mathcal{I}_\mathfrak{A}(hug_1)](a)(b)$$
$$[\mathcal{I}(hug)](b)(a) = [\mathcal{I}(hug_1)](a)(b) \tag{21a}$$

$$[\text{den}_\mathfrak{A}(hug)](b)(a) = [\text{den}_\mathfrak{A}(hug_1)](a)(b)$$
$$[\text{den}(hug)](b)(a) = [\text{den}(hug_1)](a)(b) \tag{21b}$$

$\therefore$ for all $K, M \in$ Terms$_{\tilde{e}}$, $K, M : \tilde{e} \equiv (\text{s} \to \text{e})$
$$\text{den}_\mathfrak{A}(hug(M)(K)) = \text{den}_\mathfrak{A}(hug_1(K)(M)) \tag{21c}$$
$$\text{den}(hug(M)(K)) = \text{den}(hug_1(K)(M))$$

Therefore, the following denotational equalities (22a)–(22d) hold in $\mathfrak{A}$, and a class of $\mathfrak{A}$ with such interpretations for $hug, hug_1 \in$ Consts:

$$\text{den}_\mathfrak{A}(hug) \tag{22a}$$
$$= \text{den}_\mathfrak{A}(\lambda x \, (\lambda y \, hug(x)(y))) = \text{den}_\mathfrak{A}(\lambda x_m \, (\lambda y_k \, hug(x_m)(y_k))) \tag{22b}$$
$$= \text{den}_\mathfrak{A}(\lambda x \, (\lambda y \, hug_1(y)(x))) = \text{den}_\mathfrak{A}(\lambda x_m \, (\lambda y_k \, hug_1(y_k)(x_m))) \tag{22c}$$
$$\neq \text{den}_\mathfrak{A}(hug_1) \tag{22d}$$
$$= \text{den}_\mathfrak{A}(\lambda y \, (\lambda x \, hug_1(y)(x))) = \text{den}_\mathfrak{A}(\lambda y_k \, (\lambda x_m \, hug_1(y_k)(x_m))) \tag{22e}$$

Another alternative rendering of the same sample sentence Φ is given in (23a)–(23b), by syntax-semantics interface that fills up at first the subject NP and then the complement NP:

$$\Phi \xrightarrow{\text{render}} C \equiv [\lambda y \, (\lambda x \, hug_1(y)(x))](kim)(maja) \tag{23a}$$
$$\Rightarrow_{\text{cf}} \text{cf}(C) \equiv [\lambda y \, (\lambda x \, hug_1(y)(x))](k)(m) \text{ where } \{$$
$$m := maja, k := kim \} \tag{23b}$$
$$\approx \text{cf}(B) \equiv [hug_1(k)](m) \text{ where } \{ m := maja, k := kim \} \tag{23c}$$

The formal language of acyclic recursion allows various alternative renderings of sentences that have syntactic structure similar to that of Φ, in (18a)–(18c).

For example, the same sentence Φ, in (18a)–(18c), can be rendered to one of the terms in (24a)–(24b), in syntax-semantics interface that fills up at first the complement NP and then the subject NP, similarly to the rendering in (18a)–(18c):

$$\Phi \xrightarrow{\text{render}} D \equiv [\lambda x \, (\lambda y \, hug_1(y)(x))](maja)(kim) \tag{24a}$$

$$\Rightarrow_{\text{cf}} \text{cf}(D) \equiv [\lambda x \, (\lambda y \, hug_1(y)(x))](m)(k) \text{ where } \{ \atop m := maja, \, k := kim \} \tag{24b}$$

Note 2. *In* (18a)–(18c)*, the word "hugs" is rendered to a constant* hug*. While in* (24a)–(24b)*, the same word "hugs" is rendered to a term* $\lambda x \, (\lambda y \, hug_1(y)(x))$*, which is a syntactic* λ*-construct, using the constant* hug_1*.*

Corollary 1. *Assume a class of semantic structures* $\mathfrak{A}$ *with interpretations for* hug, hug_1 $\in$ Consts*, as given in* (20a), (21a)*. Then, for* $A, B, C, D \in$ Terms*. respectively, in* (18a)–(18c), (19a)–(19c), (23a)–(23b)*, and* (24a)–(24b)*, the following algorithmic equivalences hold:*

$$\text{cf}(A) \approx_{\mathfrak{A}} \text{cf}(B), \quad \text{for } \mathfrak{A}, \text{ as in } (20a) \tag{25a}$$

$$\text{cf}(B) \approx \text{cf}(C), \quad \text{cf}(C) \approx \text{cf}(D), \tag{25b}$$
$$\text{in general, for all semantic structures}$$

$$\therefore \, A \approx_{\mathfrak{A}} B \approx C \approx D \tag{25c}$$

Proof. The algorithmic equivalences follow from the Referential Synonymy Theorem (3), because their corresponding canonical forms are algorithmically equivalent.

Note: It is important that these canonical forms are obtained by reductions using the rules in Sect. 3, especially the added (γ) rule. $\quad\square$

Note 3. *In general,* $A, B \in$ Terms*, given in* (18a)–(18c), (19a)–(19c)*, and similar ones, are not algorithmically equivalent, in all semantic* structures. That is:

$$\mathsf{cf}(A) \not\approx_{\mathfrak{A}'} \mathsf{cf}(B), \quad \textit{for some semantic structures } \mathfrak{A}' \qquad (26)$$

Counterexamples are provided by semantic structures $\mathfrak{A}$ that do not satisfy (20a)–(20c), and respectively (21a)–(21c).

Corollary 2.

(1) For every $C \in$ Terms that is explicit, irreducible, and proper, the λ-abstraction, in the order of the application in C, preserves the algorithmic equivalence:

$$C \approx \lambda(u)\,[\lambda(v)\,(C(u)(v))] \qquad (27a)$$

(2) Therefore (as a special case of C), for every $c \in$ Consts, such that $c : \tilde{e} \to (\tilde{e} \to \tilde{t})$:

$$c \approx \lambda y\,(\lambda x\, c(y)(x)) \approx \lambda y_k\,(\lambda x_m\, c(y_k)(x_m)) \qquad (28a)$$

(3) For $hug, hug_1 \in$ Consts, such that $hug, hug_1 : \tilde{e} \to (\tilde{e} \to \tilde{t})$:

$$hug_1 \approx \lambda y\,(\lambda x\, hug_1(y)(x)) \approx \lambda y_k\,(\lambda x_m\, hug_1(y_k)(x_m)) \qquad (29a)$$
$$hug_1 \not\approx \lambda x\,(\lambda y\, hug_1(y)(x)) \approx \lambda x_m\,(\lambda y_k\, hug_1(y_k)(x_m)) \qquad (29b)$$

(4) There are some proper, explicit, irreducible terms $C \in$ Terms, such that (30):

$$C \not\approx \lambda(v)\,[\lambda(u)\,(C(u)(v))] \qquad (30)$$

Proof. The algorithmic equivalences (1)–(3) follow from the Referential Synonymy Theorem (3).

The abcence of algorithmic equivalences in (3)–(4) follow also from the Referential Synonymy Theorem (3), by examples of specific semantic structures $\mathfrak{A}'$, where there are entities, for which the denotational equality is not valid.

$$\text{For some } \mathfrak{A}, \text{ and some } a, b \in \mathbb{T}_{\tilde{e}}: \qquad (31a)$$

303

$$[\mathcal{I}_{\mathfrak{A}}(hug_1)](a)(b) = 1 \tag{31b}$$
$$[\mathcal{I}_{\mathfrak{A}}(hug_1)](b)(a) = 0 \tag{31c}$$
$$[\mathrm{den}_{\mathfrak{A}}(hug_1)](a)(b) \neq [\mathrm{den}_{\mathfrak{A}}(hug_1)](b)(a) \tag{31d}$$

□

5 Restricted Algorithmic Beta-Conversion

Theorem 4. *Assume that* $u, v \in \mathsf{PureV}$ *and* $C \in \mathsf{Terms}$. *The restricted* β-conversion (32a) *is valid provided that* (32b)–(32e) *hold:*

$$\big(\lambda(u)(C)\big)(v) \approx C\{u :\equiv v\} \tag{32a}$$

$$\text{the replacement } C\{u :\equiv v\} \text{ is free} \tag{32b}$$
$$C \text{ is explicit, irreducible term} \tag{32c}$$
$$C \text{ is a proper term} \tag{32d}$$
$$v \in \mathsf{FreeV}(C\{u :\equiv v\}) \tag{32e}$$
$$\therefore \mathsf{FreeV}((\lambda(u)\,C)(v)) = \mathsf{FreeV}(C\{u :\equiv v\}) \tag{32f}$$

Proof. The term $(\lambda(u)(C))(v)$ is proper, even if C were immediate. Thus, by (32d), it follows that both terms C and $(\lambda(u)(C))(v)$ are proper terms.

The proof is by structural induction on the formation rules of the term C, in Def. 1 and using the Algorithmic Equivalence Theorem 3 and the definition of the denotation function, see it, e.g., in Loukanova [21, 20]. For sake of space, we do not provide a detailed proof in this paper. □

Loukanova [9] demonstrates that the formal language and type theory $\mathrm{L}_{\mathrm{ar}}^\lambda$ have a rich choices of alternative terms for rendering sentences. For example, the sentence of natural language in (33a) has an occurrence of non-reflexive pronoun that has context dependent interpretations. The semantic representation of such sentences can be left underspecified, e.g., as in (33a)–(33b):

$$\Phi \equiv \text{Kim hugs her.} \xrightarrow{\text{render}} A \equiv hug(m)(kim) \tag{33a}$$

$$\Rightarrow_{\mathsf{cf}} \mathsf{cf}(A) \equiv hug(m)(k) \text{ where } \{\, k := kim \,\} \tag{33b}$$

The non-reflexive pronouns are in contrast with reflexive pronouns, which are restricted by co-indexing. Both reflexive and non-reflexive pronouns obtain more adequate semantic representations in L_{ar}^{λ} that are not available in Montague's IL.

The typical λ-*co-indexing* from IL can be used in L_{ar}^{λ}, e.g., as in the term in the rendering (34a), while the term (34c), which is in canonical form, is not available in IL.

$$[\text{Mary}]_m \text{ likes } [\text{herself}]_m \xrightarrow{\text{render}} H_1 \tag{34a}$$

$$H_1 \equiv [\lambda x\,(like(x)(x))](mary) \qquad \text{(via } \lambda\text{-co-index)} \tag{34b}$$

$$\Rightarrow_{\mathsf{cf}} [\lambda x\,(like(x)(x))](m) \text{ where } \{\, m := mary \,\} \\ \text{by (ap)} \tag{34c}$$

$$\therefore H_1 \equiv [\lambda x\,(like(x)(x))](mary) \\ \approx [\lambda x\,(like(x)(x))](m) \text{ where } \{\, m := mary \,\} \tag{34d}$$

The terms in (34a)–(34b) and (36c), are algorithmically equivalent. That is, (34d) follows directly from the reduction (34a)–(34c), by the Algorithmic Equivalence Theorem 3.

Renderings can be done directly into terms in canonical forms, which are available in L_{ar}^{λ}, e.g., as in (35):

$$[\text{Mary}]_m \text{ likes } [\text{herself}]_m \xrightarrow{\text{render}} \mathsf{cf}(H_1) \\ \mathsf{cf}(H_1) \equiv [\lambda x\,(like(x)(x))](m) \text{ where } \{\, m := mary \,\} \tag{35}$$

Alternatively, in L_{ar}^{λ}, co-indexing of sub-expressions can be represented by filling up several argument slots in a L_{ar}^{λ}-term, by the same recursion variable. Typical cases include expressions with reflexive pronouns, e.g., as in (36a)–(36b):

$$[\text{Mary}]_m \text{ likes } [\text{herself}]_m \xrightarrow{\text{render}} H_2 \tag{36a}$$

$$H_2 \equiv like(m)(m) \text{ where } \{m := mary\} \quad \text{(recursion co-index)} \tag{36b}$$

$$\approx [\lambda x\,(like(x)(x))](m) \text{ where } \{\, m := mary \,\} \quad (\lambda\text{-co-index}) \tag{36c}$$

$$\approx [\lambda x \, (like(x)(x))](mary) \qquad\qquad\qquad \text{(λ-co-index)}$$
$$\tag{36d}$$

The terms in (36a)–(36d) are algorithmically equivalent, which follows, as a special case, from Corollary 3.

Corollary 3. *For any explicit, irreducible terms $D, M \in$ Terms, of the types $D : (\sigma \to (\sigma \to \tau))$, $M : \sigma$, and any $m \in$ RecV$_\sigma$, $(\sigma, \tau \in$ Types) such that (37):*

$$x \notin \mathsf{FreeV}(D) \tag{37}$$

the algorithmic equivalence (38a)–(38b) holds:

$$[\lambda x \, (D(x)(x))](m) \ \text{ where } \ \{m := M\} \tag{38a}$$
$$\approx D(m)(m) \text{ where } \{m := M\} \tag{38b}$$

Proof. The statement of the algorithmic equivalence (38a)–(38b) follows from the Referential Synonymy Theorem (3). $\qquad\square$

Note 4. *The algorithmic equivalences (38a)–(38b) does not follow from using β-conversion. Algorithmic β-conversion is not valid in general for the algorithmic equivalence in* L$_{\mathrm{ar}}^\lambda$, *except for some simple cases. See Loukanova [9].*

6 Conclusion and Future Work

In this paper, I highlight that β-reduction is valid for the denotational semantics in L$_{\mathrm{ar}}^\lambda$. I show that the general β-rule does not hold for algorithmic equivalence, except for simple spacial cases. For more on the topic, see also Moschovakis [26] and Loukanova [9].

An immediate task on extending the work in this paper is to employ the generalised γ^*-reduction, Loukanova [20, 21]. For example, reduction of recursion terms employing recursion assignments of the form $d := \lambda(\overrightarrow{y})D$, which designate constant functions of multiple arguments, is an interesting development of the type-theory of acyclic recursion.

References

[1] Marie Duží. Extensional logic of hyperintensions. In *Conceptual modelling and its theoretical foundations*, pages 268–290. Springer, 2012.

[2] Marie Duží. If structured propositions are logical procedures then how are procedures individuated? *Synthese*, 196(4):1249–1283, 2019.

[3] Marie Duží and Bjørn Jespersen. Procedural isomorphism, analytic information and β-conversion by value. *Logic Journal of the IGPL*, 21(2):291–308, 12 2012.

[4] Marie Duží and Bjørn Jespersen. Transparent quantification into hyperintensional objectual attitudes. *Synthese*, 192:635–677, 2015.

[5] Marie Duží and Miloš Kosterec. A valid rule of β-conversion for the logic of partial functions. *Organon F*, 24(1):10–36, 2017.

[6] Daniel Gallin. *Intensional and Higher-Order Modal Logic: With Applications to Montague Semantics*. North-Holland Publishing Company, Amsterdam and Oxford, and American Elsevier Publishing Company, 1975.

[7] Antonius J. C. Hurkens, Monica McArthur, Yiannis N Moschovakis, Lawrence S Moss, and Glen T Whitney. The logic of recursive equations. *Journal of Symbolic Logic*, 63(2):451–478, 1998.

[8] Bjørn Jespersen and Marie Duží. Transparent quantification into hyperpropositional attitudes de dicto. *Linguistics and Philosophy*, 2022.

[9] Roussanka Loukanova. Reference, Co-reference and Antecedent-anaphora in the Type Theory of Acyclic Recursion. In Gemma Bel-Enguix and María Dolores Jiménez-López, editors, *Bio-Inspired Models for Natural and Formal Languages*, pages 81–102. Cambridge Scholars Publishing, 2011.

[10] Roussanka Loukanova. Semantics with the Language of Acyclic Recursion in Constraint-Based Grammar. In Gemma Bel-Enguix and María Dolores Jiménez-López, editors, *Bio-Inspired Models for Natural and Formal Languages*, pages 103–134. Cambridge Scholars Publishing, 2011.

[11] Roussanka Loukanova. Syntax-Semantics Interface for Lexical Inflection with the Language of Acyclic Recursion. In Gemma Bel-Enguix, Veronica Dahl, and María Dolores Jiménez-López, editors, *Biology, Computation and Linguistics*, volume 228 of *Frontiers in Artificial Intelligence and Applications*, pages 215–236. IOS Press, Amsterdam; Berlin; Tokyo; Washington, DC, 2011.

[12] Roussanka Loukanova. Algorithmic semantics of ambiguous modifiers with the type theory of acyclic recursion. In *Web Intelligence and Intelligent Agent Technology, IEEE/WIC/ACM International Conference*, volume 3, pages 117–121, Los Alamitos, CA, USA, dec 2012. IEEE Computer Society.

[13] Roussanka Loukanova. Semantic Information with Type Theory of Acyclic Recursion. In Runhe Huang, Ali A. Ghorbani, Gabriella Pasi, Takahira Yamaguchi, Neil Y. Yen, and Beijing Jin, editors, *Active Media Technology. AMT 2012*, volume 7669 of *Lecture Notes in Computer Science*, pages 387–398. Springer, Berlin, Heidelberg, 2012.

[14] Roussanka Loukanova. Algorithmic Granularity with Constraints. In Kazayuki Imamura, Shiro Usui, Tomoaki Shirao, Takuji Kasamatsu, Lars Schwabe, and Ning Zhong, editors, *Brain and Health Informatics. BHI 2013*, volume 8211 of *Lecture Notes in Computer Science*, pages 399–408. Springer, Cham, 2013.

[15] Roussanka Loukanova. Algorithmic Semantics for Processing Pronominal Verbal Phrases. In Henrik Legind Larsen, Maria J. Martin-Bautista, María Amparo Vila, Troels Andreasen, and Henning Christiansen, editors, *Flexible Query Answering Systems. FQAS 2013*, volume 8132 of *Lecture Notes in Computer Science*, pages 164–175. Springer, Berlin, Heidelberg, 2013.

[16] Roussanka Loukanova. A Predicative Operator and Underspecification by the Type Theory of Acyclic Recursion. In Denys Duchier and Yannick Parmentier, editors, *Constraint Solving and Language Processing. CSLP 2012*, volume 8114 of *Lecture Notes in Computer Science*, pages 108–132. Springer, Berlin, Heidelberg, 2013.

[17] Roussanka Loukanova. Relationships between Specified and Underspecified Quantification by the Theory of Acyclic Recursion. *ADCAIJ: Advances in Distributed Computing and Artificial Intelligence Journal*, 5(4):19–42, 2016.

[18] Roussanka Loukanova. An Approach to Functional Formal Models of Constraint-Based Lexicalized Grammar (CBLG). *Fundamenta Informaticae*, 152(4):341–372, 2017.

[19] Roussanka Loukanova. Binding operators in type-theory of algorithms for algorithmic binding of functional neuro-receptors. In M. Ganzha, L. Maciaszek, and M. Paprzycki, editors, *2017 Federated Conference on Computer Science and Information Systems*, volume 11 of *Annals of Computer Science and Information Systems*, pages 57–66. IEEE (Digital); Polish Information Processing Society, 2017.

[20] Roussanka Loukanova. Gamma-Reduction in Type Theory of Acyclic Recursion. *Fundamenta Informaticae*, 170(4):367–411, 2019.

[21] Roussanka Loukanova. Gamma-Star Canonical Forms in the Type-Theory of Acyclic Algorithms. In Jaap van den Herik and Ana Paula Rocha, editors, *Agents and Artificial Intelligence. ICAART 2018*, volume 11352 of *Lecture Notes in Computer Science*, pages 383–407. Springer International Publishing, Cham, 2019.

[22] Duží Marie. Logic of dynamic discourse; anaphora resolution. *Information Modelling and Knowledge Bases XXIX*, 301:263–279, 2018.

[23] Marek Menšík, Marie Duží, and Jakub Kermaschek. The role of beta conversion in functional programming. *Information Modelling and Knowledge Bases XXIX*, 301:280–298, 2018.

[24] Yiannis N. Moschovakis. The formal language of recursion. *Journal of Symbolic Logic*, 54(4):1216–1252, 1989.

[25] Yiannis N. Moschovakis. The logic of functional recursion. In M. L. Dalla Chiara, K. Doets, D. Mundici, and J. van Benthem, editors, *Logic and Scientific Methods*, volume 259, pages 179–207. Springer, Dordrecht, 1997.

[26] Yiannis N. Moschovakis. A Logical Calculus of Meaning and Synonymy. *Linguistics and Philosophy*, 29(1):27–89, 2006.

[27] Reinhard Muskens. *Meaning and Partiality*. Studies in Logic, Language and Information. CSLI Publications, Stanford, California, 1995.

[28] Reinhard Muskens. Sense and the computation of reference. *Linguistics and Philosophy*, 28(4):473–504, Aug 2005.

[29] Richmond H. Thomason, editor. *Formal Philosophy: Selected Papers of Richard Montague*. Yale University Press, New Haven, Connecticut, 1974.

[30] Pavel Tichý. The foundations of Frege's logic. In *The Foundations of Frege's Logic*. Berlin: De Gruyter, 1988.

[31] Pavel Tichý. Constructions as the subject matter of mathematics. In Werner Depauli-Schimanovich, Eckehart Köhler, and Friedrich Stadler, editors, *The Foundational Debate: Complexity and Constructivity in Mathematics and Physics*, pages 175–185. Springer Netherlands, Dordrecht, 1995.

[32] Michiel van Lambalgen and Fritz Hamm. *The Proper Treatment of Events*. Explorations in Semantics. Wiley-Blackwell, Oxford, 2005.

HYPERINTENSIONS FOR PROBABILISTIC COMPUTATIONS

GIUSEPPE PRIMIERO

Logic, Uncertainty, Computation and Information Group
University of Milan, Italy
`giuseppe.primiero@unimi.it`

Abstract

HTLC, for Hyperintensional Typed Lambda Calculus, is an extension of the typed λ-calculus with hyperintensions and related rules, introduced in [11]. This contribution introduces $HTLC_p$, which adapts the former to hyperintensions for nondeterministic processes. We formulate appropriate conditions for reasoning with such objects, discuss its metatheory and conclude with some observation comparing it with the semantics of Transparent Intensional Logic.

1 Introduction

Hyperintensions, as concepts which distinguish between necessarily equivalent contents, are enjoying a large philosophical success. Since their introduction in [3], a number of notions have been identified as hyperintensional up to some extent: information, belief, knowledge, meaning, explanation, to name just a few of the most common ones. This success does not extend to the representation of computations in the physical and digital domains, where hyperintensions are still little

This research has been funded by the Project PRIN2020 BRIO - Bias, Risk and Opacity in AI (2020SSKZ7R) awarded by the Italian Ministry of University and Research (MUR).

explored, possibly with the exception of computational semantics for natural language [15].

A formal model of hyperintensions for the typed λ-calculus HTLC is introduced in [11]. The system is designed to reason with expressions for extensional, intensional and hyperintensional entities. Under the Curry-Howard isomorphism, expressions of HTLC are interpreted computationally, with a term t a computational process (e.g. the application of the successor function to define a natural number) and its type T its output (type $\mathbb{N}$ or value 4); an hyperintensional term t^* of type $*^T$ can be seen as denoting computational processes t, t' which are identical with respect to their outputs T, but for which structural identity fails (e.g. t might denote the successor function application, while t' might denote composition of the successor function and the summation function). Hence, distinct notations or modes of presentation or execution for the computation of a given natural number, or distinct operations to compute a given value of any other type, would be denoted by the same hyperintensional term.

HTLC offers an alternative, proof-theoretic view to TIL hyperintensions [8], the latter being especially suitable for modelling cases from natural language [10] because of its ability to express partial functions, the former being constrained to total functions. A recent conjecture presented in [7] states that the non-hyperintensional fragment of total functions, without modalities (quantifying over possible worlds and times) of TIL is Henkin complete. For HTLC with only total functions and proper constructions (no modalities quantifying over possible worlds and times are available), we have shown that any consistent set of closed well-formed formulas Λ is satisfiable in a model in which the domain of basic and function types is denumerable, and the domain for hyperintensional types is non-denumerable but strongly reducible to a model with a denumerable domain for extensional and intensional types. There is no form of partiality or non-determinism in HTLC.

While degrees of beliefs have long been interpreted in the form of subjective probabilities [6, 13], the link between probabilities and hyperintensions is little explored. In fact, in the domain of computations with (sharp) probabilistic outputs, hyperintensional terms and associated attitudes (such as belief and trust) become extremely natural. Standardly, types are interpreted as propositions or sets, their terms being respectively proofs and elements of the set. In the case of terms as computational processes, types are normally interpreted as their output types, e.g. the result of an addition process on natural numbers will be a natural number, thereby expressing the signature of the computation $\mathbb{N} \to \mathbb{N}$. An alternative, more characteristic interpretation is to look at input and output values respectively, so that the computational terms are natural numbers and the type is their output value. For simplicity, we focus on the latter. Let us consider an example:

> *With a fair die, the theoretical probability to obtain a 4 is*
> 0.16666666666.

One way to express this more precisely is to say: under the probability distribution corresponding to a fair die, throwing the die will produce 4 with a theoretical probability of 0.16666666666. Note that here the process of throwing the die can be seen as a computational process with 4 as its output. It is now quite natural to see computations with probabilistic outputs as possible denotations of hyperintensional terms:

1. the claim that

   the theoretical probability to obtain 4 is 0.16666666666

   can be seen as the subjective probability 1 measuring the degree of belief that the die is fair at least with respect to output 4;

2. the claim assigns a probability value valid under a number of distinct initial probability distributions expressing either a fair or a biased die; hence, an assertion of identity valid for two processes observed for a given probabilistic output, may fail for different outputs; in other words, one may believe that "the theoretical probability to obtain a 4 is 0.16666666666", without necessarily believing to be using a fair die.

To see this in more detail, consider rolling two dice, where d_1 is fair, and d_2 is biased (although we might not know that): d_1 (theoretically) shows a probability 0.16666666666 of outputting 4; d_2, may be biased in a way that the probability of outputting 4 is still the same, while the probability of outputting 5 is 0.2 and the probability of outputting any of 1, 2, 3, 6 is 0.158333333335. As their actual probability distribution is unknown, the two dice represent two systems which are behaviourally identical with respect to a given output, but structurally different otherwise. As we might not know what their actual structure is, biased or fair, they are epistemically opaque in the sense of an explanation of their working not being transparently accessible. An abstract procedure denoting the throw of a die with a probability 0.16666666666 of outputting 4 maybe actually refer to both d_1 and d_2.

In [4], further extended in [5], a natural deduction system for probabilistic computations is offered, with the above sentence formalised as follows:

$$\{x_d{:}1_{1/6}, x_d{:}2_{1/6}, x_d{:}3_{1/6}, x_d{:}4_{1/6}, x_d{:}5_{1/6}, x_d{:}6_{1/6}\} \vdash d{:}4_{0.16666666666}$$

This formula has: on the left-hand side of the derivability sign a probability distribution in which a random variable x associated with the process d receives uniformly probability 1/6 for each of its possible outputs $1, \ldots, 6$; under such distribution, one derives on the right-hand side the expected probability of 0.16666666666 to get output 4 from process d. A major task in [5] is to allow reasoning under an unknown initial distribution to infer trustworthiness of computations by comparing the frequency of a given observed output with its expected probability. Evaluating trustworthiness clearly seems a natural task for a language like HTLC which includes hyperintensional terms denoting processes which are identical for some output but are structurally distinct, as it is the case for probabilistic computations.

In this paper we sketch the extension of HTLC to probabilistic outputs in a language dubbed HTLC_p, identifying appropriate conditions for reasoning from probabilistic computations to their hyperintensional counteparts and back. We offer an overview of meta-theoretical results, and conclude with some comparison with the notion of semantic computation holding for Transparent Intensional Logic.

2 The Logic HTLC$_p$

2.1 Language

The syntax of HTLC$_p$ is a typed λ-calculus extended with probabilistic types and a type for hyperintensional terms denoting them:

Definition 1 (Grammar).

$$T_p ::= T_p \mid (T_p \; T_{p_1}^1 \ldots T_{p_n}^n) \mid (T + T')_p \mid *^{T_p}$$

$$t ::= x_t \mid [\lambda x_{t_1} \ldots x_{t_n}.t] \mid [t \; t_1 \ldots t_n] \mid t^*$$

In the present context, we are especially interested in an interpretation of types following the Curry-Howard-De Bruijn isomorphism: terms are computations and types are their (probabilistic) outputs. As mentioned above, probabilitic outputs can be seen both as types of outputs and as output values, and we focus here on the latter interpretation. Hence the first significant difference from HTLC in this grammar is the addition of a probabilistic index p to types.

The type syntax includes extensional entities T_p. Among these we might include probabilistic truth values, individuals with a certain probability to belong to a given universe, but also output values in those domains, e.g. a random number generator and its output 4 in the domain $\mathbb{N}$ with a given probability to be produced. The latter will be the most natural interpretation for HTLC$_p$.

Next, we add intensional entities with a given probability p of outputting type T when the probabilities of their input type is respectively p_1 for T_1 up to p_n for T_n). While in HTLC these are constrained to total functions, by having functions with probabilistic arguments $T_{p_1}^1 \ldots T_{p_n}^n$ we obtain an output T_p, with $p = \Pi(p_1 \cdot \ldots \cdot p_n)$ which implements a form of indeterminacy. We simplify the arrow notation of multi-argument functions $(T_1 \to \cdots \to T_n \to T)$ with the pair notation $(((T\ T_n)\dots)T_1)$. As in standard typed λ-calculi we use association to the left when dealing with function types, so the curried version can be rewritten as $(T\ T_n\ \dots\ T_1)$. We can build higher-order functions that take functions as arguments and return a function as value. Specific to this calculus is the case where all the input types to a function express the probability distribution of outputs under which a given one can be obtained: $(\lambda x)t : (T_p\ T'_{1-p})$ is thus a specific form of a function type term in which the probability p of t to get output T depends on the probability $1-p$ of t to get output T'. Returning to our example from the previous section, consider the expression $d : (4_{1/6}\ 1_{1/6}\dots 6_{1/6})$, which expresses the function type denoting a fair die such that if there is a probability $1/6$ to get any of its possible outputs, then there is a probability $1/6$ to get output 4.

Moreover, and precisely to obtain such functional interpretation, we add a disjunction type, through which we can express exclusive probabilistic outputs by the same computation, summing up to probability $p = 1$. Hence, for our fair die we can induce the expression $\vdash d : (1+\cdots+6)_1$ saying that the probability of getting one of outputs $1\dots 6$ from die d is 1.

Finally, we include hyperintensional entities of type $*^{T_p}$, where a term of this type denotes an entity with a given probability p of having an extensional or intensional type T. Hyperintensional types are thus interpreted as procedures denoting computations with a given probabilistic output. We explain this further below referring to a term of such type.

Terms of the language are thus computations. As in HTLC we include variables, abstraction terms denoting functions and application terms denoting values of functions on given arguments. Here another notable difference from HTLC is that variables get indexed with the term (or computation) for which they act as a random variable, asserting the probability of its output. Hence, the list of variables on the left-hand side of an expression provides the probability distribution for a term on its right-hand side. To follow up on the example above,

$$\{x_d : 1_{1/6}, \ldots, x_d : 6_{1/6}\} \vdash d : 4_{1/6}$$

denotes the expression that states that under the probability distribution of a fair die d, there is a probability $1/6$ to have output 4.

Hyperintensional terms denote abstract procedures for computations of any of the above forms, with a given probability for the corresponding output: the two procedures denoted by the same hyperintensional term have necessarily identical probabilities for some of their value (and in turn necessarily identical beliefs are attributed to them for those values), while they might differ for probabilities assigned to other output values (and in turn different beliefs are attributed to them for those values). This is possible for example because there is no explicit formulation of the internal structure of such computations, or such structure is not accessible (hence, technically they might be black boxes, opaque in the sense used above). For example, the intuition is that an expression $t^* : *^{4_{0.16666666666}}$ denotes any process, including any rolling die (fair or not), which may return 4 with probability 0.16666666666. Currently, we do not distinguish in the syntax the theoretical probability of the fair die from the expected probability or frequency of a die observed rolling (something we explicitly do in TPTND), but will allow ourselves to consider such distinctions informally.

To sum up, a judgement of HTLC_p is a formula of the form

$$\{x_1 : T_{p_1}^1, \ldots, x_n : T_{p_n}^n\} \vdash t : T_p$$

saying that a computation t outputs T with probability p, provided that x_1 outputs T^1 with probability p_1, up to x_n outputs T^n with probability p_n.

2.2 Rules

Inference rules for computational terms of arbitrary types are adapted from [5] and further extended with rules to define the meaning of

317

hyperintensional terms denoting probabilistic computations, see Figure 1. Note that in this preliminary formulation we merge what in TPTND [5] is strictly separated, namely reasoning about random variables, single experiments and rules for sampling.

The Assumption rule allows the derivation of a typed random variable from its own assumption. The Abstraction rule allows to construct a λ-term for a function type that takes inputs $T^1, \ldots, T^n$ respectively with probabilities $p_1, \ldots, p_n$ (possibly with more assumptions encoded in Γ), returns output T with probability p. The Application rule creates a term of type T denoting the value of a function, expressed by a computation t on the n-tuple of arguments expressed by computational terms $t_1, \ldots, t_n$, where the probability of T is computed as the product of the probabilities of all its inputs. We further formulate rules for the additivity of exclusive probabilistic outputs. Given a probability distribution encoded in Γ and process t giving output T with probability p and output T' with probability p', process t will give either output with a probability $p + p'$, with such probabilities adding up to 1. Taken the disjunction of all possible outputs $T + T'$ of process t and the probability p of one such output T, the probability of the remaining output T' will be $1 - p$.

Next we add rules for hyperintensional terms. The Trivialization rule defines the process of going from a given computation t which outputs T with probability p to the hyperintensional term t^* which denotes a construction of the computation denoted by t, i.e. t^* denotes any of the computations with the same output as t. When the trivialized computation t is of type T_p, Trivialization allows to shift from a an extensional computation to a hyperintensional one producing it. When the trivialized computation t is of type $(T_p \, T_{p_1} \ldots T_{p_n})$, Trivialization allows to shift from a functional computation to a hyperintensional one producing it. We do not consider higher-order triviliased terms here. The criterion for selecting a procedure t^* denoting any computation with output T_p is fidelity, which can be evaluated by the corresponding elimination rule. Execution proceeds from a hyperintensional type $*^{T_p}$ to its intended denoted extensional or intensional computational term with output T with probability p. In order to

318

do so, i.e. to identify a computation t' that satisfies the hyperintensionally denoted output T_p, we exploit the trustworthiness criteria defined for TPTND in [5]. Given a procedure t^* of type $*^{T_p}$, Execution returns a computation t' denoting its output, that is a term of type T with probability p', provided the absolute difference between the intended and the observed probability of the outputs is below a given critical threshold, typically the confidence interval for the intended probability parametrized over the number n of times one has observed t' outputting T over the overall number of occurrences of t' with any other output. Similarly so for a term t^* of type $*^{(T\ T_1...T_n)}$, where Execution returns the functional computation t' that denotes an intensional entity of type $(T_p\ T_{p_1}^1 \ldots T_{p_n}^n)$.

Note that Execution requires canonical terms, i.e. terminating computations in normal form by β-reduction rules, see Figure 2. In these evaluation rules, we state: evaluation by argument substitution in abstraction; evaluation to equivalent abstracted terms; evaluation to equivalent applied terms; evaluation to equivalent arguments in application. We use $\rightarrow_\beta$ for these evaluation steps. Further, we extend reduction rules with one specific instance to account for the reduction of terms with one equivalent among exclusive distinct outputs, see Figure 3. We use $\rightarrow_p$ for this evaluation step. We use simply $\rightarrow$ for the common evaluation relation and $\twoheadrightarrow$ for its transitive and reflexive closure. The latter evaluation rule allows us to establish two important variations from HTLC. First: subject reduction is in a weaker form as it allows some form of subtyping by establishing that if t' is a term obtained by an evaluation rule (including $\rightarrow_p$) from term t, its type T' must be included in the type T of the latter, up to identity, see Figure 4. Second: term identity is defined in a stronger form, under reduction for all possible output types, see Figure 5.

Coming back to our example, consider two expressions: $\Gamma \vdash d_1 : 4_{0.16666666666}$ and $\Delta \vdash d_2 : 4_{0.16666666666}$ denote respectively that the first and the second die show a behaviourally identical probability to output 4 (assuming here the same number of throws), under the (possibly unknown) distribution of their outputs encoded respectively in Γ and Δ. Applying Trivialization to each, will return the same term

$$\frac{}{x_t : T_p \vdash x_t : T_p} \; \text{Assumption}$$

$$\frac{\Gamma, x_1 : T^1_{p_1}, \ldots, x_n : T^n_{p_n} \vdash t : T_p}{\Gamma \vdash [\lambda x_1 \ldots x_n . t] : (T_p \; T^1_{p_1}, \ldots, T^n_{p_n})} \; \text{Abstraction}$$

$$\frac{\Gamma \vdash t : (T_p \; T^1_{p_1}, \ldots, T^n_{p_n}) \quad \Gamma_1 \vdash t_1 : T^1_{p_1} \ldots \Gamma_n \vdash t_n : T^n_{p_n}}{\Gamma, \Gamma_1, \ldots \Gamma_n \vdash [t \; t_1 \ldots t_n] : T_{p=p_1 \cdots p_n}} \; \text{Application}$$

$$\frac{x_t : T_p, x_t : T'_{p'}, \vdash t : T_p \quad x_t : T_p, x_t : T'_{p'}, \vdash t : T'_{p'}}{x_t : T_p, x_t : T'_{p'}, \vdash t : (T + T')_{p+p' \leq 1}} \; \text{+-I}$$

$$\frac{x_t : T_p, x_t : T'_{p'}, \vdash t : (T + T')_1 \quad x_t : T_p, x_t : T'_{p'}, \vdash t : T_p}{x_t : T_p, x_t : T'_{p'}, \vdash t : T'_{1-p}} \; \text{+-E}$$

$$\frac{\Gamma \vdash t : T_p}{\Gamma \vdash t^* : *^{T_p}} \; \text{Trivialization}$$

$$\frac{\Gamma \vdash t^* : *^{T_p} \quad \Delta \vdash t' : T_{p'} \quad |p - p'| \leq \epsilon(n)}{\Gamma, \Delta \vdash t' : T_{p'}} \; \text{Execution}$$

Figure 1: HTLC$_p$ Rules System

$t^* : *^{4_{0.16666666666}}$. Hence, given a term $t^* : *^{4_{0.16666666666}}$, Execution would return either d_1 or d_2. But assuming d_1 fair and d_2 biased, e.g. such that the probability of outputting 4 is still the same, while the probabilities of outputting 5 is 0.2 and the probability of outputting any of $1, 2, 3, 6$ is 0.158333333335, the two dice are identical with respect to a given output (namely 4), but structurally different otherwise. Hence, Subject Reduction will be satisfied for at least one evaluation (namely of the term type with output $4_{0.16666666666}$), but not for all such outputs, making term identity fail. The desirable property to infer the identity of the two dice with respect to output

$$\Gamma \vdash [[\lambda x_1 \ldots x_n . t] \; t_1 \ldots t_n] \to_\beta t[x_1/t_1 \ldots x_n/t_n] : T_p$$

$$\frac{\Gamma, x_1 : T_{p_1}^1, \ldots, x_n : T_{p_n}^n \vdash t \to_\beta t' : T_p}{\Gamma \vdash [\lambda x_1 \ldots x_n . t] \to_\beta [\lambda x_1 \ldots x_n . t'] : (T_p \; T_{p_1}^1 \ldots T_{p_n}^n)} \; \beta\text{-Abstr}$$

$$\frac{\Gamma \vdash t \to_\beta t' : (T_p \; T_{p_1}^1 \ldots T_{p_n}^n) \quad \Gamma_1 \vdash t_1 : T_{p_1}^1 \ldots \Gamma_n \vdash t_n : T_{p_n}^n}{\Gamma, \Gamma_1 \ldots \Gamma_n \vdash [t \; t_1 \ldots t_n] \to_\beta [t' \; t_1 \ldots t_n] : T_{p=p_1 \cdots p_n}} \; \beta\text{-App}$$

$$\frac{\Gamma \vdash t : (T_p \; T_{p_1}^1 \ldots T_{p_i}^i \ldots T_{p_n}^n) \quad \Gamma_1 \vdash t_1 : T_{p_1}^1 \ldots \Gamma_n \vdash t_n : T_{p_n}^n \quad \Gamma_i \vdash t_i \to_\beta t_i' : T_{p_i}^i}{\Gamma, \Gamma_1 \ldots \Gamma_n \vdash [t \; t_1 \ldots t_i \ldots t_n] \to_\beta [t \; t_1 \ldots t_i' \ldots t_n] : T_{p=p_1 \cdots p_i' \cdots p_n}} \; \beta\text{-App}$$

Figure 2: HTLC_p: β-rules

$$\frac{\Gamma \vdash t \to_\beta t : (T_p + T_{p'}')_1}{\Gamma \vdash t \to_p t : T_{p=1-p'}} \; \to_p$$

Figure 3: HTLC_p: Probabilistic reduction

4 is granted by a successful instance of the Execution rule, where the first premise is induced by Trivializion on e.g. d_1 and the second premise is an instance of d_2:

$$\frac{\Gamma \vdash t^* : *4_{0.1666666666} \quad \Gamma, \Delta \vdash d_2 : 4_{0.1666666666} \quad |\, 4_{0.1666666666} - 4_{0.1666666666}\,| = 0}{\Gamma, \Delta \vdash d_2 : 4_{0.1666666666}} \; \text{Execution}$$

Similarly, distinguishing among the two dice with respect to the outputs $1, 2, 3, 5, 6$ is possible by a failing instance of the Execution rule using any of the other outputs:

$$\frac{\Gamma \vdash t^* : *1^{0.1666666666} \quad \Gamma, \Delta \vdash d_2 : 5_{0.2} \quad |\, 4_{0.1666666666} - 4_{0.1666666666}\,| > \epsilon(n)}{\Gamma, \Delta \nvdash d_2 : 4_{0.1666666666}} \; \text{Execution}$$

2.3 Structural Properties

Structurally, the behaviour of HTLC_p differs from its deterministic counterpart and it is inspired by TPTND in [5], see Figure 6.

$$\frac{\Gamma \vdash t : T_p \qquad t \twoheadrightarrow t'}{\Gamma \vdash t' : T'_{p'} \subseteq T_p}$$

Figure 4: HTLC: Subject reduction

$$\frac{\Gamma \vdash t : T^i_{p_i}, \forall T^i_{p_i} \in \Gamma \mid \sum^i_{1...n} p_i = 1 \qquad t \twoheadrightarrow t'}{\Gamma \vdash t = t'}$$

Figure 5: HTLC: Term Identity

By the Weakening rule, given the probability of a computation t to have output T with probability p, provided the value assigned to some variable x_1 is T^1 with probability p_1; if the latter assignment is independent ($\perp\!\!\!\perp$) of a distribution Δ from which some process $t_2 : T^2_{p_2}$ follows, then the starting conditional probability can be extended by additional assumptions Δ and the probability p is left unaltered. In other words: any given probability distribution can be extended at will without inferring new probabilities, as long as no new dependencies are introduced.

The Contraction rule expresses a form of plausibility test on hypotheses. Given more than one distinct hypothesis on the theoretical probability p of some process with output T^1, on which computation t with output T with expected probability p depends, a contraction is a function $f[0,1]$ which extracts one value $x : T^1_{p_i}$ from the hypotheses $p_1, \ldots, p_n$. The function f can be chosen at will, e.g. the Maximum Likelihood function.

3 Normalization

In this section we adapt the Normalization strategy shown for HTLC in [11] and provide an appropriate novel interpretation for the role of hyperintensional terms of probabilistic computations.

Applying Execution is possible to derive distinct equivalent com-

$$\frac{\Gamma, x_1 : T^1_{p_1} \vdash t : T_p \qquad \Delta \vdash t_2 : T^2_{p_2} \qquad x_1 : T^1_{p_1} \perp\!\!\!\perp \Delta}{\Gamma, x_1 : T^1_{p_1}, \Delta \vdash t : T_p} \text{ Weakening}$$

$$\frac{\Gamma, x_1 : T^1_{p_1}, \ldots, x_1 : T^1_{p_n} \vdash t : T_p}{\Gamma, x_1 : T^1_{f(p_1,\ldots,p_n)} \vdash t : T_p} \text{ Contraction}$$

Figure 6: Structural Rules

putations (i.e. terms for which the symmetric closure of $\twoheadrightarrow$ holds) of a base type or of a function type, and hence they return the same output value with the same probability when executed. When such reduction includes disjunction types, evaluation may lead to terms that satisfy only some of their subtypes. This is the classical example of failing identity for hyperintensions of probabilistic computations. Normalization grants the identity of processes with outputs with the same probability, when these are obtained by Execution on the common hyperintensional term. In particular, an application of Trivialization on distinct but reducible terms t_1, t_2 (denoting e.g. respectively the fair and the biased die) may induce distinct hyperintensional terms t^*_1, t^*_2 (two distinct representation of the distributions) of the same type. We would like to show that they may be indistinguishable from the point of view of output 4 when Execution is applied, but distinguishable for other outputs, as illustrated above. To show the first case, we need to restrict evaluation to β-reductions:

Lemma 1 (Trivialized Diamond). *Let $\Gamma \vdash t_1 : T_p$, $\Gamma \vdash t_2 : T_p$ and $t_1 \to_\beta t_2$. Let, moreover, $t^*_1 : *^{T_p}$ be obtained from $t_1 : T_p$ by Trivialization, and $t^*_2 : *^{T_p}$ be obtained from $t_2 : T_p$ by Trivialization. Then $t^*_1 \twoheadrightarrow_\beta t^*_2$.*

Proof. Identical to the proof for HTLC, see [11]. □

Theorem 1 (Church-Rosser). *If $\Gamma \vdash t : T_p$, $t \twoheadrightarrow_\beta t'$ and $t \twoheadrightarrow_\beta t''$ then there is a term u such that $t' \twoheadrightarrow_\beta u$ and $t'' \twoheadrightarrow_\beta u$ and $\Gamma \vdash u : T_p$.*

323

Proof. By induction on t, t', t'', and u using Subject Reduction and the Diamond Lemma if the term u is of the form t'^* (and thus t''^*). □

Note that the above results do not hold in general for $\twoheadrightarrow$, i.e. when evaluation includes the reduction from disjunction types to any of their subtypes, as the same premise may induce different conclusions (i.e. differently probabilistically indexed outputs).

To illustrate this further, let us come back to our example. Consider two dice, for which it can be proven that if $d_1 : 4_{0.16666666666}$ and $d_2 : 4_{0.16666666666}$, then the hyperintensional term d_1^* will also reduce to d_2^*, i.e they will have the same type (behaviourally: the same output) by Subject Reduction. Then for d_1, d_2 there is also a term u to which they both reduce and for which $u^* : *^{4_{0.16666666666}}$ can be obtained by Trivialization. Note, however, that such reasoning cannot be performed for all other outputs of d_1, d_2 as requested by Term Identity. Hence, when $\twoheadrightarrow_p$ is involved different types may be obtained and identity of terms resulting from Execution fails.

4 On diverging semantic computations

The reading of hyperintensions as terms denoting computational processes with probabilistic outputs allows to maintain a distinction natural in TIL, see [14]:

- syntactic computation (also dubbed "construction transformation") corresponds to a computational step obtained by a β-reduction;

- semantic computation (or "v-constructing") correlates to a computational step returning a specific interpretation or evaluation of a λ-term.

Note that for construction transformation it is assumed that β-reduction preserves strict equivalence. From this perspective, in our notation $\twoheadrightarrow_\beta$ is syntactic computation, while typing expresses semantic computation, which includes $\twoheadrightarrow_p$. Then Term Identity and Normalization express syntactic and semantic equivalence. However, probabilistic computations with our weaker notion of Subject reduction introduce diverging semantic computations by non-reducible syntactic transformations:

Proposition 1. *Consider* t_1, t_2 *such that* $t_1 : T_p$ *and* $t_2 : T_p$, *then* $t_1 \twoheadrightarrow_\beta t_2$. *Consider further* $t_1 : U_{1-(p-n)}$ *and* $t_2 : U'_{1-(p-n')}$ *but it is not the case that* $t_1 \twoheadrightarrow t_2$ *because in particular it is not the case that* $t_1 \twoheadrightarrow_p t_2$. *Then* $t_1 \neq t_2$.

In the present paper we do not further investigate the relation of HTLC_p hyperintensions with other relevant aspects of TIL, e.g. the different kinds of β-conversions possible when considering the contrast between syntactic and semantic transformations, or the connection to Church-Rosser, see [9, 12]

5 Conclusions

The system HTLC_p introduced in this paper is a typed λ-calculus with hyperintensions for probabilistic computations. Its semantics is defined by rules for expressions in which contexts are representations of probability distributions, terms are computational processes and types are their outputs (values or types). Under this reading, hyperintensional terms are useful to denote computations with some identical probabilistic outputs, but obtained under possibly different probability distributions. The ability to make such distinctions is granted by a strong form of term identity and a Normalization result by appropriate versions of the Diamond Lemma and the Church-Rosser Theorem holding for β-reducible terms in a system that allows further reducibility of probabilistic terms.

A natural domain of application for logics such as HTLC_p is the verification of AI systems for which only certain behavioural properties are accessible, while their inner structure may remain opaque. In such cases, the ability to evaluate identity with respect to transparent models is essential to establish safety and liveness properties, see

[1, 2], or more generally accuracy and trustworthiness, as done e.g. in [16].

Extensions of HTLC$_p$ concern in particular: its relational semantics, further study of its meta-theoretical properties and applications to bias identification for labeling methods in AI.

References

[1] Nicola Angius and Giuseppe Primiero. The logic of identity and copy for computational artefacts. *J. Log. Comput.*, 28(6):1293–1322, 2018.

[2] Nicola Angius and Giuseppe Primiero. Copying safety and liveness properties of computational artefacts. *Journal of Logic and Computation*, 08 2022. exac053.

[3] M. J. Cresswell. Hyperintensional logic. *Studia Logica*, 34(1):25–38, 1975.

[4] Fabio Aurelio D'Asaro and Giuseppe Primiero. Probabilistic typed natural deduction for trustworthy computations. In Dongxia Wang, Rino Falcone, and Jie Zhang, editors, *Proceedings of the 22nd International Workshop on Trust in Agent Societies (TRUST 2021) Co-located with the 20th International Conferences on Autonomous Agents and Multiagent Systems (AAMAS 2021), London, UK, May 3-7, 2021*, volume 3022 of *CEUR Workshop Proceedings*. CEUR-WS.org, 2021.

[5] Fabio Aurelio D'Asaro and Giuseppe Primiero. Checking trustworthiness of probabilistic computations in a typed natural deduction system, 2022. https://arxiv.org/abs/2206.12934.

[6] Bruno de Finetti. *Theory of Probability: A Critical Introductory Treatment*. New York: John Wiley, 1970.

[7] M. Duží. Hyperintensions as abstract procedures. Presented at Congress on Logic Methodology and Philosophy of Science and Technology 2019, 2019.

[8] M. Duží, B. Jespersen, and P. Materna. *Procedural Semantics for Hyperintensional Logic – Foundations and Applications of Transparent Intensional Logic*, volume 17 of *Logic, Epistemology, and the Unity of Science*. Springer, 2010.

[9] Marie Duží and Miloš Kosterec. A valid rule of β-conversion for the logic of partial functions. *Organon F*, 24(1):10–36, 2017.

[10] Marie Duzí and Marek Mensík. Inferring knowledge from textual data by natural deduction. *Computación y Sistemas*, 24(1), 2020.

[11] Michal Fait and Giuseppe Primiero. HTLC: hyperintensional typed lambda calculus. *FLAP*, 8(2):469–496, 2021.

[12] Milos Kosterec. Substitution contradiction, its resolution and the church-rosser theorem in TIL. *J. Philos. Log.*, 49(1):121–133, 2020.

[13] Henry Ely Kyburg. *Studies in Subjective Probability*. Krieger, 1964.

[14] Ivo Pezlar. On two notions of computation in transparent intensional logic. *Axiomathes*, 29(2):189–205, 2018.

[15] C. Pollard. Hyperintensions. *Journal of Logic and Computation*, 18(2):257–282, 2008.

[16] Alberto Termine, Giuseppe Primiero, and Fabio Aurelio D'Asaro. Modelling accuracy and trustworthiness of explaining agents. In Sujata Ghosh and Thomas Icard, editors, *Logic, Rationality, and Interaction - 8th International Workshop, LORI 2021, Xi'ian, China, October 16-18, 2021, Proceedings*, volume 13039 of *Lecture Notes in Computer Science*, pages 232–245. Springer, 2021.

THE ZINFANDEL/PRIMITIVO PUZZLE

BJØRN JESPERSEN
Faculty of Philosophy, University of Groningen, and Department of Computer Science, VŠB-TU Ostrava
bjornjespersen@gmail.com

What does Transparent Intensional Logic make of a case like "Lærke likes Zinfandel, but dislikes Primitivo"? Does it perhaps fit the template of "Lærke adores woodchucks, but detests groundhogs", which TIL likens to a Mates-style puzzle, which is then explained away as a non-puzzle? Or maybe rather the template of "Lærke is seeking an abominable snowman, but not a yeti", which is equally well-understood in TIL? Or perhaps something altogether different? I have argued elsewhere that popular examples in the vein of "Lærke adores woodchucks, but detests groundhogs" fail to motivate hyperintensional distinctions among attitude complements and the invocation of opacity. Pairs of synonymous predicates like {'is a woodchuck', 'is a groundhog'} are freely intersubstitutable in non-quotational intensional and hyperintensional attitude reports in TIL, as long as our only concern is to preserve the identity of the agent's attitude. Poetic, rhetorical or pragmatic factors may tell against substitution, though. I will argue below that the pair of predicates {'is a Zinfandel grape', 'is a Primitivo grape'} is importantly different from the pair {'is a woodchuck', 'is a groundhog'}, and so are not freely intersubstitutable. The distinction between these two cases is drawn while observing compositionality and transparency.

Grant of SGS No. SP2022/123 'Application of Formal Methods in Knowledge Modelling and Software Engineering IV'. I wish to thank Miloš Kosterec and Marie Duží for comments — though Marie may not have known at the time where the fruits of our discussion would end up.

This essay is dedicated to Marie in deep appreciation of our friendship and cooperation over the last two decades — with much more to come!

There is much to like about this passage from the entry in the Stanford Encyclopedia of Philosophy on hyperintensionality:

> One logically sophisticated, neo-Fregean structured account, offering a systematic analysis of a range of hyperintensional phenomena, is the *Transparent Intensional Logic* approach. The view, pioneered by Tichý [...], treats the meanings of expressions as given by structural procedures, called *constructions*, built out of entities that are somewhat like Fregean senses. In particular, different names and different predicates, even if they necessarily co-designate, may be associated with different senses, so the meanings of 'Robin Hood' and 'Robin of Locksley', or of 'furze' and 'gorse', may be distinct even if those meanings are not built up out of others. This gives the system resources to handle many hyperintensional contexts straightforwardly. In particular, it manages to give a powerful compositional account where other approaches have to resort to pragmatics. [1]

But as a card-carrying TILian, I must take issue with two claims made there. One is that the pair of names {'Robin Hood', 'Robin of Locksley'} and the pair of predicates {'is furze', 'is gorse'} would semantically be on the same page; they are not. This pair of names is like the pair {'Mark Twain', 'Samuel Langhorne Clemens'}, where one is a pen name and the other a given name plus surname, and they do not have the same meaning. In fact, they do not even necessarily co-designate, or co-designate at all, because the pen name designates an intension and the given name-plus-surname designates an individual.[1]

---

[1]See Duží et al. [5, 298, 3.2]. At some empirical indices, the pen name and the given name will share the same *reference*, namely the individual that is both the denotation of the given name-plus-surname and the value of the intension denoted

The other issue is that the pair {'is furze', 'is gorse'} not only necessarily co-designate, but are synonymous and as such semantically indistinguishable. In fact, the pair is an instance of lexical synonymy. The two nouns 'furze', 'gorse' are lexemes, i.e., lexically simple units, that are related by cognitive synonymy or descriptive meaning.[2] The fact that these two predicates are semantically indistinguishable makes one of them semantically redundant. Pairs, or even triples, of lexical synonyms are abundant in English. Standard examples include {'is a puma', 'is a cougar'}, and {'is a whistle-pig', 'is a groundhog', 'is a woodchuck'}. It is clear what TIL makes of such examples, and this stance deviates from what appears to be the received view. Our stance is that these examples are instances of syntactic distinctions without a semantic difference. There may be lots of reasons for favouring one predicate over another, but none of them is of cognitive or semantic or logical import. Rather such reasons are informed by considerations as to which terms fit together lyrically or rhetorically with which other terms, or which terms the target audience is likely to know, or by the perlocutionary effect the speaker seeks to elicit. So-called absolute synonymy is most likely not attainable.[3]

Below I will go through a few of the templates for analysis TIL

---

by the pen name. But the category of reference falls outside the perimeter of semantics and is instead an empirical category. See (*ibid.*, 13-14). The denotation of a term with the semantics of a proper name coincides extensionally with its reference.

[2]Or so I am assuming. I am not involved in field linguistics, so I do not know for a fact that they are. Strictly speaking, I am working from the assumption that *if* a pair of lexemes are identical in point of descriptive (but perhaps not expressive) meaning *then* they are semantically and inferentially indistinguishable and, *therefore*, intersubstitutable in any non-quotational context.

[3]As for the calibration of synonymy I am aiming for, consider substitution of synonyms in poetry. There, the bar for substitutability has gone up (synonymy being necessary but not sufficient for substitutability). See [7]. See also [13, 195] on the hyperintensionality of poems being such that 'replacing an expression with its synonym changes [the] meaning [of the poem]', offering as an example the pair {'sheen', 'lustre'} as occurring in a particular stanza. Contrast this with Frege's example in [6, 37] of four words for *horse*. No distinction among them is relevant to any *Gedanke* expressed by using any of these four predicates. This is the measure of synonymy I have in mind.

makes available for establishing whether or not one term (such as a
name or predicate) or a longer phrase can be validly substituted for
another within an ascription of a hyperintensional attitude to a given
agent.[4] The objective is to assign the right template to a specific
case. I will also explain a constraint that informs the correct use of
"$F = G$", which expresses identity of properties.

1

Here is the case I want to take a closer look at in this paper: the
pair of predicates {'is a Zinfandel grape', 'is a Primitivo grape'}. The
backstory is the following. It is a key metaphysical assumption that
types or races of wine grapes are individuated by their DNA:

> Zinfandel is a grape primarily grown in California. Prim-
> itivo is a grape primarily grown in Italy [Apulia]. But
> these grapes are actually the same. And even more, 'Prim-
> itivo' and 'Zinfandel' were never the original names for
> this grape. The grapes are originally from Croatia, where
> they're called 'Tribidrag' and sometimes 'Crljenak kašte-
> lanski'. While we take this information for granted in 2015,
> we didn't know Zinfandel and Primitivo were the same
> grape until [1968]. Furthermore, we didn't trace the roots
> of these grapes to Tribidrag until 2001.[5]

Two data need to be reconciled:

- that Zinfandel and Primitivo are genetically identical grapes;

- that this identity had to be established by scientific rather than
  analytic or semantic means.

---

[4]I am presupposing some familiarity with the ideography of TIL. I will be
providing references to relevant sources throughout this essay.

[5]https://www.vivino.com/wine-news/the-origin-of-zinfandel-and-
primitivo. Unfortunately, it seems that this site is no longer accessible. The
reader might be referred here instead: https://www.city-vino.com/blogs/blog/
primitivo-vs-zinfandel-vs-tribigdrag/

Those philosophers who are still taken with Kripke's category of necessity a posteriori will have a ready answer: Zinfandel and Primitivo are necessarily identical, but this identity is neither logical nor analytic nor nomic but metaphysical, and must be established empirically/scientifically. However, TIL argues that the Marcus-Kripke schema "if $a = b$ then $\Box(a = b)$" reduces to this triviality: if $a = a$ then $\Box(a = a)$. The necessitation in the consequent just makes explicit what is implicit in the antecedent, namely that the true instances of "$a = b$" conforming to the Marcus-Kripke schema are instances of self-identity (or strict identity). Hence, the Kripkean cannot preserve the Zinfandel/Primitivo case as a Frege case, because Frege cases are obviously not about establishing self-identity by empirical/scientific means.

2

The first template concerns contingent co-extensionality of two different intensions. For an example, let us take the best-known Frege case. TIL offers this analysis of "Hesperus is Phosphorus":

$$\lambda w \lambda t [^0{=}\ ^0H_{wt}\ ^0Ph_{wt}]$$

Types: $= /(o\iota\iota); H, Ph/\iota_{\tau\omega}; w/*_1 \to \omega; t/*_1 \to \tau$. The semantics is that 'Hesperus' and 'Phosphorus' both denote *offices*. 'Hesperus' denotes the office of the individual that is the brightest non-lunar celestial body in the evening sky; 'Phosphorus' denotes the office of the individual that is the brightest non-lunar celestial body in the morning sky. These two offices have been extensionalized so as to descend from the type $\iota_{\tau\omega}$ to the type ι. The Closure above produces the proposition/$o_{\tau\omega}$, or empirical truth-condition, that these two offices are co-occupied by the same individual. Or more technically, the Closure produces a function from worlds to a partial function from times to truth-values that returns the truth-value **T** whenever these two different intensions are co-extensional. The actual world and the present moment are among the worlds and times that satisfy this condition. It counts as a worthwhile piece of astronomical knowledge to know about this co-occupation, because this co-occupation is a matter of nomic, hence a posteriori, and not analytic, mathematical or logical, hence a priori, necessity.[6] It is an additional piece of astronomical knowledge, and one that cannot be coaxed from this instance of co-occupation, that the shared occupant is Venus/ι. This template is suitable for cases involving two definite descriptions, or syntactic names with the semantics of definite descriptions, and two offices. The formula above readily generalizes to properties; just replace $H, Ph/\iota_{\tau\omega}$ by $F, G/(o\iota)_{\tau\omega}$:

$$\lambda w \lambda t[^0=' \ ^0F_{wt} \ ^0G_{wt}]$$

Type: $=' /(o(o\iota)(o\iota))$. This Closure presents the proposition that is satisfied at those worlds and times, at which the extension (a set) of F is identical to the extension of G; for instance, when it is the case that everybody is cynophile iff they are oenophobic. These two properties are not internally related in any way; their co-extensionality is as contingent as anything. This template is appropriate for cases involving two predicates and two properties.

The second template addresses instances of one property and several predicates.[7] I will consider the property of being a woodchuck and the property of being a groundhog.[8] First of all, I am going to affirm that the *Trivialization* of the property of being a woodchuck is identical to the *Trivialization* of the property of being a groundhog.[9]

---

[6]It cannot be read off of the formula that the logically and analytically contingent co-extensionality of H, Ph is a case of nomic necessity; i.e., their co-extensionality is nomically necessary. This fact is one I am adding in prose. On how to capture nomic necessity in TIL, see Duží et al. [5, §4.5] for a sketch.

[7]The exposition and discussion of this second template draws on material from my (ms.).

[8]Cf. https://en.wikipedia.org/wiki/Groundhog.

[9]The principle of individuation regulating meaning in TIL is *procedural isomorphism*. This principle identifies any pair of structured meanings as procedurally isomorphic, as soon as their divergence in structure is semantically irrelevant. Such structured meanings, though perhaps procedurally distinct, are semantically indistinguishable. They are also indistinguishable in point of cognitive value. As a limiting case, a pair of structured meanings will be identical, and not merely isomorphic, procedures. This is realized when a pair of structurally atomic meanings (i.e., procedures of one step) are one and the same meaning. There is no

Hence, $=''$ is of type $(o*_1 *_1)$. I will use infix notation for '$=$' for better readability, and leave out the Trivialization of the identity relation.

$$[^{00} Woodchuck\ =''\ ^{00} Groundhog]$$

This one Trivialization is encoded in two different ways, '$^0 Woodchuck$' and '$^0 Groundhog$'. In a regimented ideography purged of notational redundancy, only one of these two terms would survive. Had I instead written

$$[^0 Woodchuck\ ='''\ ^0 Groundhog]$$

I would have expressed that the *property* of being a woodchuck is identical to the *property* of being a groundhog. This is also true, as per assumption, though it is quite another claim. One of the two predicates involved is redundant in a regimented ideography. *Type:* $='''/(o(o\iota)_{\tau\omega}(o\iota)_{\tau\omega})$.

3

TIL is a broadly Fregean framework. For instance, the SEP entry on hyperintensionality is right that our hyperintensions (which Tichý dubbed *constructions* and we tend to call *procedures* these days) are created in the image of Frege's notion of *Sinn*. And the notion of *Funktion* — both understood as a mapping (function-in-extension) and as a computational procedure (function-in-intension) for generating a mapping — is a cornerstone of TIL. But TIL also deviates from the historical Frege. Not only does TIL come with a typed universe, but Frege's distinction between opaque and transparent contexts is not replicated in TIL. In TIL, all non-quotational contexts are transparent, because TIL adheres to an invariantist semantics.[10] The meaning

---

structural divergence between them, and the semantic and cognitive processing is the same for both elements of the respective pairs. See [3, 9, 11] for explanation of procedural isomorphism.

[10]Quotational or mixed contexts – e.g., "Lærke thinks that 'røgede ørreder og jordbærgrød er de skønneste sommerretter' ", "Vår thinks that smoked trout and 'jordbærgrød' are the most delicious summer dishes" — are complicated, because

and the denotation of a given term remains invariant across contextual embedding. And compositionality is respected throughout. But when a given term or phrase is embedded within an extensional or an intensional or a hyperintensional context, the granularity profile of the context dictates which semantic aspect of the term becomes salient.[11]

Here is a way to schematize the difference between transparency and opacity.[12] The type of $=$ is $(o\alpha\alpha)$, α an arbitrary type for generality. Let substitutability (itself presupposing the validity of Leibniz's Law) be encapsulated by this implication:

Substitutability. $a = b \rightarrow (\varphi \leftrightarrow \varphi[b/a])$

Then opacity is defined as a negated instance of substitutability:

Opacity. $a = b \wedge \neg(\varphi \leftrightarrow \varphi[b/a])$

By the lights of TIL, this conjunction is necessarily false, because the conjuncts are mutually exclusive. Instead, this is how TIL construes transparency, regardless of whether the implication is applied to extensional or intensional or hyperintensional contexts:

---

quoted words and phrases occur simultaneously used and mentioned. The attitude logic of TIL is geared toward used occurrences only. This is also seen from the fact that while TIL can quantify into any hyperintensional attitude context TIL cannot quantify into quotational or the linguistic portions of mixed contexts occurring as attitude complements.

[11] See [11, §3.6]. Non-extensional contexts are typically, though not necessarily, agent-involving. But extensional contexts can be agent-involving, too. Thus, "Lærke is kicking Vår" simply records physical interaction between two human bodies, so the agency being ascribed to Lærke amounts to her initiating a causal chain and does not include any intent or purpose or anything else of an intellectual nature, as found in appetitive (e.g., *seeking*) or contemplative (e.g., *doubting*) attitudes. I should add that "Lærke is kicking Vår" induces a different sort of extensional context than does "$1 + 2 = 3$". The former does not obtain throughout, or independently of, logical space, but only within a portion of it. Since *kicking* is a relation-in-intension, the sentence gets formalized thus: $\lambda w \lambda t[^0 Kick_{wt} \; ^0 L \; ^0 V]$. Every context comes with a modal profile in TIL, and the modal profile of "Lærke is kicking Vår" is contingency. The modal profile of "$1 + 2 = 3$" is necessity (and the modal profile of "$1+2 = 4$" is impossibility) due to being independent of logical space, which explains why these are two different kinds of extensional context.

[12] This exposition relies on Jespersen and Duží [11, §2.3], which in turn engages with Caie et al. [2]. See also Lederman [12].

336

Transparency. $a = b \rightarrow (\chi\varphi a = \chi\varphi b)$

χ is a placeholder for an operator with an arbitrary granularity profile. Substitutability within the scope of χ is a necessary condition for the identity of a and b outside the scope of $\chi : a = b \rightarrow (\chi\varphi a \leftrightarrow \chi\varphi[b/a])$. This raises the bar for true instances of "$a = b$". Hence, this sort of conjunction will consist of mutually exclusive conjuncts:

$$\varphi a = \varphi b \wedge (\chi\varphi a \wedge \neg\chi\varphi b)$$

What constrains our use of '$=$' is that substitution of identicals must be valid, and that if 'a', 'b' are constants with the same compositional semantic value then "$a = b$" must be true. Opacity, on the other hand, helps the opacitist to a true conjunction and the preservation of the Frege puzzles of cognitive significance. The complications that the adoption of opacity incurs — developing two logics, one for transparency and one for opacity; maintaining a system of double bookkeeping, one book for transparent contexts and another book for opaque contexts; pointless problems with coordination between transparent and opaque contexts; etc. — serve the purpose of maintaining a fairly simple semantics for "$a = b$", "φa", etc., in transparent contexts. Transparentists, for their (our) part, will have to discard the first conjunct in order to preserve the Frege puzzles. To see this, the conjunction

$$a = b \wedge (\chi\varphi a \wedge \neg\chi\varphi b)$$

comes out necessarily false, whereas the conjunction

$$a = b \wedge (\chi\varphi a \wedge \chi\varphi b)$$

comes out necessarily true, with the conjuncts of "$\chi\varphi a \wedge \chi\varphi b$" being mere notational variants. In order to preserve both transparency and the non-triviality of both conjuncts, $a = b$ and $(\chi\varphi a \wedge \chi\varphi b)$, the transparentist develops a more elaborate semantics for "$a = b$", "φa" and "$\varphi a = \varphi b$" that remains the same whether occurring within the scope of χ or not.[13]

---

[13]See Jespersen and Duží [11, §2.3] for further details, including an account in prose of a transparentist analysis of knowing that Hesperus is visible at night while not knowing that Phosphorus is visible at night.

Turning to the pair of predicates (F, G) specifically and interpreting χ as belief, the transparency constraint on identity outside the scope of the operator amounts to this:

$$F = G \rightarrow (B_a F_b = B_a G_b)$$

If being a woodchuck is identical to being a groundhog, then believing that b is a woodchuck is identical to believing that b is a groundhog. The identity of properties is a sufficient condition for the identity of beliefs (though note the caveat in fn. 14). The identity of beliefs is a necessary condition for the identity of properties. Thus, any fine-graining found in the consequent will be carried through to the antecedent. There is no Fregean puzzle involved, and two-way 'Millian' substitution of 'F' and 'G' in belief reports comes out valid.[14]

Therefore, the implication above is not right for the purposes of the Zinfandel/Primitivo example. The Zinfandel/Primitivo case is different from the groundhog/woodchuck case. It is one thing to believe, or know, that the wine in the glass before you is a Zinfandel and another thing to believe, or know, that the wine in the glass before you is a Primitivo. Yet the wine in both glasses has been squeezed from genetically identical grapes. But we have barred ourselves from simply writing "Zinfandel = Primitivo". Neither predicate is redundant.[15] So, what is the right move to make at this point?

---

[14]The implication above does come with one caveat, though. Empirical properties are individuated up to co-intensionality (i.e., necessary co-extensionality) so the property of being half-full comes out identical to the property of being half-empty. Yet I want to allow that it is one thing to believe that the pint before you is half-empty and another that it is half-empty. This is feasible, as soon as the belief is hyperpropositional. There are infinitely many hyperintensional modes-of-presentation of this one property. Importantly, these are the respective meanings of different predicates, such as 'is half-empty', 'is half-full', sporting different constituents or structures (see [10, 332]). The values of F, G are restricted to properties that are presented in identical manners and denoted by synonymous predicates, such as {'is furze', 'is gorse'} or {'is a cougar', 'is a puma'}. This is also why the *paradox of analysis*, which applies to pairs such as {'is a bachelor', 'is a single male'}, does not apply to the thus restricted values of F, G.

[15]In my (ms.) I offer "Boris Johnson = Alexander de Pfeffel" as an example of what I take to be an instance of a trivial identity, 'Boris Johnson' and 'Alexander de Pfeffel' arguably being semantically indistinguishable. Linguistic incompetence,

4

TIL generates its invariantist semantics by generalizing from the hard-est case.[16] Thus, in order to offer a semantics for "Zinfandel is Prim-itivo", we need to first offer a semantics for a pair of attitude ascrip-tions in the vein of {"Lærke believes that Zinfandel is tasty", "Lærke believes that Primitivo is tasty"}. If these two ascriptions were syn-onymous, then "Zinfandel is Primitivo" would be simply using two different predicates to express that a particular grape is self-identical. For background, I will quickly go through my account of attitude ascriptions involving 'is a woodchuck' and 'is a groundhog' to show why it is not suitable for attitude ascriptions involving 'Primitivo' and 'Zinfandel'.[17]

Consider:

(1) "Lærke doubts that woodchucks are groundhogs".

(1.1) $\lambda w \lambda t [^0 Doubt_{wt} \, {}^0 L\!\ae rke \, {}^0 [\lambda w \lambda t [^0 \forall \lambda x [[^0 Woodchuck_{wt} \, x] \supset {}^0 Groundhog_{wt} \, x]]]]]$

Type: $Doubt/(o\iota^*_n)_{\tau\omega}$: a relation-in-intension between an individual and an n-order procedure, in this case, a hyperproposition producing a truth-condition (proposition); that is, $[\lambda w \lambda t [^0 \forall \lambda x [[^0 Woodchuck_{wt} x] \supset {}^0 Groundhog_{wt} x]]]] \to o_{\tau\omega}$.

Alternatively:

(1.2) $\lambda w \lambda t [^0 Doubt_{wt} \, {}^0 L\!\ae rke \, {}^0 [[\lambda w \lambda t [\lambda x [^0 Woodchuck_{wt} \, x]]] = ''' [\lambda w \lambda t [\lambda x [^0 Groundhog_{wt} \, x]]]]]$

---

such as knowing one but not another name for a particular individual, is irrelevant to semantic indistinguishability.

[16]But what is the hardest case? We know that very fine-grained contexts are hard, so the semantics we devise for those is the one we extend to all other contexts as well. Yet a kind of context harder than the one we are currently generalizing from may well come upon our radar, in which case we will have to generalize from that one instead. In any event, TIL extends a semantics designed for hyperinten-sional contexts to intensional and extensional contexts. See also Duží et al. [5, 11] on how to obtain universal transparency.

[17]This exposition draws on material from my (ms.).

Or after η-reduction:[18]

(1.3) $\lambda w \lambda t [^0 Doubt'_{wt} \, {}^0 L\ae rke \; {}^0[^0 Woodchuck =''' \, {}^0 Groundhog]]$

Type: $Doubt'/((o\iota_n^*)_{\tau\omega}$: a relation-in-intension between an agent and an n-order procedure, in this case, a hyperproposition producing a truth-value; that is, $[^0 Woodchuck =''' \, {}^0 Groundhog] \to o$.

The alternative analyses reflect the fact that "All woodchucks are groundhogs" is ambiguous between predication and identity. Are we predicating the property of being a woodchuck of each groundhog, or are we identifying a property with a property? The philosophical point I wish to make is unaffected by which analysis one picks. Formulae (1.1), (1.2) express that Lærke doubts that the same triviality is true. Therefore, we can substitute $^0 Woodchuck$ for $^0 Groundhog$, or the other way around.[19]

Next, consider this objectual (i.e., non-propositional) attitude:

(2) "Lærke adores woodchucks, but detests groundhogs."

Its logical analysis:

(2.1) $\lambda w \lambda t [[^0 Adore_{wt} \, {}^0 L\ae rke \, {}^{00} Woodchuck] \wedge$
$[^0 Detest_{wt} \, {}^0 L\ae rke \, {}^{00} Groundhog]]$

It is kindred to, but importantly different from, this attitude:

(3) "Lærke is seeking an abominable snowman, but not a yeti."

Its logical analysis:

(3.1) $\lambda w \lambda t [[^0 Seek_{wt} \, {}^0 L\ae rke \; {}^0[^0 Abom \, {}^0 Snowman]] \wedge$
$\neg [^0 Seek_{wt} \, {}^0 L\ae rke \, {}^{00} Yeti]]$

---

[18]This is η-conversion in the standard λ-calculus: $\lambda x (fx) \equiv_\eta f$, provided x does not occur free in f. I am presupposing, not exactly unproblematically, that η-conversion does not upset substitutability within hyperintensional contexts. η-converts differ in point of logical processing. I am adding the η-converted alternatives merely to make the formulas easier to read. The identity of properties is what matters for present purposes.

[19]See [11, Defs. 10, 11, 12] on the logic of substitution.

Types: Yeti, Snowman/ $(o\iota)_{\tau\omega}$; *Abom*$/((o\iota)_{\tau\omega}(o\iota)_{\tau\omega})$: a property modifier; $[^{0}Abom\ ^{0}Snowman] \rightarrow (o\iota)_{\tau\omega}$: this Composition produces an entity typed as a property; *Seek*$/(o\iota_{n}^{*})_{\tau\omega}$: a relation-in-intension between an individual and an *n*-order procedure.

Lærke enters twice into a relation-in-intension of type $(o\iota_{n}^{*})_{\tau\omega}$. First time around, she is related to the Composition $[^{0}Abom$ $^{0}Snowman]$, which explains why the Composition is Trivialized so as to display it. Had the Composition not been Trivialized it would have occurred in executed mode instead, and Lærke would have been intentionally related to a property. The appropriate type would have been *Seek*$'/(o\iota(o\iota)_{\tau\omega})_{\tau\omega}$ then. Second time around, Lærke is related to the Trivialization $^{0}Yeti$, which explains why the Trivialization is itself Trivialized so as to display it. Had it not been, Lærke would have been related to what $^{0}Yeti$ produces, to wit, the property of being a yeti. The procedures $[^{0}Abom\ ^{0}Snowman]$ and $^{0}Yeti$ are not procedurally isomorphic, hence not substitutable within hyperintensional contexts. It is, thus, irrelevant that they are equivalent, in that they co-produce the same property. TIL is able to make perfectly good sense of (3).[20] No inconsistent or confused attitude is being attributed to Lærke. Instead, the attitude being attributed to her is that she conceptualizes a particular property in one way, but not another.

The analysis that applies to (3) does not apply to (2). The reason is simply that $^{0}Woodchuck$ and $^{0}Groundhog$ are procedurally isomorphic (in fact, even trivially so). Thus, though not apparent from (2.1), this formalization attributes an inconsistent and confused attitude to Lærke.[21]

---

[20]See [4, §5.2].

[21]Of course, there should be room for psychological subtlety in day-to-day attitude ascriptions. It is inherent to *la condition humaine* that we both love and hate (or less drastically, are both attracted to and repelled by) the same thing or the same person. But I am assuming that Lærke's attitude is that she entirely and exclusively adores, or entirely and exclusively detests, whistle-pigs. This leaves no room for psychological subtlety, nor should it, because it would distract from the combination of two predicates with one property and two attitudes that are polar opposites.

Now consider the centerpiece of this essay:

(4)   "Lærke likes Zinfandel, but dislikes Primitivo."

Which, if any, of the templates above is the right one for this ascription? The next section provides the answer.

5

The austere transparency-preserving constraint TIL imposes upon itself was presented above: $F = G \rightarrow (B_a F_b = B_a G_b)$. Consequently, if it is true that Lærke likes Zinfandel and dislikes Primitivo then it must be false that Zinfandel and Primitivo are identical grapes. And there is room for this to be true, because the second data point leaves room for an agent to believe, without being conceptually confused, that Zinfandel has a property that Primitivo lacks, or vice versa. So, how should we accommodate the first data point, that Zinfandel and Primitivo are genetically identical grapes?

If I went no further than acknowledging that Zinfandel and Primitivo are genetically identical grapes, then 'is a Zinfandel grape' and 'is a Primitivo grape' would be reduced to notational variants, and we would be unable to express or generate the Frege case required. So, there has to be at least one property that sets Zinfandel and Primitivo grapes apart. The quotation above providing the two data points also provides the clue: "Zinfandel is a grape primarily grown in California. Primitivo is a grape primarily grown in Italy [Apulia]." The *discovery* is that the genetic code of Zinfandel is identical to the genetic code of Primitivo. Call this genetic code *DNA*. In simple terms, the discovery is this: "These grapes right here have *DNA*; the grapes over there have *DNA* — so, it is the same DNA in both cases! Who would have thought?!" (This is akin to discovering that the position of the brightest object in the evening sky and the position of the brightest object in the morning sky is the same *position*.) In fact, there is an additional discovery involved, namely *which* grape is their common ancestor; Tribidrag, as it happens. These are *DNA*-encoded grapes

342

cultivated in Dalmatia.[22]

We are looking at a case of two predicates and two properties, though with the important qualification that these two properties — *being a Primitivo grape, being a Zinfandel grape* — share a good many properties that are internal to them. One of them is sharing the genetic code of *DNA*. Another is where they are grown, and in this respect Primitivo and Zinfandel grapes must differ (as must Tribidrag grapes). On my analysis, it is internal to Zinfandel grapes to be grown in California and for Primitivo grapes to be grown in Apulia (and for Tribidrag grapes to be grown in Dalmatia).

TIL has a method for handling internal (or if you like, essential) properties, which is to invoke the *requisite* relation. The basic idea is to take an intensional entity, such as a property, and stack other intensional entities upon it in such a way that anything that instantiates the intension at the base must also instantiate all the other intensions stacked upon it. The stacked intensions are what is known as the requisites of the intension at the base.[23]

Formally, I define the property of being a Zinfandel grape as the property with the two requisite properties of having *DNA* and being grown in California (*Calif*), and I define the property of being a Primitivo grape as the property with the two requisite properties of having *DNA* and being grown in Apulia (*Apulia*):

$$^0Zinfandel =_{df}^{\prime\prime\prime} [^0\lambda p[[^0Req\ ^0DNA\ p] \wedge [^0Req\ ^0Calif\ p]]]$$
$$^0Primitivo =_{df}^{\prime\prime\prime} [^0\lambda p[[^0Req\ ^0DNA\ p] \wedge [^0Req\ ^0Apulia\ p]]]$$

Types: $p/^*_1 \rightarrow (o\iota)_{\tau\omega}$; $Req/(o(o\iota)_{\tau\omega}(o\iota)_{\tau\omega})$; $\imath/((o\iota)_{\tau\omega}(o(o\iota)_{\tau\omega}))$.

The upshot is two distinct properties, with an overlap of requisite properties. These two (so-called ontological) definitions can be introduced only after the empirical discovery of the genetic code of Zinfandel and Primitivo grapes on pain of preempting the outcome of the empirical inquiry into their genetic make-up. The definitions serve to enshrine definitionally or analytically what we now know constitutes

---

[22] *That* they have a common ancestor is trivial.

[23] See[5, 361] for the requisite relation between properties of individuals.

the genetic code of Zinfandel and Primitivo grapes. Genetically, Zinfandel and Primitivo are Tribidrag grapes. But location matters. The grammatical proper name 'Zinfandel' serves to denote *DNA*-encoded grapes grown in California, while the grammatical proper name 'Primitivo' serves to denote *DNA*-encoded grapes grown in Apulia. It will amount to abuse of language (and conceivably also a violation of copyright laws) to label or market Primitivo as 'Zinfandel' and Zinfandel as 'Primitivo'.

We now have what we need in order to obtain the desired result. It is perfectly possible for Lærke to like Zinfandel and dislike Primitivo, because *DNA*-encoded grapes grown in California are to her liking while *DNA*-encoded grapes grown in Apulia are not. Vår may hold the opposite view, while Sørine finds them equally tasty. The kind of account I do not want to give says that Zinfandel and Primitivo are identical, full stop, i.e., *Zinfandel = Primitivo*, but that Lærke may fail to know that 'Zinfandel' and 'Primitivo' are just two names for the same property. This sort of account treats a Frege puzzle as a Mates puzzle. But Lærke's actual ignorance bears on something objectual rather than something linguistic. It is my opinion that if someone has persuaded themselves that (4) induces an opaque context then it becomes very tempting to treat a Frege puzzle as though it were a Mates puzzle.

6

When I was outlining above how transparency is secured, I invoked hyperintensional attitudes. This does not imply, however, that a case like (4) must be framed as a hyperintensional attitude. In fact, a good old-fashioned intensional attitude will work just fine.[24] It would not for the yeti/abominable-snowman case (which can be solved only by means of hyperintensional, though not procedurally isomorphic attitudes) or the woodchuck/groundhog example (which needs to be dissolved, rather than solved, by explaining why it fails to justify

---

[24]See the discussion of "Tama fears that some woodchucks are poisonous" in my (ms.)

hyperintensional distinctions). The intuitive idea underlying (4) is that F and G are distinct properties and that Lærke likes instances of F and dislikes instances of G. The fact that F and G share one of the two requisite properties explicitly mentioned in the two definitions above does not impinge on Lærke's attitude. The logical analysis of "Lærke likes Zinfandel, but dislikes Primitivo" is this:

(4.1) $\lambda w \lambda t [^0 Like_{wt} \, ^0L\!\ae rke \, ^0Zinfandel \wedge \neg [^0 Like_{wt} \, ^0L\!\ae rke \, ^0Primitivo]]$

Type: $Like/(o\iota(o\iota)_{\tau\omega})_{\tau\omega}$: a relation-in-intension between an individual and a property, just like *Seek'*. We are a far cry from being permitted to substitute 'Zinfandel' and 'Primitivo' for one another in an attitude ascription, because they do not even co-denote. The template I am invoking is the first one, $\lambda w \lambda t [^0 = \, ^0F_{wt} \, ^0G_{wt}]$. For sure, there is an internal link between F and G in the case of *Zinfandel/Primitivo*: for any world and any time, F/*Zinfandel* and G/*Primitivo* share the property that they are genetically identical in virtue of *DNA*. This is so thanks to the requisite relation. This does not entail, however, that they are necessarily co-instantiated: there are worlds and times blessed with Zinfandel grapes but deprived of Primitivo grapes, and other worlds and times blessed with Primitivo but deprived of Zinfandel, and yet other worlds and times blessed with both of them or bereft of both of them. Which of the scenarios happens to obtain depends on the viticultural facts on, and in, the ground in California and Apulia.

7

TIL is renowned for its acute attention to minute differences in meaning. Yet there are cases where TIL is pulling in the opposite direction. Whenever a pair of terms or phrases are semantically indistinguishable, we draw no further distinctions. Any further ones would be syntactic or pragmatic or poetic or rhetorical distinctions without a semantic difference. In the absence of a semantic difference, there can be no logical, or inferential, difference, either. Any differences beyond semantics and logic are beyond the enterprise of logical analysis of natural language. Therefore, pairs of contexts like "χ ...groundhog ...", "χ ...woodchuck ...", or "χ ...furze ...", "χ ...gorse ...", are semantically and logically indiscernible. Such pairs, in and by themselves, fail to motivate the introduction of hyperintensional distinctions among predicates or their meanings, and they certainly fail to motivate the introduction of any sort of opacity. These pairs are merely pairs of predicates that are notational variants of one another and as such freely intersubstitutable in any fine-grained, non-quotational attitude report. They are too feeble to constitute Frege cases, which thrive on congruence or equivalence, but are undermined by synonymy.

But then we come across a case like the pair "χ ...Zinfandel ...", "χ ...Primitivo ...", which is ostensibly cut from the same cloth as the previous two pairs. Only it cannot be, provided we want (as we should) to preserve it as a Frege case. This particular example forces us to combine the identity of Zinfandel and Primitivo grapes with the fact that this identity could only be established by scientific, or more broadly empirical, means. The solution turned out to consist in defining *being a Zinfandel grape* as a distinct, if closely related, property from *being a Primitivo grape*. Zinfandel and Primitivo are genetically identical grapes, to be sure, but they are grown in two different locations. The discovery, in simple terms, was that the grapes grown over here are the same as the grapes grown over there.

Just like with woodchuck/groundhog, or furze/gorse, the Zinfandel/Primitivo puzzle is one whose solution does not require, hence nor motivates, going hyperintensional. But the reasons are different. In the former case, the puzzles generated from them are dissolved by pointing out that they are a matter of synonymy, not of equivalence or congruence, and as such misdiagnosed as Frege cases. In the latter case, the puzzle is dispelled by pointing out that there is more to Zinfandel and Primitivo grapes than their genetic code. So, we are confronted with an instance of two properties and two predicates, not one property and two predicates. Intensional distinctions suffice for solving the Zinfandel/Primitivo puzzle.

In general, it is required of a theoretical framework geared toward logical analysis of natural language that it must be able to provide both intensional and hyperintensional solutions. But there are cases where opting for a hyperintensional solution, or pushing for fine-grained distinctions, is too much of a good thing. What is not required of such a theoretical framework is that it should be able to provide both transparent and opaque solutions, sometimes obeying and sometimes violating constraints such as Leibniz's Law and compositionality. Opacity is best left to wither on the vine.

References

[1] Berto, F., D. Nolan (2021), 'Hyperintensionality', *The Stanford Encyclopedia of Philosophy* (Summer 2021 Edition), Edward N. Zalta (ed.), https://plato.stanford.edu/archives/sum2021/entries/hyperintensionality/.

[2] Caie, M., J. Goodman, H. Lederman (2020), 'Classical opacity', *Philosophy and Phenomenological Research* 101, 524-566.

[3] Duží, M. (2019), 'If structured propositions are logical procedures then how are procedures individuated?', *Synthese* 196, 1249-843.

[4] Duží, M., B. Jespersen (2017), 'Transparent quantification into hyperintensional objectual attitudes', *Synthese* 192 (2015), 635-77.

[5] Duží, M., B. Jespersen, P. Materna (2010), *Procedural Semantics for Hyperintensional Logic*, LEUS vol. 17, Springer.

[6] Frege, G. (1918), 'Der Gedanke', *Beiträge zur Philosophie des deutschen Idealismus* 2, 58-77.

[7] Glavaničová, D. and M. Kosterec (2021), 'The fine-grainedness of poetry: a new argument against the received view', *Analysis* 81, 224-31.

[8] Jespersen, B. (ms.), 'A misdiagnosed conundrum about *woodchuck* and 'groundhog" (in submission).

[9] Jespersen, B. (2021), 'First among equals: co-hyperintensionality for structured propositions', *Synthese* 199, 4483—97.

[10] Jespersen, B. (2015), 'Structured lexical concepts, property modifiers, and Transparent Intensional Logic', *Philosophical Studies* 172, 321-45.

[11] Jespersen, B., M. Duží (2022), 'Transparent quantification into hyperpropositional attitudes de dicto', *Linguistics and Philosophy*, 45, 1119–64.

[12] Lederman, H. (2022), 'Fregeanism, sententialism, and scope', *Linguistics and Philosophy*, 45, 1235–1275.

[13] Lepore, E. (2009), 'The heresy of paraphrase: when the medium really is the message', *Midwest Studies in Philosophy* 33, 177-97.